# Chest Atlas

# M.L. Durizch

ASSISTANT PROFESSOR
DEPARTMENT OF RADIOLOGY
COLLEGE OF MEDICINE
UNIVERSITY OF SOUTH ALABAMA

# J.T. Littleton

EMERITUS PROFESSOR
DEPARTMENT OF RADIOLOGY
ADJUNCT PROFESSOR
DEPARTMENT OF STRUCTURAL
AND CELLULAR BIOLOGY
COLLEGE OF MEDICINE
UNIVERSITY OF SOUTH ALABAMA

# Chest Atlas

## Radiographically Correlated Thin-Section Anatomy in Five Planes

With 616 Figures

Contributing Author:
Wan C. Lim
Consulting Author:
W.P. Callahan

Springer-Verlag

**J.T. Littleton, M.D., F.A.C.R.**
Emeritus Professor
Department of Radiology
Adjunct Professor
Department of Structural
and Cellular Biology

**M.L. Durizch, R.T., B.S., M.B.A.**
Assistant Professor
Department of Radiology

**Wan C. Lim, Ph.D.**
Instructor
Department of Structural
and Cellular Biology

**W.P. Callahan, Ph.D.**
Associate Professor (Ret)
Department of Structural
and Cellular Biology

All University of South Alabama
College of Medicine
Mobile, Alabama

**Library of Congress Cataloging-in-Publication Data**
Littleton, Jesse T. (Jesse Talbot), 1917–
    Chest atlas : radiographically correlated thin-section anatomy in five planes /
J.T. Littleton, M.L. Durizch ; contributing author, Wan C. Lim;
consulting author, W.P. Callahan.
        p.   cm.
    Includes bibliographical references and index.
    ISBN 0-387-97928-X.—ISBN 3-540-97928-X
    1. Chest—Tomography—Atlases.   I. Durizch, Mary Lou.    II. Title.
    [DNLM: 1. Thoracic Radiography—atlases.   2. Thorax—anatomy &
histology—atlases.   3. Tomography, X-Ray—atlases. WF 17 L781c]
    4. Computed tomography, multiplanar, direct acquisition
    QM541.L54   1993
    611'.94—dc20
    DNLM/DLC
    for Library of Congress                                                92-49474

Printed on acid-free paper.

Production supervised by Bob Hollander and managed by Karen Phillips;
manufacturing supervised by Rhea Talbert.
Typeset by Best-set Typesetter, Hong Kong.
Printed and bound by Walsworth Publishing Company, Inc., Marceline, MO.
Printed in the United States of America.

9 8 7 6 5 4 3 2 1

ISBN 0-387-97928-X Springer-Verlag   New York   Berlin   Heidelberg
ISBN 3-540-97928-X Springer-Verlag   Berlin   Heidelberg   New York

# Preface

*"Knowledge of anatomic detail enables the radiologist to achieve correct evaluation of clinical manifestations."*

Luzsa (26)

The beginnings of the development of sectional anatomy occurred in the early part of the sixteenth century. Carter et al. (7) have lucidly recounted the chronological history of sectional anatomy since the early descriptions of Leonardo da Vinci, who pictured median sagittal sections of the bodies of both the male and the female. The interested reader will find this comprehensive historical review of great interest.

A renaissance of scientific interest in sectional anatomy has occurred in the last two decades in response to the "explosion" of new sectional imaging methods in radiology, particularly computed tomography (CT) and, more recently, magnetic resonance (MR). Prior to CT, the most concentrated attention to sectional anatomy came from radiological interest in the sectional imaging of conventional pluridirectional tomography (CPT).

This atlas is a comprehensive, thin-section anatomic display of the chest in five planes (axial, coronal, sagittal, left and right posterior oblique). A contact radiograph of each anatomic section and clinical images of MR, CT, and CPT are correlated to the pristine anatomy of cadavers preserved only by freezing. The lungs were reexpanded to reestablish normal spatial relations. Natural color and normal radiographic density have been preserved as nearly as possible.

The atlas is intended to offer anatomic information for clinical application, primarily to radiologists, but also to those clinicians who must view and understand sectional images, such as surgeons, pulmonologists, cardiologists, and other specialists. Additionally, it is intended to be a reference source for medical students, residents, anatomists, and others concerned with a detailed knowledge of anatomy of the thorax.

The incentive to develop an anatomic–roentgenologic correlative text of the thorax has been incubating in the senior author's mind during a 31-year experience with multiplanar sectional imaging of all body systems. Clinical interpretation of sectional images in multiple planes without a multiplanar sectional anatomic reference often made this precise science time-consuming and tedious. Cognitive reconstruction of the shape and location of anatomic objects, seen in a single sectional plane or in a sequence of planes, into a three-dimensional impression had to be made with the aid of displays and descriptions in conventional anatomic texts. The delivery of roentgenologic support to good medical care was often unnecessarily prolonged, less accurate, and consequently not as effective as it might have been. Objectivity passed unnecessarily into subjectivity. The need can be exemplified by an experience of watching a chief resident patiently trying to convince his mentor—an experienced pulmonary radiologist—that a mass in the left upper lobe of the lung on a coronal sectional image at the level of the vertebral bodies was indeed a carcinoma rather than a pulmonary vessel, as insisted by the senior radiologist. A suitable sectional anatomic reference would have quickly shown that no large pulmonary vessels appear at such a posterior plane in this pulmonary segment.

The initial planning for this atlas began in 1977. The 15-year delay in its gestation primarily involved the acquisition of suitable anatomic material. Nineteen bodies were donated to this project, but only five were acceptable—the last one in 1989. During the first five years, nine bodies were examined, but none were acceptable!

Since 1977, 22 sectional anatomic texts which are confined to, or include, the thorax, have come to our attention (2–8, 10–12, 14, 17, 20–22, 27, 28, 38, 39, 43, 44, 47). All have the expressed intent to provide an anatomic base to the new imaging techniques of CT, MR, and ultrasound; none relate to conventional pluridirectional tomography.

The improved soft-tissue discrimination of CT, its favorable axial orientation of many anatomic structures, and its operational simplicity soon resulted in the acceptance of CT as the optimal sectional imaging method in the thorax. The authors, however, still find the multiplanar images of trispiral tomography of the chest to be an indispensable adjunctive procedure. Trispiral tomography is especially useful to image the tracheobronchial segments, the fine detail of pulmonary nodules, vascular patterns, and numerous other instances where nonaxial projections provide better orientation of the pathological process. In the authors' experience, the coronal plane has been found to be most useful; however, in numerous instances the orthogonal sagittal plane and, less frequently, the oblique planes will be required.

Many of the modern texts cited earlier are limited to axial projection, but some have detailed the coronal and sagittal planes. None, however, include sectional anatomy in right and left oblique planes. The oblique projections were first described in 1937 by Peirce and Stocking (36), who presented line drawings and black-and-white photographs of thick tissue slabs taken from embalmed, diseased cadavers. These images were sparsely labeled and can no longer correlate with modern clinical sectional images.

The authors feel that an increasing need for oblique imaging already exists in MR, especially in the heart and vascular structures. Future sectional imaging techniques (multiplanar CT and tomosynthesis, for example, as alluded to in the Appendix), will be enhanced by anatomic support from the multiple planes presented in this atlas.

The atlas is structured in six chapters, beginning with a detailed recounting of the preparation of the anatomic specimens, specimen radiographs, and correlated clinical images: trispiral tomography, CT, and MR (Chapter 1, Methods). In the following five chapters, which detail sectional anatomic–radiographic correlations, reader convenience was of highest priority in the design of each page. The anatomic and radiographic images appear on facing pages with extensive labeling of the anatomic specimen to include minor anatomic elements. The same label designation appears for the same anatomic part throughout the atlas, with a small capital or lowercase letter to indicate the anatomic part—arteries, bronchi, lobes, and so forth—and a larger number to indicate the unit part within these systems. The master anatomic list follows the Acknowledgments.

Each of the five projections (axial, coronal, sagittal, left posterior oblique, and right posterior oblique) is treated in a separate chapter. Page layout for the sections in each chapter is designed as a flow pattern with a colored photograph of the anatomic specimen on the top left of two facing pages, the specimen radiograph on the right-hand page, and the correlative clinical radiographs on the lower part of the pages. The anatomic specimen is as completely labeled as space will permit. Unfortunately, images in clinical practice do not emerge from processing fully labeled; hence, we felt it would be of greater reference value if the specimen radiographs and comparative clinical images were not obstructed by overlying labels and reference lines.

Where crowding of labels on the anatomic section would have obscured pertinent anatomy, appropriate labels appear on adjacent sections to provide continuity of anatomic labeling throughout each series. We regret any reader inconvenience occasioned by the "skip" technique, but it seemed to be a necessary trade-off.

The anatomic labeling key is positioned with the colored anatomic specimen for quick reference. The radiograph of the anatomic specimen serves as a highly detailed reference of radiographic anatomy for the clinical images. The ultimate goal of sectional imaging is the *absolute reproduction of anatomic detail in clinical images*. This goal may never be attained, but it can be projected from the specimen radiographs in this atlas, where finite detail of fresh anatomic material preserved in a natural state by freezing has been reproduced as a sectional radiographic image. Some variance will be noted between the structures seen in the colored anatomic specimen and the specimen radiograph. The former is a photographic image of only one surface of a 3- to 10-mm-thick section of tissue, whereas the radiograph records the three-dimensional detail contained between the two surfaces. The high resolution and tissue density discrimination of the specimen radiograph is an absolute radiographic anatomic reference. A "better than hoped for" correlation resulted between the specimen radiograph and the clinical images of the trispiral tomograms and CT. The clinical images in each of the five planes were made from the respective cadavers prior to sectioning. Even the MR images, which were taken from size-related volunteers, exhibit relatively good anatomic correlation.

The level of each section can be quickly identified from the section plan illustration at the beginning of Chapters 2–6 and the reference key which appears at the top right of the facing page layout for every section. In the latter, the bar across the miniature line drawing identifies the precise level of section, while the relative width of the bar and its color denote the thickness of that section.

Textual material does not appear with the correlative section displays. The highlights of anatomic structures and occasional clinical references are contained in a short introduction to each chapter. The tables appearing with each chapter are intended to be a quick reference to the optimal bronchial imaging in each plane.

## Epilogue

A comparative sectional anatomic reference such as this atlas has a useful life which can be reasonably estimated to be equal to the state-of-the-art of the new clinical images presented. This premise becomes strikingly evident from the rapid obsolescence of the majority of the correlated references that have appeared since this atlas was planned in 1977. Early CT and MR images no longer have comparative value to the clinician. The quality of CPT and CT images has probably reached a mature state; however, MR is still in its infancy, and current MR images can be expected to change significantly.

The true comparison base in this atlas, the colored anatomic specimen and its high-resolution companion radiograph, will not change significantly in the reasonable future. These anatomic illustrations are representative of living tissue, since they were taken from fresh-frozen, unembalmed normal tissue with postmortem inflation of the lungs. The manner of presentation of this information via video display and other new techniques may improve the ease with which the clinician can digest the detail and assimilate it into the interpretation of clinical images.

J.T.L.
M.L.D.

# Acknowledgments

The authors are indebted to many who contributed their support and talents in the development of this atlas. We wish to extend our appreciation especially to the following supporters, the order of whose listing bears no relationship to the magnitude of their contributions. We hope that we have acknowledged everyone, but if there are omissions we humbly apologize—your omission is an oversight, not a measure of your contribution. Unless otherwise designated, the individuals listed were or are all associated with the University of South Alabama.

**Leroy Riddick, M.D., Director, Mobile Laboratory of the Alabama Department of Forensic Sciences, and Clinical Professor, Department of Pathology, University of South Alabama**, for essential advice throughout the entire experience and also for assistance in procuring cadaver material.

**Sara Jo Daniel, M.D., formerly Associate Professor of Pathology, Department of Pathology**, for assistance in the preparation of bodies, especially at 2:00 A.M.

**Byron G. Brogdon, M.D., University Distinguished Professor and Chairman, and Arvin E. Robinson, M.D., former Chairman, Department of Radiology**, for sage council and cooperative forbearance by granting uncounted requests in favor of "the book."

**Steven R. Goodman, Ph.D., Professor and Chairman, Lemoyne Yielding, M.D., former Professor and Chairman, and Glenn L. Wilson, Ph.D., Professor, Department of Cellular and Structural Biology**, gave essential support from the section of gross anatomy.

**William P. Callahan, Ph.D., former Associate Professor, Department of Structural and Cellular Biology**, for assistance during the development of sectioning techniques and for enthusiastic help in the preparation of the cadavers and sawing of the anatomic sections for this atlas.

**Wan C. Lim, Ph.D., Instructor, Department of Cellular and Structural Biology**, joined us late but contributed immeasurably by her diligent review of the anatomic labeling.

**Michael Carmichael, Walter Beckham, Jr., Jon Traudt, Tollef Tollefsen, and Frank Vogtner, Jr., University of South Alabama; and Malcolm Yunker and Bob Jones, Professional Photographers, Mobile, Alabama**, who gave early advice in the development of photographic and illustrative displays.

**Edward D. Sellers, D.V.M., Mobile, Alabama**, made greyhound dog material available to us and assisted in the preparation of these laboratory animals.

**Mr. Bill Brady, Professional Photographer, Mobile, Alabama**, gave unselfishly of his time and talents in printing all of the colored anatomic and some black-and-white prints. We especially wish to recognize his ability to maintain uniform color balance throughout the multiple series of colored prints.

**Mrs. Edda Gilbert, Medical Photographer**, who with professional grace and good humor clearly demonstrated that more than usual talent is required in the ebbing art of black-and-white photography.

**Families of Donors**: We are especially indebted to the families of those who made donations to the State of Alabama Anatomical Board. Without their generosity and personal understanding, this study would not have been possible.

**Ralph H. Ewing, Jr., M.D., Department of Radiology, Medical Center Clinic, P.A., Pensacola, Florida**, identified important considerations in the making of the correlated MR images early in the planning process.

**Dietbert Hahn, M.D., Professor, Department of Radiology, University of Munich, Munich, Germany**, was an essential and indispensable consultant in the selection of MR parameters.

**Liston Orr, M.D., Herbert Hamilton, M.D., and Raymond LaRue, M.D., Residents in Radiology, University of South Alabama**, unselfishly volunteered to spend long, uncomfortable hours in the scanner serving as "models" to make the size-related correlative MR images possible.

**Mrs. Pat DeWitt**, our long-suffering secretary, typed and retyped, filed and refiled, did multiple Medline searches, ran errands, and in every way fulfilled her responsibilities in the development of the manuscript. She took more weekends from her family than she will admit to help us make publishing deadlines.

**CT Imaging Specialists: Jean LaFontaine, R.T., Melanie Bedsole, R.T., and Christopher Ainsworth, R.T.**, always found time from busy clinical schedules to make the CT images for the atlas.

**MR Imaging Specialists: Patrick Brady, R.T., and Jerry Walker, R.T.**, took personal time from busy schedules to make all of the MR images for the atlas.

**Siemens Medical Systems, Inc., Iselin, New Jersey, and Picker International, Inc., Highland Heights, Ohio**, responded with highly effective advisory personnel during the making of the CT cadaver scans and the MR images.

**The Staff at Springer-Verlag, New York**, never wavered from a cooperative, enthusiastic spirit throughout the long development of this book. Repeated extensions of manuscript delivery dates were always forthcoming and seemed to become part of the process.

# Contents

# Anatomic Key List

## Abbreviations for anatomic structures

| | |
|---|---|
| A | ARTERIES |
| B | BRONCHI |
| D | DUCT, THORACIC |
| E | ESOPHAGUS |
| F | FASCIA |
| f | FAT |
| F | FISSURES |
| G | GLANDS |
| H | HEART |
| L | LOBES AND BRONCHOPULMONARY SEGMENTS (LUNGS) |
| M | MUSCLES |
| n | NEURAL STRUCTURES |
| N | NODES (LYMPH NODES) |
| O | ORGANS |
| P | PLEURAL STRUCTURES |
| S | SKELETAL STRUCTURES |
| T | TRACHEA AND LARYNX |
| V | VEINS |
| | NONANATOMIC |

**ARTERIES, Coronary**
see Heart

**ARTERIES, Pulmonary (variations are frequent)**

A 1   pulmonary trunk
A 2   main pulmonary, right
A 2*  main pulmonary, right (interlobar part)
A 3   right upper lobe (RUL)
A 4       RUL, apical branch
A 5       RUL, posterior branch
A 6       RUL, anterior branch
A 7   right middle lobe (RML)
A 8       RML, lateral branch
A 9       RML, medial branch
A 10  right lower lobe (RLL)
A 11      RLL, superior branch
A 12      RLL, medial basal branch
A 13      RLL, anterior basal branch
A 14      RLL, lateral basal branch
A 15      RLL, posterior basal branch
A 16  main pulmonary, left
A 17  left upper lobe (LUL)
A 18      LUL, apicoposterior branch
A 19      LUL, anterior branch
A 20      LUL, lingula
A 21      LUL, superior lingular branch
A 22      LUL, inferior lingular branch
A 23  left lower lobe (LLL)
A 24      LLL, superior branch
A 25      LLL, anteromedial basal branch
A 26      LLL, lateral basal branch
A 27      LLL, posterior basal branch

**ARTERIES, Systemic**

A 28  aorta
A 29  aortic arch
A 30  aorta, ascending
A 31  aorta, descending
A 32  axillary, left
A 33  axillary, right
A 34  brachiocephalic (innominate) trunk
A 35  bronchial, left
A 36  bronchial, right
A 37  carotid, common, left
A 38  carotid, common, right
A 39  costocervical trunk, left
A 40  costocervical trunk, right

A 41  intercostal, highest (superior), left
A 42  intercostal, highest (superior), right
A 43  intercostal, anterior, left
A 44  intercostal, anterior, right
A 45  intercostal, posterior, left
A 46  intercostal, posterior, right
A 47  internal thoracic (mammary), left
A 48  internal thoracic (mammary), right
A 49  subclavian, left
A 50  subclavian, right
A 51  subscapular, left
A 52  subscapular, right
A 53  suprascapular, left
A 54  suprascapular, right
A 55  thoracic, lateral, left
A 56  thoracic, lateral, right
A 57  thoracoacromial, left
A 58  thoracoacromial, right
A 59  thyrocervical trunk, left
A 60  thyrocervical trunk, right
A 61  thyroid, inferior, left
A 62  thyroid, inferior, right
A 63  transverse cervical, left
A 64  transverse cervical, right
A 65  vertebral, left
A 66  vertebral, right

**BRONCHI**

B 1   main, right
B 2   right upper lobe (RUL)
B 3       RUL, apical segment
B 4       RUL, posterior segment
B 5       RUL, anterior segment
B 6   intermediate, right
B 7   right middle lobe (RML)
B 8       RML, lateral segment
B 9       RML, medial segment
B 10  right lower lobe (RLL)
B 11      RLL, superior segment
B 12      RLL, medial basal segment
B 13      RLL, anterior-basal segment
B 14      RLL, lateral basal segment
B 15      RLL, posterior basal segment
B 16  main, left
B 17  left upper lobe (LUL)
B 18      LUL, apicoposterior segment
B 19      LUL, anterior segment
B 20      LUL, lingula
B 21      LUL, superior lingular segment
B 22      LUL, inferior lingular segment

B 23  left lower lobe (LLL)
B 24  LLL, superior segment
B 25  LLL, anteromedial basal segment
B 26  LLL, lateral basal segment
B 27  LLL, posterior basal segment

**DUCT**

D 1  thoracic

**ESOPHAGUS**

E 1  esophagus
E 2  esophagogastric junction

**FASCIA**

F 5  infradiaphragmatic (transversalis)
F 6  perivisceral
F 7  prevertebral
F 8  Sibson's
F 9  superficial (investing)
F 10  supradiaphragmatic

**FAT**

f 1  axillary
f 2  epicardial
f 3  extrapleural
f 4  mediastinal
f 5  pericardial
f 6  perirenal
f 7  subcutaneous

**FISSURES**

F 1  left major (oblique)
F 2  right major (oblique)
F 3  right minor (horizontal)
F 4  aberrant (LUL)

**GLANDS**

G1  adrenal
G2  pancreas
G3  thymus
G4  thyroid

**HEART**

H 0  arteriosum, ligamentum
H 1  atrium, left
H 2  atrium, right
H 3  auricle, left
H 4  auricle, right
H 5  cardiac vein, great
H 6  cardiac vein, middle
H 7  cardiac vein, small
H 8  coronary artery, circumflex
H 9  coronary artery, descending
H 10  coronary artery, left
H 11  coronary artery, right
H 12  coronary sinus
H 13  pericardium, parietal serous
H 14  pericardium, visceral serous
H 15  septum, interatrial
H 16  septum, interventricular
H 17  valve, tricuspid
H 18  valve, pulmonary
H 19  valve, mitral
H 20  valve, aortic
H 21  ventricle, left
H 22  ventricle, right

**LARYNX,**
    see TRACHEA

**LOBES AND BRONCHOPULMONARY
SEGMENTS (LUNGS)**

L 1  right upper lobe (RUL)
L 2  RUL, apical segment
L 3  RUL, posterior segment
L 4  RUL, anterior segment
L 5  right middle lobe (RML)
L 6  RML, lateral segment
L 7  RML, medial segment
L 8  right lower lobe (RLL)

L 9  RLL, superior segment
L 10  RLL, medial basal segment
L 11  RLL, anterior basal segment
L 12  RLL, lateral basal segment
L 13  RLL, posterior basal segment
L 14  left upper lobe (LUL)
L 15  LUL, apicoposterior segment
L 16  LUL, anterior segment
L 17  LUL, lingula
L 18  LUL, superior lingular segment
L 19  LUL, inferior lingular segment
L 20  left lower lobe (LLL)
L 21  LLL, superior segment
L 22  LLL, anteromedial basal segment
L 23  LLL, lateral basal segment
L 24  LLL, posterior basal segment

**MUSCLES**—alphabetical list (a superior to
inferior list follows on page xiii)

M 1  abdominal, external oblique
M 2  abdominal, internal oblique
M 3  abdominis, rectus
M 4  abdominis, transversus
M 5  biceps brachii, long head
M 6  biceps brachii, short head
M 7  coracobrachialis
M 8  deltoid
M 9  diaphragm
M 10  diaphragm, central tendon
M 11  diaphragm, crus, left
M 12  diaphragm, crus, right
M 13  erector spinae
        (see also iliocostalis, longissimus,
        spinalis)
M 14  iliocostalis, thoracis
M 15  infraspinatus
M 16  intercostal
        intercostal, external
        intercostal, innermost
        intercostal, internal
M 17  interspinalis, cervicis
M 18  interspinalis, thoracis
M 19  intertransversarius, cervicis
M 20  intertransversarius, thoracis
M 21  latissimus dorsi
M 22  levator costarum
M 23  levator scapulae
M 24  linea alba (not labeled)
M 25  longissimus, capitis (not labeled)
M 26  longissimus, cervicis
M 27  longissimus, thoracis
M 28  longus colli
M 29  multifidus
M 30  omohyoid
M 31  pectoralis, major
M 32  pectoralis, minor
M 33  platysma
M 34  psoas
M 35  rhomboid, major
M 36  rhomboid, minor
M 37  rotator
M 38  scalene, anterior
M 39  scalene, middle
M 40  scalene, posterior
M 41  semispinalis, capitis
M 42  semispinalis, cervicis
M 43  semispinalis, thoracis
M 44  serratus, anterior
M 45  serratus, posterior inferior
M 46  serratus, posterior superior
M 47  spinalis, capitis (not labeled)
M 48  spinalis, cervicis
M 49  spinalis, thoracis
M 50  splenius, capitis
M 51  splenius, cervicis
M 52  sternocleidomastoid
M 53  sternohyoid
M 54  sternothyroid
M 55  subclavius

M 56  subcostal
M 57  subscapularis
M 58  supraspinatus
M 59  teres major
M 60  teres minor
M 61  transversospinalis
        (see also rotator, multifidus,
        semispinalis)
M 62  transversus thoracis
M 63  trapezius

**NEURAL STRUCTURES**

n 1  brachial plexus
n 2  intercostal nerve
n 3  phrenic nerve
n 4  recurrent laryngeal nerve
n 5  spinal cord
n 6  spinal dura
n 7  spinal nerve
n 8  splanchnic nerve
n 9  sympathetic nerve chain
n 10  vagus nerve

**NODES** (Lymph Nodes)

N 1  axillary
N 2  bronchopulmonary (hilar)
N 3  diaphragmatic
N 4  intercostal (not labeled)
N 5  intrapulmonary
N 6  mediastinal, anterior
N 7  mediastinal, posterior
N 8  parasternal (internal thoracic)
N 9  paratracheal
N 10  supraclavicular
N 11  tracheobronchial, inferior (subcarinal)
N 12  tracheobronchial, superior

**ORGANS**

o1  colon
o2  kidneys, left and right
o3  liver
o4  small bowel
o5  spleen
o6  stomach

**PLEURAL STRUCTURES**

P 1  anterior junction line
P 2  costal (parietal)
P 3  costodiaphragmatic recess
P 4  diaphragmatic (parietal)
P 5  mediastinal (parietal)
—  pericardial (see Heart)
P 6  visceral

**SKELETAL STRUCTURES**

s 1  clavicle
s 2  humerus
s 3  rib
s 4  rib, body, anterior
s 5  rib, body, lateral
s 6  rib, body, posterior
s 7  rib, costal cartilage
s 8  rib, head and neck
s 9  rib, tubercle
s 10  scapula
s 11  scapula, acromion
s 12  scapula, coracoid process
s 13  scapula, glenoid cavity
s 14  scapula, neck
s 15  scapula, spine
s 16  sternum
s 17  sternum, body
s 18  sternum, manubrium
s 19  sternum, xiphoid

**SKELETAL, Spine (spinal cord, dura,
nerve—see NEURAL STRUCTURES)**

s 20  intervertebral disc
s 21  vertebra, body

s 22 vertebra, costal facet
s 23 vertebra, inferior articular process
s 24 vertebra, intervertebral foramen
s 25 vertebra, lamina
s 26 vertebra, pedicle
s 27 vertebra, spinous process
s 28 vertebra, superior articular process
s 29 vertebra, transverse process

## TRACHEA AND LARYNX

T 1 trachea
T 2 carina (bifurcation of trachea)
T 3 larynx

## VEINS, Coronary

see Heart

## VEINS, Pulmonary

v 0 pulmonary, superior, RUL
v 1 pulmonary, superior, RML
v 2 pulmonary, superior, LUL
v 3 pulmonary, superior, LUL, lingula
v 4 pulmonary, inferior, RLL
v 5 pulmonary, inferior, LLL

## PULMONARY VEIN SUBDIVISIONS NOT LABELED ON PRINTS, BUT LISTED

Subdivisions of the superior and inferior pulmonary veins are too small to label in these sectional images. The segmental distribution of veins in the lung differs from that of the pulmonary arterial branches. The following is a list of segmental pulmonary vein classifications by Lusza (26); however, he notes there can be many variations in their branching patterns.

v 0 pulmonary, superior, RUL
　　apical
　　posterior
　　anterior
v 1 pulmonary, superior, RML
　　lateral
　　medial
v 2 pulmonary, superior, LUL
　　apicoposterior
　　anterior
v 3 pulmonary, superior, LUL, lingula
　　superior lingula
　　inferior lingula
v 4 pulmonary, inferior, RLL
　　apical (superior)
　　superior basal
　　inferior basal
　　common basal
v 5 pulmonary, inferior, LLL
　　apical (superior)
　　common basal

## VEINS, Systemic

v 6 axillary (left or right)
v 7 azygos
v 8 brachiocephalic (innominate), left
v 9 brachiocephalic (innominate), right
v 10 cephalic, left

v 11 cephalic, right
v 12 hemiazygos
v 13 hemiazygos, accessory
v 14 intercostal, highest (superior), left
v 15 intercostal, highest (superior), right
v 16 intercostal, anterior, left
v 17 intercostal, anterior, right
v 18 intercostal, posterior, left
v 19 intercostal, posterior, right
v 20 internal thoracic (mammary), left
v 21 internal thoracic (mammary), right
v 22 jugular, anterior, left
v 23 jugular, anterior, right
v 24 jugular, external, left
v 25 jugular, external, right
v 26 jugular, internal, left
v 27 jugular, internal, right
v 28 lateral thoracic, left
v 29 lateral thoracic, right
— mediastinal (see venous plexus)
v 30 subclavian, left
v 31 subclavian, right
v 32 subscapular (left or right)
v 33 suprascapular (left or right)
v 34 thoracoacromial (left or right)
v 35 thyroid, inferior (left or right)
v 36 transverse cervical, left
v 37 transverse cervical, right
v 38 vena cava, inferior
v 39 vena cava, superior
v 40 venous plexus, mediastinal
v 41 venous plexus, vertebral
v 42 vertebral, left
v 43 vertebral, right

## NONANATOMIC

* antemortem trauma
NB nylon bolt
† pacemaker site
‡ postmortem trauma
PC pulmonary cyst

MUSCLES—anatomical, superior to inferior list, with same muscle number as in alphabetical list above

### Muscles Connecting Upper Extremity to Vertebral Column

M63 trapezius
M21 latissimus dorsi
M35 rhomboid, major
M36 rhomboid, minor
M23 levator scapulae

### Muscles Connecting Upper Extremity to Thoracic Walls

M31 pectoralis, major
M32 pectoralis, minor
M55 subclavius
M44 serratus, anterior

### Muscles of the Shoulder

M 8 deltoid
M57 subscapularis
M58 supraspinatus
M15 infraspinatus

M59 teres major
M60 teres minor

### Muscles of the Arm

M7 coracobrachialis
M5 biceps brachii, long head
M6 biceps brachii, short head

### Muscles of the Neck

M33 platysma
M52 sternocleidomastoid
M53 sternohyoid
M54 sternothyroid
M30 omohyoid

### Anterior and Lateral Vertebral Neck Muscles

M28 longus colli
M38 scalene, anterior
M39 scalene, middle
M40 scalene, posterior

### Deep Muscles of the Back

M50 splenius, capitis
M51 splenius, cervicis
M13 Erector Spinae Group (Sacrospinalis Group)
M14 　iliocostalis (cervicis & thoracis)
M25 　longissimus, capitis (not labeled)
M26 　longissimus, cervicis
M27 　longissimus, thoracis
M47 　spinalis, capitis (not labeled)
M48 　spinalis, cervicis
M49 　spinalis, thoracis
M61 Transversospinalis Group
M41 　semispinalis, capitis
M42 　semispinalis, cervicis
M43 　semispinalis, thoracis
M29 　multifidus
M37 　rotator
M17 interspinalis, cervicis
M18 interspinalis, thoracis
M19 intertransversarius, cervicis
M20 intertransversarius, thoracis

### Muscles of the Thorax

M16 intercostal (external, internal, innermost)
M56 subcostal
M62 transversus thoracis
M22 levator costarum
M44 serratus, anterior
M46 serratus, posterior superior
M45 serratus, posterior inferior
M 9 diaphragm
M10 diaphragm, central tendon
M11 diaphragm, crus, left
M12 diaphragm, crus, right

### Muscles of the Abdomen

M 1 abdominal, external oblique
M 2 abdominal, internal oblique
M 3 abdominis, rectus
M 4 abdominis, transversus
M24 linea alba (not labeled)
M34 psoas

# Chapter 1

# Methods

## Introduction

In life, and so in death, the chest is the most dynamic physiologic organ center of the human body. Delicate pulmonary structures are extremely difficult to preserve in a fresh state that will permit the lung and mediastinal structures to be sectioned and presented to clinicians and anatomists in realistic dimensions in true color. Structural relationships become distorted in embalmed specimens where partial to complete color change takes place, blood is washed from the heart and vascular structures, lungs collapse, and extensive shrinkage occurs. The body assumes the decor of the anatomy laboratory cadaver—our learning model—which imposes a significant imaginative conversion from the anatomy laboratory to current sectional radiologic images or to living images of the operating room.

The singular objective of this atlas is to make a true sectional anatomic reference base of the thorax available to radiologists, anatomists, other clinicians, and health care professionals. It is intended to be a reference that is representative of living tissue presented from five imaging planes: axial, coronal, sagittal, right posterior oblique, and left posterior oblique. To accomplish this end, special attention to the preparation of the cadavers was required.

Human lung is extremely delicate, even when frozen. Since no previous experiences could be identified in the literature (other than those referred to in the preface), and in view of the obvious paucity of fresh-frozen human specimens, the techniques described here were first developed using racing greyhounds. These dogs have relatively generous lung capacity, and by the nature of their "profession" healthy animals are destroyed on a regular basis. The ultimate techniques we employed in sectioning unembalmed, frozen cadavers into slices as thin as 3-mm have not been previously reported, and hence they are recounted here in detail for the interested reader.

## Cadaver Preparation

Specimens were obtained through the Anatomical Gifts Program of the State of Alabama Anatomical Board, Department of Anatomy, University of South Alabama College of Medicine. Each cadaver was prepared in a similar manner using techniques previously described in part by the authors (24).

Immediately after death, as soon as the donation of the subject was legally secured (usually 1–4 hours), an endotracheal (ET) tube was passed to reinflate the lungs to estimated full inspiration. Inflation pressure must not exceed the expiratory capability of the operator blowing through an extender tube attached to the ET tube. Mechanical inflation may create higher pressures than necessary, resulting in alveolar rupture and inevitable pneumothorax. The manual inflation procedure reinflated zones of postmortem atelectasis and restored normal spatial relationships within the thorax, which is essential for image correlation. After double clamping of the ET tube, an anteroposterior (AP) radiograph was made to insure that the thorax was normal, that the lungs were fully inflated, and that the thorax was suitable for sectioning.

After acceptability of the subject was assured, both lower extremities were amputated at the junction of the upper and middle third of the thigh; vessels were securely ligated; and the specimen was carefully placed in a true supine position in a standard home freezer (2' × 4' × 3') and allowed to "cure" at −10° F for a minimum of 3 months.

In one instance, antemortem pneumonic consolidation was not evident until posterior sections were made through the coronal subject (Chapter 3). Since no essential anatomic structures were obliterated by the area of lobar pneumonia, the specimen was used in the atlas.

Careful precautions must be taken to prevent air from entering the thoracic vascular structures via an open vessel. In two cadavers the carotid arteries and jugular veins were open (as a result of a gunshot wound and a limited autopsy), permitting entry of air into both cardiac chambers, pulmonary arteries, and large thoracic vessels. Both cadavers were salvaged by selectively bivalving the chest at preplanned planes that permitted selective catheterization of cardiac chambers and great vessels under fluoroscopic control. These vascular cavities were filled with packed red cells to restore normal color and radiographic density.

Postmortem "sludging" is an unavoidable problem that appeared to various degrees throughout all five bodies. Settling of red cells into the dependent portion of the large vessels and cardiac chambers resulted in variable color shades, from the light yellow of the supernatant serum in the anterior portion of the vessels (the aortic arch, for example) to deep red to black in the dependent portions such as the lower thoracic aorta. This phenomenon also alters the radiographic density sufficiently to be seen as layering in the specimen radiographs and CT images. Corrective tinting of the color illustrations was not attempted, and it is hoped that the reader will soon become accustomed to the color change within the blood vessels. Postmortem sludging occurs within 1 to 3 hours after death and probably cannot be prevented by more rapid freezing, but it probably could be prevented by rotating the subject during the freezing process. Induced antemortem clotting, in a brain-dead patient on life support, for example, would also be ineffective since red cell settling also occurs in clotted blood. This same phenomenon accounts for the deeper red color of the dependent planes of the lungs in all five specimens. In this instance we are experiencing dependent fluid settling as well through the interstices of the inflated lungs.

### Clinical Images

#### Computed Tomography

After "curing" in the freezer, each thorax was isolated and taken to the Department of Radiology for filming. Sectional images of the specimen chosen for the axial sections were taken on a Siemens Somatom Plus CT scanner (Figure 1.1).

The axial specimen, as well as all others, was first fluoroscoped and carefully marked with skin sutures to define cutting axes and limits upon which to realign the specimens in both the Radiology Department and the Anatomy Laboratory throughout the entire process (Figure 1.1A and B). Without this precise registration, the anatomic planes cut would not have matched the clinical planes; essential correlation would have been lost. A Magic Marker pen had been used to mark the dog specimens, but the saw carried the ink through the tissue to soil the lung surface.

Axial CT scans were made from C7 to T12 with mediastinal and lung parameters. For the lung series, 2-mm-thick overlapping axial scans were made using a 2-mm slice thickness and 1-mm table movement. One-second scans were made with a high-resolution, lung algorithm at 225 mA and 137 kV. Window width (W) and center (C) settings of W 1200, C −750 were best for viewing the frozen lung tissue. The mediastinum series was done with contiguous 3-mm-thick slices using a mediastinal algorithm at 1 sec, 290 mA, and 125 kV. These scans were viewed at W 400, C 10. The CT images were subsequently matched to the colored photograph and radiograph of the gross anatomic specimen.

#### Trispiral Tomography

The remaining four specimens were tomogrammed on an Exatome device (General Electric-CGR, Milwaukee, Wisconsin). Sections were contiguous, with a layer height change of 1-mm, using a 14°/40° trispiral movement. The resulting pluridirectional tomographic sections were 1-mm thick. The entire diameter of each of the four specimens was sectioned in each of four planes: AP projection (coronal—Chapter 3); lateral projection (sagittal—Chapter 4); left posterior oblique projection (LPO—Chapter 5); and right posterior oblique projection (RPO—Chapter 6).

During filming, an aluminum trough filter, attached to the x-ray tube collimator, was used to even density gradients between the lung and mediastinum for the coronal and oblique projections. Technical factors employed were $4\frac{1}{2}$ sec, 20 mA, 75–80 kV for the coronal and oblique projections and $4\frac{1}{2}$ sec, 35 mA, 76 kV for the sagittal projection. DuPont Quanta Fast Detail screens and Cronex 7 film were used throughout.

Each specimen was positioned on the tomographic table using skin sutures and laser localizing beams to permit a return to the same precise position for the sawing procedure that was to take place later in the anatomy laboratory (Figure 1.2). Here, as with CT, the tomographic images were subsequently matched to the colored photograph and to the radiograph of the gross anatomic specimen.

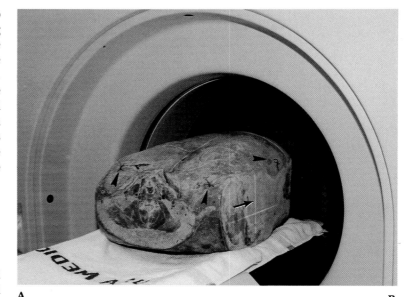

A              B

**Figure 1.1.** (A) Isolated thorax in CT gantry of a Siemens Somatom Plus (Siemens AG, Erlangen, Germany) for the axial sections (Chapter 2). Note black silk suture reference marks (arrowheads) for laser localization and alignment during radiographic imaging and sawing. Vertical laser at a midthorax axial CT level (arrow) corresponds to level of cut surface in 1.1(B). (B) Measurement from the simulated laser beam (red band) to the skin sutures ensured exact registration and realignment of the CT plane.

**Figure 1.2.** Specimen bolted to 45° plastic frame preparatory to obtaining conventional tomographic images in right posterior oblique plane. Simulated laser beam (red band) represents the tomographic plane.

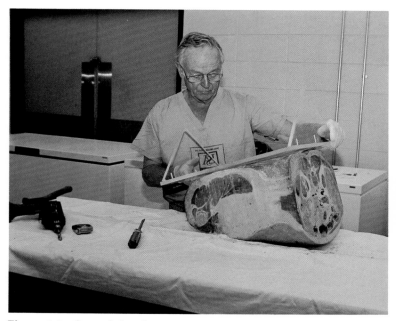

**Figure 1.3.** One of the chest specimens being bolted to plastic frame by Dr. Littleton.

**Figure 1.4.** A specially designed cutting device (A) with a heavy ($\frac{1}{2}''$) vertical plastic plate (B) was bolted to the moving platform (C) of the saw. A screw drive (arrow) for the vertical assembly to which the specimen was attached permitted accurate advancement of the specimen into the saw by 1-mm increments.

## Magnetic Resonance

The low temperature of the frozen torso ($-10°$F) reduces electron activity to such a level that no MR images of frozen specimens are possible. Accordingly, all MR images were matched from size-related, living subjects.

The MR images were T1-weighted images with cardiac gating made on a Picker Vista HPQ 1.5 tesla scanner (Picker Corporation, Cleveland, Ohio). At the present time, the image quality of thin (2–5 mm) MR images is not as detailed as that obtained with thicker slices; therefore, a compromised 7-mm slice thickness was chosen. To ensure an optimal match to the anatomic section and the specimen radiograph, three series of scans were made of each subject in all five planes. Each of the three series consisted of a single acquisition of 7-mm-thick contiguous scans, with each acquisition set to begin at a level 2-mm below the previous one. This protocol produced a sequence of overlapping MR scans from which images could be chosen to closely match to the anatomic detail of each anatomic section. The distance covered in the axial projection was too great to capture in the three overlapping series. Therefore, an additional single acquisition series (without overlapping) was made through the lower cervical area and the lung apices.

## Cutting of Anatomic Sections

A frozen body can be machined within reasonable tolerances using simple woodworking skills. Clear plastic frames were constructed to accommodate each of the five thoraces to permit sectioning in each plane. The isolated chest was drilled, tapped, and bolted to the heavy plastic frames using $\frac{1}{4}''$ and $\frac{1}{2}''$ nylon screws and bolts. This provided a permanent, securely mounted base from which the specimen was never removed. Whenever possible, bone, especially the spine, was tapped to give a solid anchor (Figure 1.3). Nylon screws and bolts could be placed in the sawing area when necessary and did not interfere with the sawing procedure. Also, when necessary, the plastic frames could be sawed with the specimen.

The plastic frame holding the mounted specimen was then securely bolted, for the duration of the cutting process, to a vertical $\frac{1}{2}''$ plastic plate. This vertical plate could easily be attached to and removed from an aluminum cutting platform specially designed and constructed for a butcher's Hobart Meat Saw (Figure 1.4). The aluminum platform was equipped with a screw drive that would permit the accurate advancement of the specimen toward the saw by 1-mm increments. Constant sawing pressure and speed was obtained by hanging 5 lb of

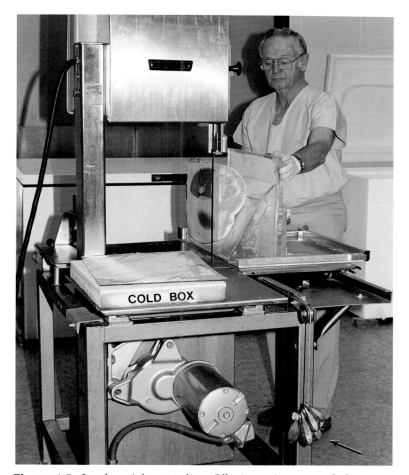

**Figure 1.5.** Lead weights totaling 5 lb (arrow), suspended over a pulley and attached to the moving platform of the saw, impart a constant sawing pressure. An aluminum "cold box" filled with dry ice is shown in position to receive a cut section.

lead weight over a pulley at the distal end of the moving platform of the saw (Figure 1.5). Constant sawing pressure was extremely important to prevent the saw blade from "wandering." After the first "machining" cut, all sections exhibited uniform thickness over the entire surface, even the 3-mm cuts. The saw kerf consumed approximately 1.5-mm of tissue.

Once cut, it was imperative that the anatomic sections be kept frozen throughout the remainder of the process. To this end, customized ($2'' \times 14'' \times 17''$) aluminum "cold boxes" were fabricated (Figure 1.5). One end of each box could be opened by loosening a single screw to allow the box to be packed with dry ice at the beginning of each sawing session.

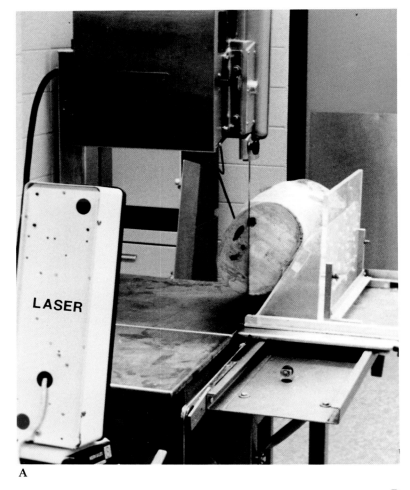

A

B

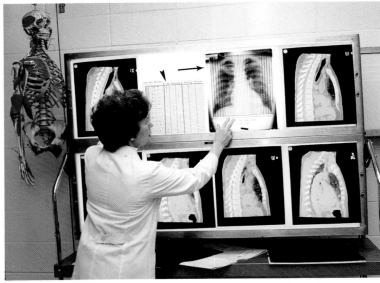

**Figure 1.7.** Daily sawing procedures were always carefully preplanned from the master section plan (arrow), the cutting log (arrowhead), and radiographs of previous sections.

Final alignment of each specimen in the saw was accomplished with a single laser beam projected in vertical axis (Figure 1.6A) and an adjustable point light source to insure the horizontal axis (Figure 1.6B). By "shimming" to these two projected light references, the alignment used during the filming of the clinical images could be exactly reproduced. At the end of each daily procedure, the specimen (still attached to the plastic plate) was removed from the cutting platform and returned to the freezer. Realignment for each succeeding sawing session was assured, since the specimen mount did not need to be disturbed.

Daily sawing procedures were always preplanned from a lined grid of the predesigned section plan overlying a radiograph of the specimen, the daily log, and radiographs of the previously cut sections (Figure 1.7). Once the planning and aligning were accomplished, three to five sections were cut and immediately placed in the freezer. The remainder of the thorax was also returned to the freezer to prevent any thawing prior to the next cutting session. The specimen block could not be kept out of the freezer for longer than 30–60 minutes without experiencing some surface thawing.

As each section came from the saw, it was caught on a cold $\frac{1}{8}''$ plastic sheet wrapped in Saran Wrap® to prevent the specimen from refreezing or sticking to it (Figure 1.8). The "catching"

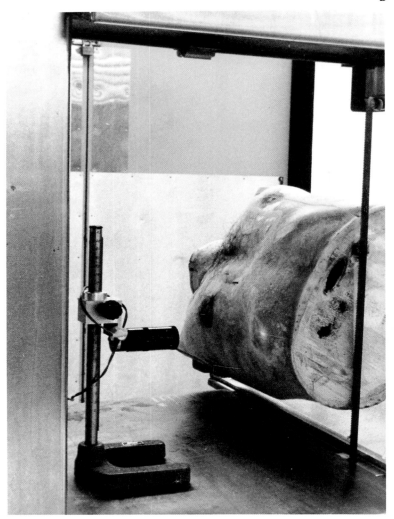

**Figure 1.6. (A)** The tomographic plane was precisely reproduced on the sawing platform by projection of an externally mounted laser beam. **(B)** The longitudinal axis of the thorax was assured by the projection of a point light source along the length of the specimen as it was advanced with the saw platform.

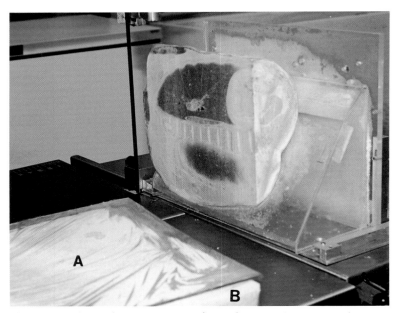

A

B

**Figure 1.8.** As each section came from the saw, it was caught on a cold sheet covered with Saran Wrap® (A), and immediately placed on a cold box (B).

A
B

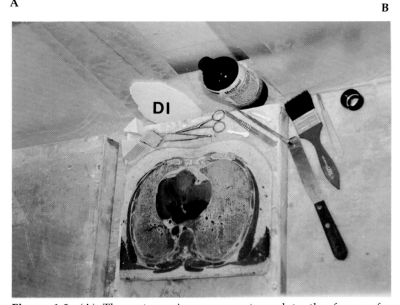

**Figure 1.9.** (**A**) The cut specimen was returned to the freezer for cleaning and was kept on a "cold box" at all times. A few minutes' exposure to room temperature without a cold box would result in irreversible thawing damage. (**B**) The implements for cleaning included assorted sponges, forceps, scissors, scalpels, makeup applicators, brushes, spatulas, probes, and methyl alcohol. A thin sheet of dry ice (DI) was kept in readiness to refreeze small areas of inadvertent thawing.

sheet was kept cold on one of the aluminum cold boxes. The freshly cut section, still on the cold box, was returned to the freezer immediately and stayed on this box throughout the cleaning, filming, and radiographic procedure.

Each section was meticulously cleaned in the freezer to remove the saw debris using methyl alcohol-soaked soft sponges, probes, brushes, scissors, scalpels, and infinite patience (Figure 1.9A and B). Cleaning both surfaces of one section required approximately 1 hour. It is difficult to describe how delicate normal lung tissue is, other than to say that it can be aptly compared to pie meringue.

## Photographic Procedures

It was necessary to build a "photographic shed" next to the sawing area that could be closed at the moment of exposure to block out all ambient light (Figure 1.10). The shed was enclosed by black cloth and paper. Photography was done using two Nikon F3 cameras with 55 mm flat copy lenses (to make duplicate series) and Kodak Kodacolor 100 ISO film. Each camera lens was covered with a polarizing filter, and floor-mounted polarized light stands were used to eliminate highlights from the wet surface glare.

When completely cleaned, the section specimen was transferred to a plate of nonglare glass overlying a background sheet of white opal plastic. Both were placed on a cold box and centered on the photographic stand. The nonglare glass eliminated any reflection from the margins. In some instances, both surfaces of the anatomic specimen were photographed. Surface anatomic detail most closely matching the detail of the specimen radiograph was chosen for publication. Careful attention was given to precise identification of section numbers (Figure 1.11).

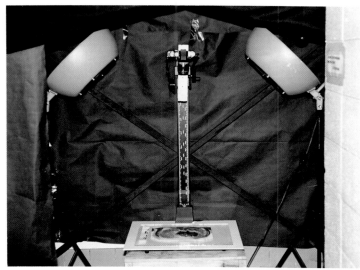

**Figure 1.10.** Photographic shed showing specimen in place for photography. A Nikon F3 camera equipped with a polarizing filter is mounted on the copy stand. Polarizing light stands are seen angled to the specimen to avoid unwanted glare from the wet surface of the specimen.

**Figure 1.11.** Final alignment of specimen on photographic platform. Careful precautions were taken to ensure that the marker (arrow) correctly displayed all of the identifying data for the section.

## Specimen Radiograph Procedures

Immediately after the photographic procedure was completed, the specimen was carefully transferred from the nonglare glass to a clear sheet of 14″ × 17″ x-ray film (for support) and then placed on a prepacked sheet of 14″ × 17″ nonscreen x-ray film

Figure 1.12. Specimen in position for contact specimen radiograph.

Figure 1.13. Specimen was carefully wrapped in Saran Wrap®, labeled, and stored in freezer for future pathologic verification of questionable structures.

in preparation for radiography (Figure 1.12). The radiographs were made with a Faxitron, a shielded cabinet x-ray system designed primarily for specimen radiography. The Faxitron was a Hewlett-Packard model No. 43807-N (Hewlett-Packard, McMinnville, Oregon) with a dual cabinet, one atop the other, having a cutout in the floor of the top unit to permit a tube–film distance sufficient to cover a 14″ × 17″ film. Exposure factors were 20 kV, 3 mA, and times of 1–3 minutes depending on the thickness of the specimen (3, 5, 10 mm). Two sets of bracketed exposures were made for each specimen, using Kodak fine-grain XTL-2 individually wrapped film packets.

As soon as the specimens had been radiographed, they were securely wrapped in Saran Wrap®, sealed, numbered, and returned to a storage freezer to await completion of the atlas (Figure 1.13). Several sections were reopened to permit pathologic verification of labeled structures. Efforts to plasticize these thin anatomic specimens so that they could be permanently retained met with no success.

## Photographic Reproductions

The colored prints of the anatomic specimens were processed in batches for each series to ensure color balance throughout. Standard color techniques were used but had to be "customized" to ensure uniformity.

The CT illustrations were printed directly from tone reversal images (negatives) made directly from the CT data by a laser printer in an effort to avoid loss of detail in the photographic process. Negatives for MR images were made on TMAX 100 Kodak film and printed with special attention to balancing the wide gray scales present.

Negatives for the specimen radiographs and all of the conventional trispiral tomograms were made on a specially designed copy stand previously designed and reported by the senior author (23). This stand (Figure 1.14) provides for a light source every square inch over a 14″ × 17″ viewing surface. Each light is wired from its own rheostat to change its brightness level from zero to maximum illumination. Light control of this order allows the operator to develop both contrast and interest areas within the image, and to select an optimal density level over the entire film. The films were photographed with a large-format (4″ × 5″) Calumet camera with a 6″ (150 mm) f4.8 lens by Ibex. Polaroid 55 professional film was used, which produced a polaroid print for immediate direct inspection and a high quality negative to be retained for final printing.

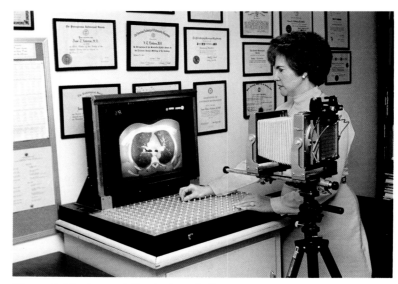

Figure 1.14. A specially designed illuminator (23) was used to copy the specimen radiographs and trispiral tomograms. Ms. Durizch is shown adjusting the rheostat panel to develop the desired light balance across the film.

## Labeling

Labeling of anatomic structures was done by the senior author (Littleton) and reexamined by Dr. Lim, Department of Structural and Cellular Biology. The terminology employed is based on the most recent edition of the Nomina Anatomica (N.A.) (6th ed., 1989, Churchill Livingstone). A common and accepted practice is to translate the Latin terms into the vernacular of the particular country concerned; the N.A. is followed, but using the anglicized form. In a few instances, terms unique to the parlance of the clinical radiologist have been included.

Variance from standard terminology occurred only when dictated by common usage among radiologists. A master anatomic list was compiled and alphabetized by anatomic categories: arteries, bronchi, fat, fissures, lobes, etc. A small capital or lower case letter precedes the number of the structure within each anatomic category: for examples, aorta = A28, spinal cord = n5. Each anatomic structure has only one number key throughout the entire atlas. This simplified, repetitive labeling technique allows the frequent user to identify the structure quickly just from seeing the label without additional reference to the anatomic key accompanying each section. The master anatomic key list is found on page xi. Identification of anatomic parts was taken heavily from available body section atlases cited in the preface, but frequent reference was made to standard text books of anatomy (1, 9, 13, 15, 16, 19, 31, 32–34, 37, 40–42, 48).

Following the master anatomic list, the muscles, which are so clearly seen in MR images, are further broken out for reader convenience by functional groups. These groups, for example, Muscles Connecting Upper Extremity to Vertebral Column, Muscles of the Neck, etc., are arranged from superior to inferior.

### Fascial Labeling

The interconnecting, ramifying, and contiguous visceral and muscular fascial planes of the neck, thorax, and abdomen are extremely difficult to identify in fresh anatomic specimens. A working knowledge of these structures, however, is essential to the understanding of the flow of tissue fluid, blood, infected material, and gas in the neck, mediastinum, and muscle planes of the thorax. Fascial structures in the neck and mediastinum are conveniently avoided in many sectional and general anatomy texts and would have been from this atlas were it not for the suggestions of Dr. Shusuke Sone (personal communication). Terminology varies from text to text and has been changing with newer references.

Anatomic detail of the planes should be treated in a separate communication, but after literature review it was elected to adapt the most simple nomenclature of Marchand (29) and Heitzman (18), and to label those fascial structures which could be identified with reasonable certainty in the axial, sagittal, and coronal sections. Labeling of fascial structures was omitted in the LPO and RPO projections.

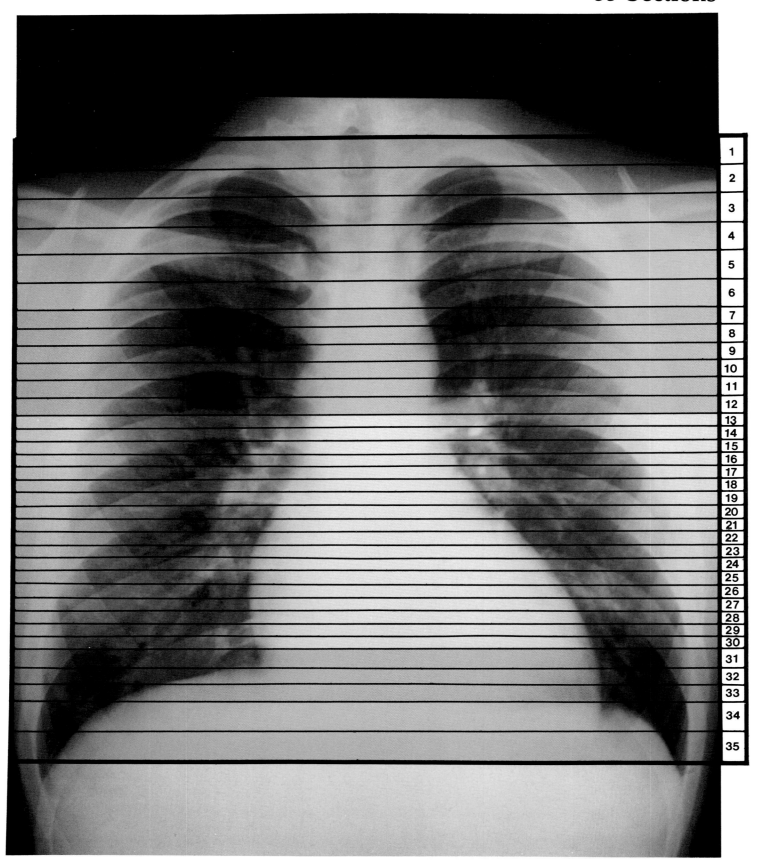

1
2
3
4
5
6
7
8
9
10
11
12
13
14
15
16
17
18
19
20
21
22
23
24
25
26
27
28
29
30
31
32
33
34
35

# Chapter 2

# Axial Plane

## The Anatomic Specimen

The cadaver for the axial sections was that of a 56-year-old white male who experienced a precipitous cardiac death. The subject had endured several years of symptomatic coronary heart disease and a pacemaker had been implanted beneath the skin below the lateral third of the left clavicle. Electrodes were present in the right ventricle. The pacemaker and its electrodes were removed after the body was frozen, leaving a defect in the anterior chest wall in superior Sections 1–3.

The lungs were reexpanded 2 hours after death. A routine chest radiograph disclosed mild cardiomegaly, but otherwise the chest and its contents appeared to be essentially normal for the patient's stated age. The cadaver was placed in the freezer in a neutral supine position and remained in the freezer for 38 months prior to isolation of the chest, removal of the pacemaker, and sectioning. As described in Chapter 1, moderate intravascular postmortem sludging and fluid settling into the dependent posterior lung segments occurred. A small amount of postmortem gas appeared in the cardiac chambers. Except for these minor pathologic and postmortem changes, the specimen was considered representative of a normal thorax.

## Section Plan

The plan for the axial sections (facing page), as shown in colored strips over a routine anteroposterior radiograph of the chest, represents a precise charting of the position of the gross sections. The sections are color-coded according to section thickness: **blue**, 10 mm; **pink**, 5 mm; **yellow**, 3 mm. The level and thickness of section is indicated throughout the chapter by the color bar overlying the miniature reference key at the top right of each double-page layout.

The sectioning process began at the level of C7 and proceeded inferiorly to the level of the vertebral body of T11 and the superior articular processes of T12. Only a very small amount of the inferior portion of each lower lobe of the lungs remained in the most inferior section. The majority of the structures at this level were intraabdominal.

Sections 1–6 are each 10 mm thick and extend from the inferior margin of C7 inferiorly to the body of T3. The next six inferior sections (Sections 7–12) are 5 mm thick and extend to the level of the transverse processes of T5 and the distal

trachea at the level of the entry of the azygos vein into the superior vena cava. Sections 13–30, with the exception of Section 13, are all approximately 3 mm thick to record detail of the major airways and the cardiac, mediastinal, and hilar structures. These sections conclude at the level of the body of T8. A 2-mm-thick section was attempted at Section 13. The sawing procedure went well, but this thin slice was too delicate to clean properly without tissue distortion. This problem can be seen in the specimen radiograph, Section 13. Sections 31–33 are again 5 mm thick extending inferiorly into the body of T10. Sections 34 and 35 are each 10 mm thick and end at the level of the inferior portion of the body of T11.

When nongeometric anatomic objects are cut, it is most probable that slightly asymmetric structures will result in a significant difference in the images of the same structures on opposite sides of the thorax. For example, it will be noted throughout this axial series that lateral skeletal structures, such as the scapulae, and some muscle groups will be visualized in different profiles. This problem could not be avoided by alignment at the time of sawing, but hopefully clarification is accomplished by labeling both sides when the structure profiles differ.

The major and minor fissures exhibit "skip" areas in some sections when the sigmoid course of the fissures coincides with the section plane.

## Orientation of Illustrations

All axial sections are viewed as though seen from below. Right-hand anatomic structures will appear on the reader's left, and left-hand anatomic structures will be on the reader's right. Two CT images are presented at each level, one with mediastinum and the other with lung window protocols. Axial chest CT is routinely performed with the arms raised alongside the head to reduce artifacts from the humeri. To be adaptable to many imaging methods (e.g., MR) the arms of the cadaver for the axial sections were not raised. Shoulder structures are presented in the standard anatomic position.

## Anatomic–Radiographic Correlation

The CT images, done prior to gross sectioning, were not carried superiorly to the level of C7; hence, there is no CT mediastinal match for Section 1. The lung fields were first encountered in Section 3; hence, there are no CT lung images for Sections 1 and 2.

The axial dimension is the latest addition to sectional imaging, occasioned by the present physical design of data acquisition in CT systems. Some anatomic structures and their relationships to adjacent structures can be seen best in axial projection. The CT images have the added advantage of improved soft tissue discrimination, which renders the images in this chapter quite different from the clinical images of trispiral tomography displayed in the coronal, sagittal, and oblique projections in Chapters 3–6. In addition, the viewer of CT images has the capability to computer manipulate the digital data to conform to the clinical problem of soft tissue, bone, etc.

High-resolution CT (HRCT) has made it possible to investigate the lung parenchyma in sufficient detail to evaluate interstitial pathologic processes with an increasing degree of reliability. In the axial sections and throughout the atlas, the interstitial architecture of the lungs is so well seen in the specimen radiographs that these images may serve as a radiographic base to normal interstitial structure, even within the pulmonary lobule.

Many authors have responded to the demand for improved anatomic reference in the interpretation of clinical images (CT, MR) in the axial plane. Twenty-two atlases of thoracic sectional anatomy in the axial projection, many with clinical correlations, have appeared since 1977, and there may be others which we have failed to identify (2–8, 10–12, 14, 17, 20–22, 27, 28, 38, 39, 43, 44, 47). It would be redundant to review all of the details of axial anatomic–radiographic correlations in this atlas. The interested reader is referred to the cited references.

### Bronchial Structures

The bronchopulmonary system consists of a complex orientation of tubes of decreasing diameter as they progress from the trachea to peripheral segmental bronchi. With few exceptions, the major and segmental bronchi are oriented in a near vertical anatomic plane and throughout the axial sections will be seen only as circular or ovoid cut surfaces rather than as longitudinal, tubular segments such as are presented in the trispiral tomographic images. All of the mainstem and segmental bronchi are seen as cross-section images at least once in axial Sections 14–26 (see Table 2.1). The trachea (T1) which is seen in cross section from C7 to the carina (T2) in Sections 1–14 is not charted in Table 2.1.

Only a few bronchi that are charted in Table 2.1 are seen in profile in the axial series to permit the optimal study of these bronchial segments in longitudinal dimension. The anatomic orientation of these bronchial segments is near parallel to the axial plane so that there are no "skip areas" between contiguous sections. Clinical assessment of these bronchi can be made with a high degree of assurance in the CT images. Bronchial segments that are optimally visualized in profile on CT are identified in Table 2.2 which follows.

**Right Lung**: In the axial projection, the posterior bronchus to the RUL (B4) is well seen in Section 14. The right mainstem (B1) and right upper lobe bronchus (B2) can be seen in Section 16, with the anterior (B5) and posterior (B4) divisions of the RUL seen well in Sections 16 and 17. The anterior bronchus to the RUL (B5) is again well seen in Section 18. The major extent of the superior bronchus to the RLL (B11) is seen especially well in CT images in Sections 22 and 23. The proximal portions of the lateral (B8) and medial (B9) divisions of the RML bronchus are seen in Section 23.

The lateral basal segmental bronchus to the RLL (B14) is seen well in Section 24, while both the medial (B12) and lateral basal (B14) divisions to the RLL are well seen in Section 25. The posterior basal bronchus to the RLL (B15) appears in Sections 25 and 26.

**Left Lung**: The left bronchial system is only sparsely seen in profile in axial projection. The anterior bronchus to the LUL (B19) is well visualized in Sections 16 and 17. The left mainstem bronchus (B16) is nicely outlined in Section 19. The superior bronchus to the LLL (B24) is seen in profile in Section 20, while

the lingular bronchus (B20) with its superior (B21) and inferior (B22) divisions are seen to excellent advantage in Section 21. The lateral basal segment to the LLL (B26) is seen in Section 25.

When the axial table (Table 2.2) is compared to Table 2.3; a composite of the tables from the coronal, sagittal, left posterior oblique, and right posterior oblique projections (Tables 3.1, 4.1, 5.1, and 6.1), the bronchial imaging potential of multiplanar projections becomes obvious. It is apparent that the axial plane does not provide a complete clinical assessment of the bronchial system. No one projection will visualize all of the bronchial segments, however, in the authors' experience using trispiral tomography, the coronal projection permits the most comprehensive review of the tracheobronchial tree with the smallest number of sections (Table 3.1).

### Fissures

The fissures are well seen in the axial anatomic sections and CT images at lung window settings but are not as well seen in the specimen radiographs. The thin pleural sheets comprising the fissures can be seen on the surface of the gross anatomic sections, but for the most part, they are oblique to the central ray of the x-ray tube in the specimen radiographs and hence do not cast a discrete shadow. In the CT images, the fissures are more completely seen because of the 360° rotation of the x-ray tube. Viewing the fissures of both lungs in sequence in CT images from Section 11 through Section 32, the reader can lucidly reconstruct the three-dimensional anatomy of the major fissures (F1 and F2) of both lungs and of the horizontal fissure (F3) of the right lung.

### Neural Structures

The spinal cord, spinal dura, and frequent spinal nerves are seen throughout the axial series in the gross anatomic specimens and specimen radiographs.

Segments of the brachial plexus (n1) are seen in Anatomic Sections 3 and 5–8 on both sides and are seen with assurance in both the CT (mediastinal technique) and MR images at these same levels. For more detailed textual description of the gross anatomy and for CT and MR images of the brachial plexus, the reader is referred to Webb (45) and Wechsler (46).

The vagus nerve (n10), confirmed by pathologic section, is seen in the mediastinum in Anatomic Section 7 and the recurrent laryngeal nerve (n4) adjacent to the esophagus in Anatomic Section 8. Neither is seen with assurance in the corresponding CT and MR images.

### Vascular Structures

Vascular structures as seen in axial projection correlate well to the clinical images of CT (mediastinal technique) and MR. The pulmonary arteries and veins are well visualized in the CT images at lung settings. Special attention is drawn to the azygos arch (V7) in Anatomic Section 13.

### Other Structures

The esophagus (E1) is seen in cross–section throughout the series. The thoracic duct (D1), identified only in the axial sections of this atlas, is seen in most of the anatomic specimen Sections 12–27. Fascial planes of significance in the mediastinum are seen in Sections 4, 15–18, 20, 26, 30, and 35.

### Magnetic Resonance Images

The MR images made from a size-related living volunteer correlate almost precisely with the cadaver images: the anatomic specimens, specimen radiographs, and CT. The muscle groups are outlined in clear detail throughout and correlate exactly to the related anatomic specimen and the specimen radiograph. Supraclavicular and axillary structures, including segments of the brachial plexus, are seen in Sections 5–8. The azygos arch (V7) is clearly seen in MR Section 12, and the internal thoracic vessels in MR Sections 9–29. The mediastinal vessels and intracardiac chambers are clearly seen throughout. Skeletal images are poorly seen, an inherent limitation of MR.

## Table 2.1
### Index to Bronchial Imaging Axial Projection
Specimen Radiograph and Computed Tomogram

Right  Bronchus  Left

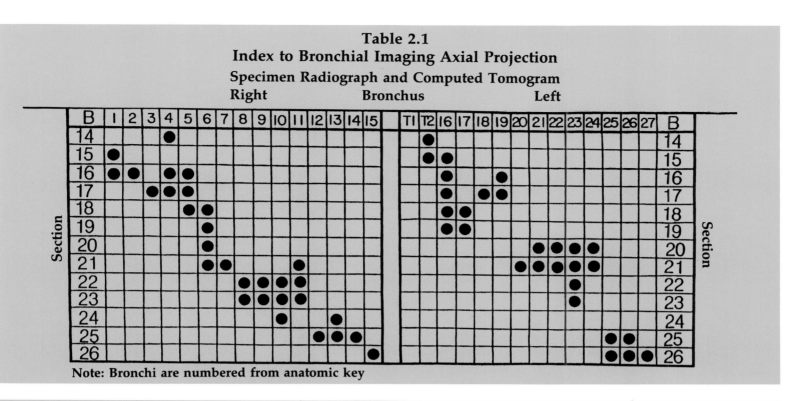

Note: Bronchi are numbered from anatomic key

## Table 2.2
### Index to Optimal Bronchial Imaging Axial Projection
Specimen Radiograph and Computed Tomogram

Right  Bronchus  Left

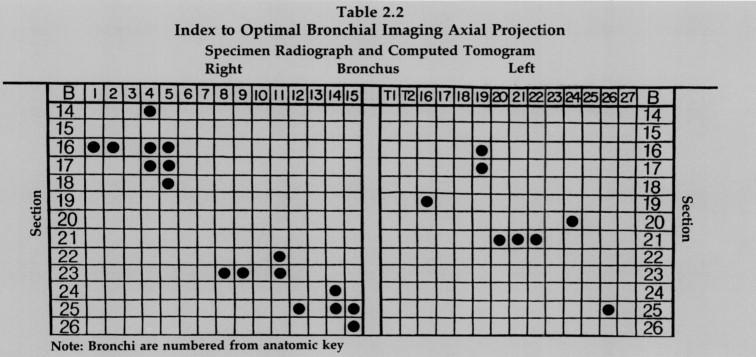

Note: Bronchi are numbered from anatomic key

## Table 2.3
### Bronchial Imaging Coronal Sagittal Oblique Projections
Specimen Radiograph and Trispiral Tomogram

Right  Bronchus  Left

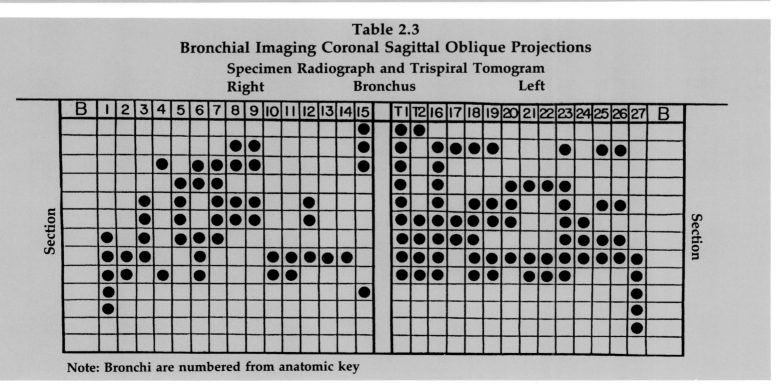

Note: Bronchi are numbered from anatomic key

# Axial

## Section 1

### Anatomic Key

**Arteries**
A 37 carotid, common, left
A 38 carotid, common, right
A 61 thyroid, inferior, left
A 65 vertebral, left

**Esophagus**
E 1 esophagus

**Glands**
G 4 thyroid

**Muscles**
M 7 coracobrachialis
M 8 deltoid
M 14 iliocostalis, thoracis
M 15 infraspinatus
M 23 levator scapulae
M 28 longus colli
M 31 pectoralis, major
M 33 platysma
M 36 rhomboid, minor
M 38 scalene, anterior
M 39 scalene, middle
M 41 semispinalis, capitis
M 42 semispinalis, cervicis
M 44 serratus, anterior
M 46 serratus, posterior superior
M 50 splenius, capitis
M 51 splenius, cervicis
M 52 sternocleidomastoid
M 53 sternohyoid
M 54 sternothyroid
M 55 subclavius
M 57 subscapularis
M 58 supraspinatus
M 61 transversospinalis
M 63 trapezius

**Neural Structures**
n 5 spinal cord
n 6 spinal dura

**Skeletal Structures**
S 1 clavicle
S 2 humerus
S 9 rib, tubercle (1st)
S 10 scapula (superior margin)
S 12 scapula, coracoid process
S 13 scapula, glenoid cavity
S 15 scapula, spine
S 20 intervertebral disc (C7–T1)
S 21 vertebra, body (C7)
S 24 vertebra, intervertebral foramen
(C7–T1)
S 25 vertebra, lamina (T1)
S 27 vertebra, spinous process (T1)
S 29 vertebra, transverse process
(T1)

**Trachea**
T 1 trachea

**Veins**
V 10 cephalic, left
V 11 cephalic, right
V 22 jugular, anterior, left
V 23 jugular, anterior, right
V 26 jugular, internal, left
V 27 jugular, internal, right
V 30 subclavian, left
V 33 suprascapular, left
V 37 transverse cervical, right
V 42 vertebral, left

**Nonanatomic**
* antemortem trauma
† pacemaker site

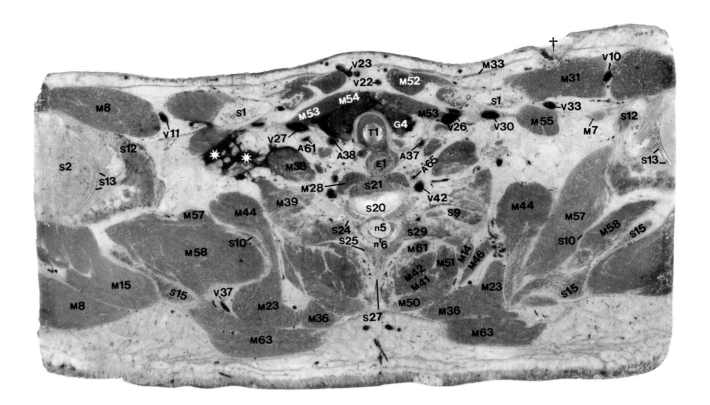

**Anatomic Specimen**

**Magnetic Resonance**

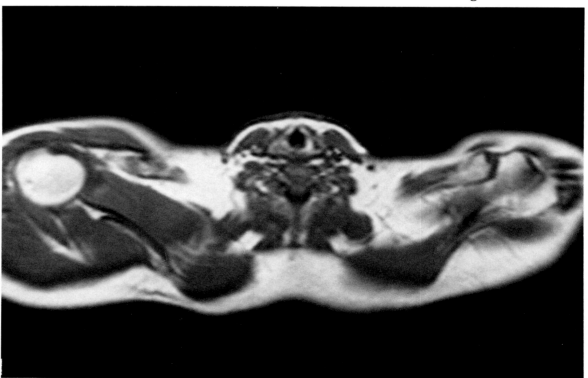

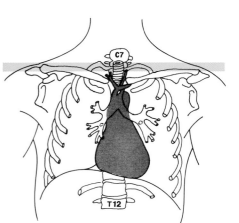

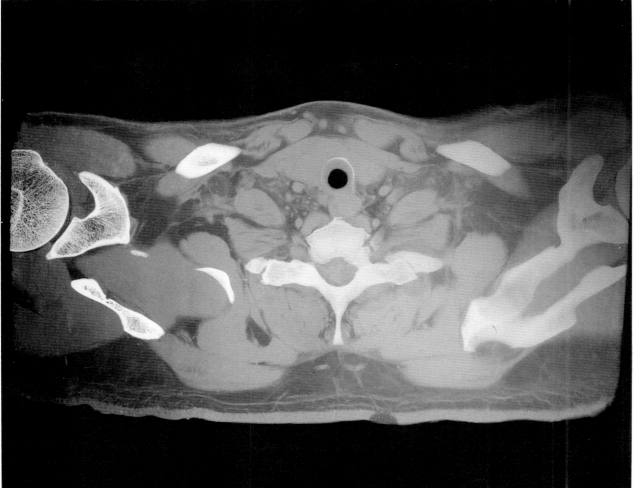

**Specimen Radiograph**

No
Computed Tomogram
Mediastinum Technique
made

No
Computed Tomogram
Lung Technique
made

# Axial

## Section 2

### Anatomic Key

**Arteries**
A 23 axillary, left
A 38 carotid, common, right
A 49 subclavian, left
A 51 subscapular, left
A 66 vertebral, right

**Esophagus**
E 1 esophagus

**Muscles**
M 7 coracobrachialis
M 8 deltoid
M 14 iliocostalis, thoracis
M 15 infraspinatus
M 23 levator scapulae
M 28 longus colli
M 31 pectoralis, major
M 32 pectoralis, minor
M 33 platysma
M 35 rhomboid, major
M 36 rhomboid, minor
M 38 scalene, anterior
M 39 scalene, middle
M 41 semispinalis, capitis
M 42 semispinalis, cervicis
M 44 serratus, anterior
M 46 serratus, posterior superior
M 48 spinalis, cervicis
M 50 splenius, capitis
M 51 splenius, cervicis
M 52 sternocleidomastoid
M 53 sternohyoid
M 54 sternothyroid
M 55 subclavius
M 57 subscapularis
M 58 supraspinatus
M 61 transversospinalis
M 63 trapezius

**Neural Structures**
n 5 spinal cord
n 6 spinal dura
n 7 spinal nerve

**Skeletal Structures**
S 1 clavicle
S 2 humerus
S 8 rib, head and neck (1st)
S 9 rib, tubercle (1st)
S 10 scapula (superior margin)
S 12 scapula, coracoid process
S 13 scapula, glenoid cavity
S 14 scapula, neck
S 15 scapula, spine
S 21 vertebra, body (T1)
S 22 vertebra, costal facet (T1)
S 25 vertebra, lamina (T1)
S 27 vertebra, spinous process (T1)
S 29 vertebra, transverse process (T1)

**Trachea**
T 1 trachea

**Veins**
V 11 cephalic, right
V 23 jugular, anterior, right
V 24 jugular, external, left
V 26 jugular, internal, left
V 27 jugular, internal, right
V 30 subclavian, left
V 33 suprascapular, left
V 35 thyroid, inferior, right
V 43 vertebral, right

**Nonanatomic**
* antemortem trauma
† pacemaker site

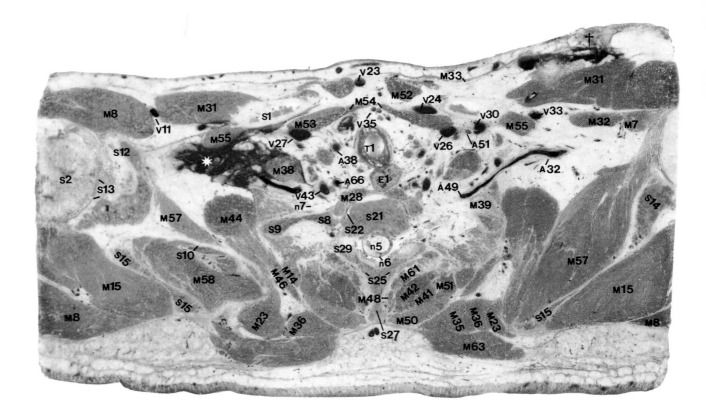

**Anatomic Specimen**

**Magnetic Resonance**

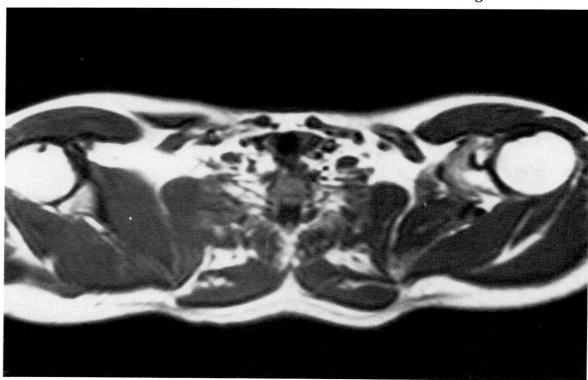

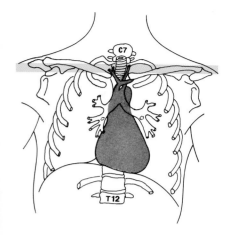

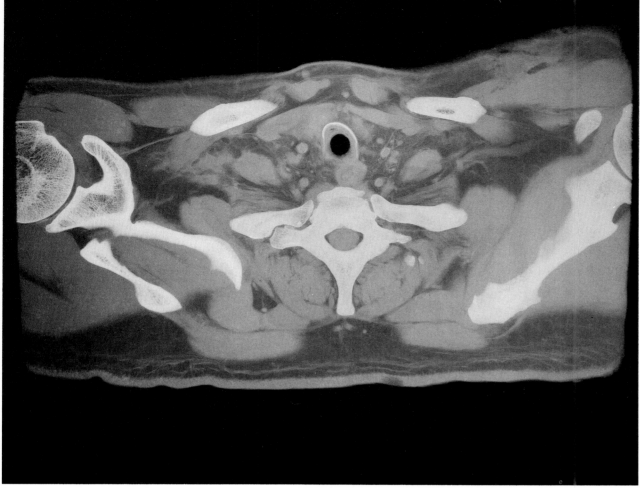

**Specimen Radiograph**

**Computed Tomogram—Mediastinum Technique**

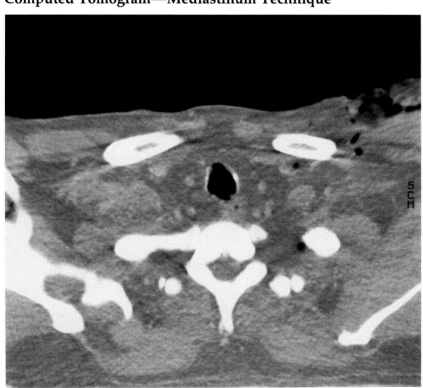

No
Computed Tomogram
Lung Technique
made

# Axial

## Section 3

### Anatomic Key

**Arteries**

A 32 axillary, left
A 37 carotid, common, left
A 38 carotid, common, right
A 49 subclavian, left
A 50 subclavian, right
A 62 thyroid, inferior, right
A 65 vertebral, left
A 66 vertebral, right

**Esophagus**

E 1 esophagus

**Fat**

f 1 axillary

**Lobes**

L 1 right upper lobe (RUL)
L 2 RUL, apical segment
L 14 left upper lobe (LUL)
L 15 LUL, apicoposterior segment

**Muscles**

M 7 coracobrachialis
M 8 deltoid
M 14 iliocostalis, thoracis
M 15 infraspinatus
M 16 intercostal
M 26 longissimus, cervicis
M 28 longus colli
M 31 pectoralis, major
M 32 pectoralis, minor
M 33 platysma
M 35 rhomboid, major
M 36 rhomboid, minor
M 38 scalene, anterior
M 42 semispinalis, cervicis
M 44 serratus, anterior
M 46 serratus, posterior superior
M 51 splenius, cervicis
M 52 sternocleidomastoid
M 53 sternohyoid
M 54 sternothyroid
M 57 subscapularis
M 61 transversospinalis
M 63 trapezius

**Neural Structures**

n 1 brachial plexus
n 5 spinal cord
n 6 spinal dura

**Nodes**

N 7 mediastinal, posterior

**Pleural Structures**

P 6 visceral

**Skeletal Structures**

S 1 clavicle
S 2 humerus
S 6 rib, body, posterior (1st)
S 9 rib, tubercle (2nd)
S 10 scapula (medial margin)
S 12 scapula, coracoid process
S 13 scapula, glenoid cavity
S 14 scapula, neck
S 21 vertebra, body (T1)
S 27 vertebra, spinous process (T1)
S 29 vertebra, transverse process (T2)

**Trachea**

T 1 trachea

**Veins**

V 23 jugular, anterior, right
V 26 jugular, internal, left
V 27 jugular, internal, right
V 30 subclavian, left
V 31 subclavian, right
V 33 suprascapular, left
V 35 thyroid, inferior, right
V 42 vertebral, left
V 43 vertebral, right

**Nonanatomic**

* antemortem trauma
† pacemaker site

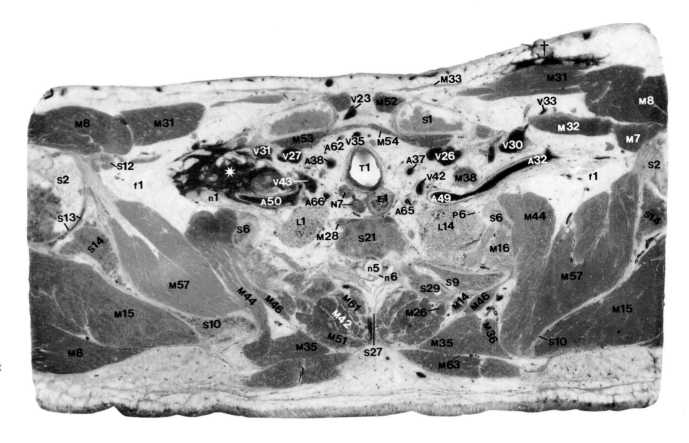

**Anatomic Specimen**

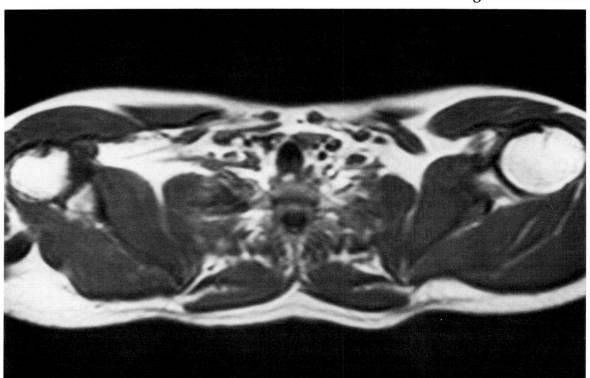

**Magnetic Resonance**

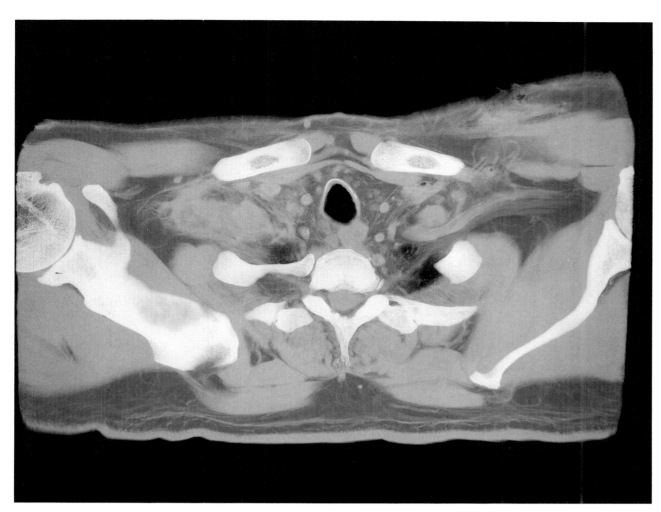

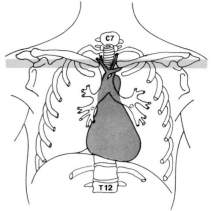

**Specimen Radiograph**

**Computed Tomogram—Mediastinum Technique**

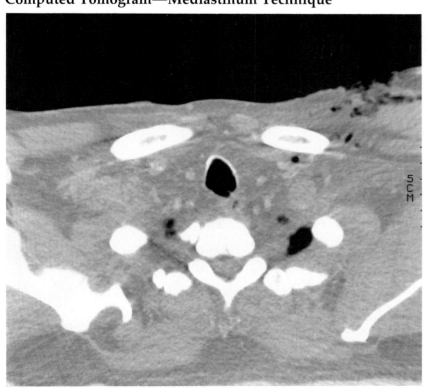

**Computed Tomogram—Lung Technique**

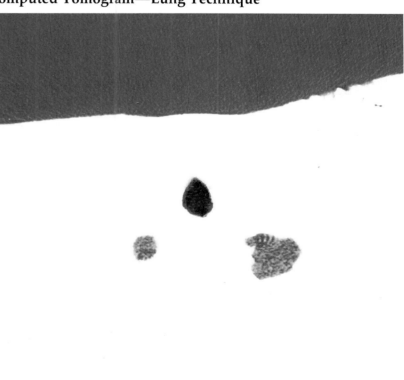

# Axial

## Section 4

## Anatomic Key

### Arteries
A 32 axillary, left
A 33 axillary, right
A 37 carotid, common, left
A 38 carotid, common, right
A 49 subclavian, left
A 50 subclavian, right
A 65 vertebral, left

### Esophagus
E 1 esophagus

### Fascia
F 9 superficial (investing)

### Fat
f 1 axillary

### Lobes
L 2 RUL, apical segment
L 15 LUL, apicoposterior segment

### Muscles
M 7 coracobrachialis
M 8 deltoid
M 14 iliocostalis, thoracis
M 15 infraspinatus
M 16 intercostal
M 31 pectoralis, major
M 32 pectoralis, minor
M 33 platysma
M 35 rhomboid, major
M 42 semispinalis, cervicis
M 44 serratus, anterior
M 46 serratus, posterior superior
M 51 splenius, cervicis
M 54 sternothyroid
M 57 subscapularis
M 61 transversospinalis
M 63 trapezius

### Neural Structures
n 5 spinal cord
n 6 spinal dura

### Nodes
N 7 mediastinal, posterior
N 9 paratracheal

### Pleural Structures
P 2 costal (parietal)
P 6 visceral

### Skeletal Structures
s 1 clavicle
s 2 humerus
s 5 rib, body, lateral (1st)
s 6 rib, body, posterior (2nd)
s 8 rib, head and neck (2nd)
s 9 rib, tubercle (2nd)
s 10 scapula (medial margin)
s 13 scapula, glenoid cavity
s 14 scapula, neck
s 20 intervertebral disc (T2–T3)
s 21 vertebra, body (T2)
s 22 vertebra, costal facet (T2)
s 25 vertebra, lamina (T2)
s 26 vertebra, pedicle (T2)
s 27 vertebra, spinous process (T2)
s 29 vertebra, transverse process (T2)

### Trachea
T 1 trachea

### Veins
V 6 axillary, left and right
V 8 brachiocephalic (innominate), left
V 9 brachiocephalic (innominate), right
V 31 subclavian, right
V 35 thyroid, inferior, right
V 41 venous plexus, vertebral (external)

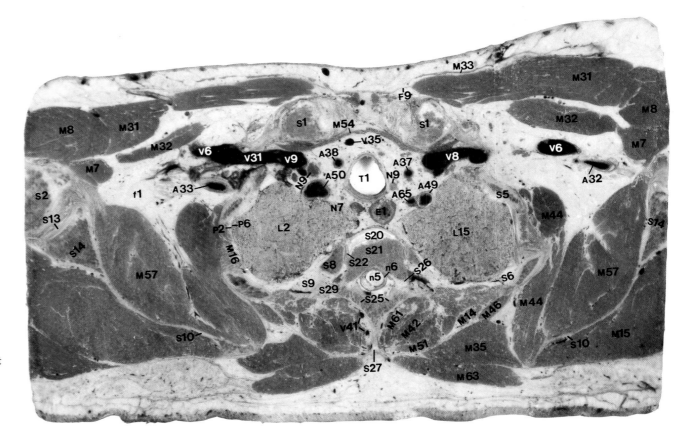

**Anatomic Specimen**

**Magnetic Resonance**

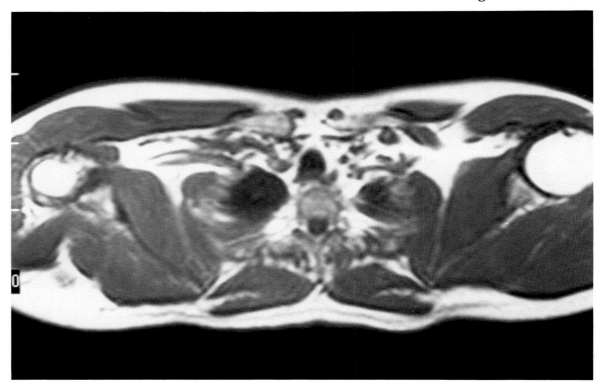

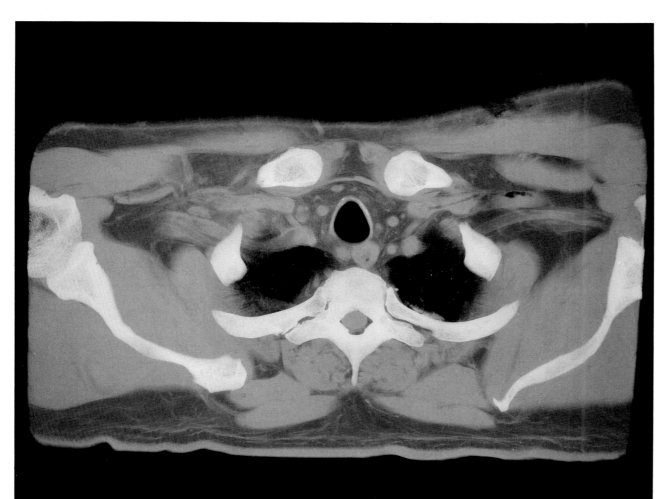

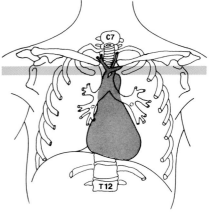

**Specimen Radiograph**

**Computed Tomogram—Mediastinum Technique**

**Computed Tomogram—Lung Technique**

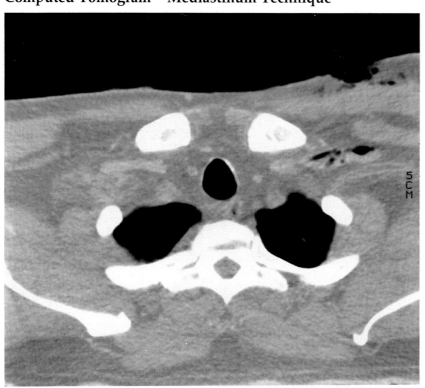

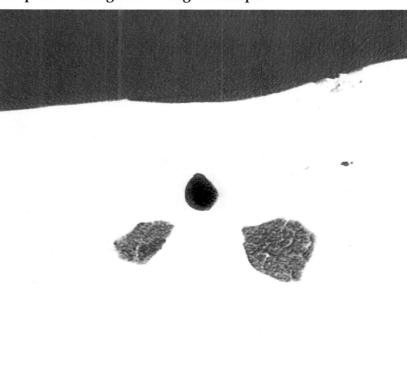

# Axial

## Section 5

### Anatomic Key

**Arteries**

A 32 axillary, left
A 33 axillary, right
A 37 carotid, common, left
A 38 carotid, common, right
A 47 internal thoracic (mammary), left
A 48 internal thoracic (mammary), right
A 49 subclavian, left
A 50 subclavian, right

**Esophagus**

E 1 esophagus

**Fat**

f 1 axillary

**Lobes**

L 2 RUL, apical segment
L 15 LUL, apicoposterior segment

**Muscles**

M 6 biceps brachii, short head
M 7 coracobrachialis
M 8 deltoid
M 14 iliocostalis, thoracis
M 15 infraspinatus
M 16 intercostal
M 31 pectoralis, major
M 32 pectoralis, minor
M 33 platysma
M 35 rhomboid, major
M 42 semispinalis, cervicis
M 44 serratus, anterior
M 46 serratus, posterior superior
M 51 splenius, cervicis
M 54 sternothyroid
M 57 subscapularis
M 61 transversospinalis
M 63 trapezius

**Neural Structures**

n 1 brachial plexus
n 5 spinal cord
n 6 spinal dura

**Nodes**

N 6 mediastinal, anterior
N 9 paratracheal

**Skeletal Structures**

S 1 clavicle
S 2 humerus
S 4 rib, body, anterior (1st)
S 6 rib, body, posterior (2nd)
S 7 rib, costal cartilage (1st)
S 9 rib, tubercle (3rd)
S 10 scapula (medial margin)
S 18 sternum, manubrium
S 21 vertebra, body (T2)
S 27 vertebra, spinous process (T2)
S 29 vertebra, transverse process (T3)

**Trachea**

T 1 trachea

**Veins**

V 6 axillary, left and right
V 8 brachiocephalic (innominate), left
V 9 brachiocephalic (innominate), right
V 35 thyroid, inferior, right
V 41 venous plexus, vertebral (external)

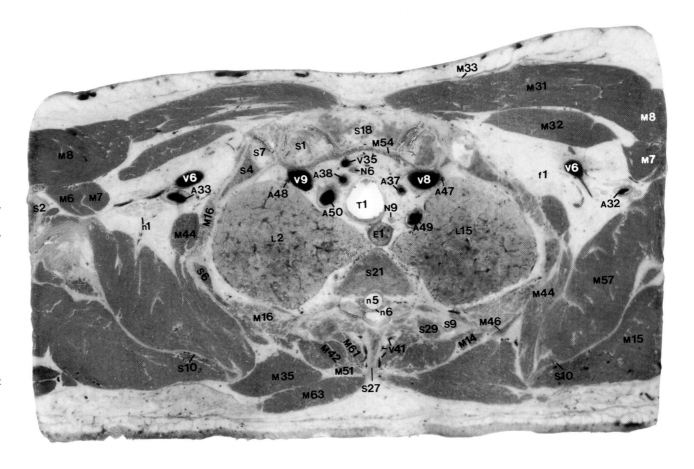

**Anatomic Specimen**

**Magnetic Resonance**

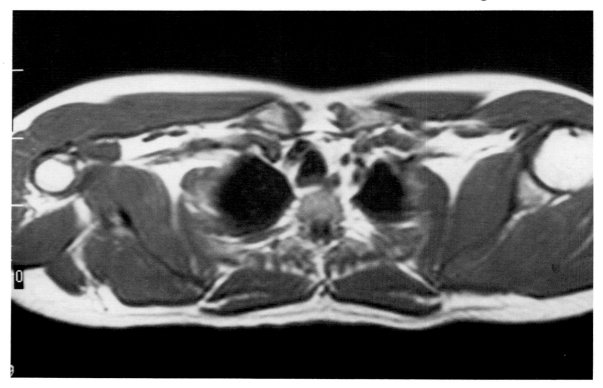

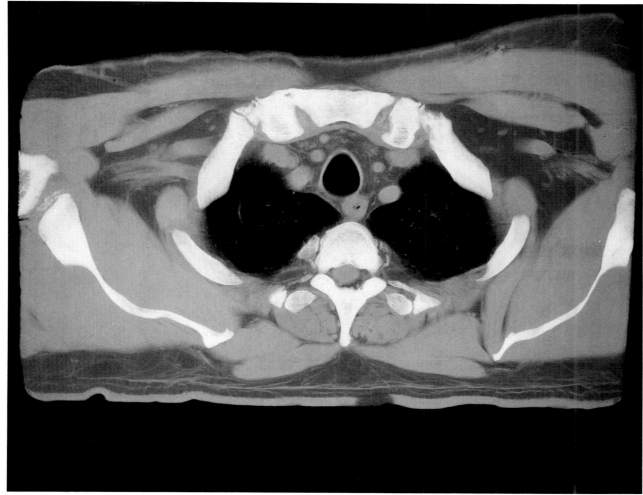

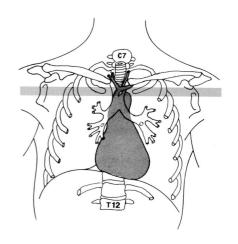

Specimen Radiograph

Computed Tomogram—Mediastinum Technique

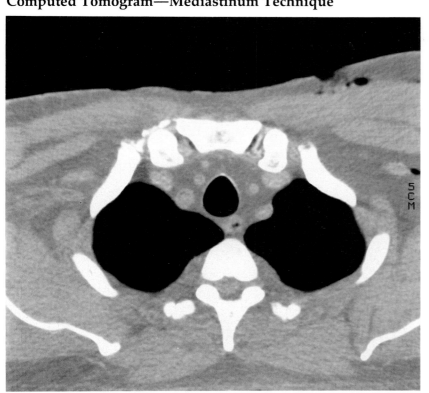

Computed Tomogram—Lung Technique

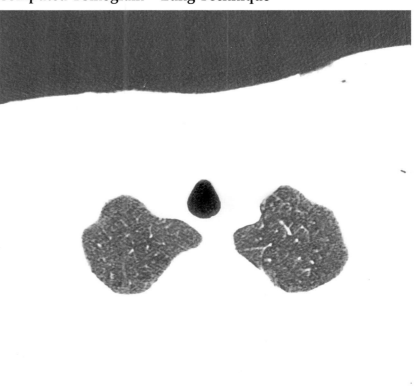

# Axial

## Section 6

### Anatomic Key

**Arteries**

A 32 axillary, left
A 33 axillary, right
A 37 carotid, common, left
A 38 carotid, common, right
A 47 internal thoracic (mammary), left
A 48 internal thoracic (mammary), right
A 49 subclavian, left
A 50 subclavian, right

**Esophagus**

E 1 esophagus

**Fat**

f 1 axillary

**Lobes**

L 2 RUL, apical segment
L 15 LUL, apicoposterior segment

**Muscles**

M 6 biceps brachii, short head
M 7 coracobrachialis
M 8 deltoid
M 14 iliocostalis, thoracis
M 15 infraspinatus
M 16 intercostal
M 31 pectoralis, major
M 32 pectoralis, minor
M 33 platysma
M 35 rhomboid, major
M 42 semispinalis, cervicis
M 44 serratus, anterior
M 46 serratus, posterior superior
M 51 splenius, cervicis
M 54 sternothyroid
M 57 subscapularis
M 61 transversospinalis
M 63 trapezius

**Neural Structures**

n 1 brachial plexus
n 5 spinal cord
n 6 spinal dura

**Nodes**

N 6 mediastinal, anterior
N 9 paratracheal

**Skeletal Structures**

s 1 clavicle
s 5 rib, body, lateral (2nd)
s 7 rib, costal cartilage (1st)
s 8 rib, head and neck (3rd)
s 9 rib, tubercle (3rd)
s 10 scapula (medial margin)
s 18 sternum, manubrium
s 21 vertebra, body (T3)
s 22 vertebra, costal facet (T3)
s 25 vertebra, lamina (T3)
s 27 vertebra, spinous process (T3)
s 29 vertebra, transverse process (T3)

**Trachea**

T 1 trachea

**Veins**

V 6 axillary, left and right
V 8 brachiocephalic (innominate), left
V 9 brachiocephalic (innominate), right
V 35 thyroid, inferior, right
V 41 venous plexus, vertebral (external)

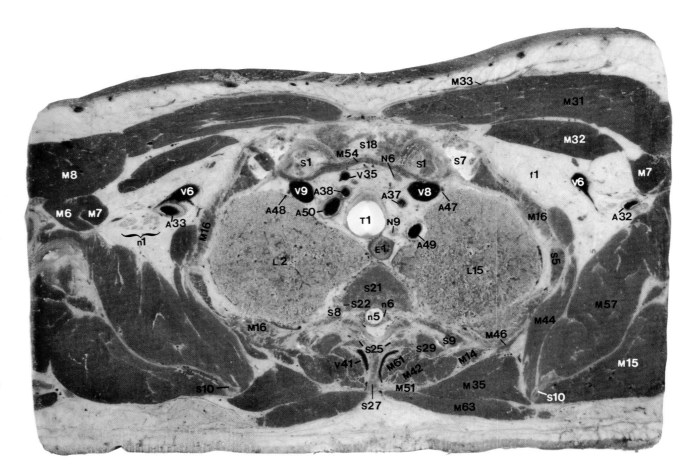

**Anatomic Specimen**

**Magnetic Resonance**

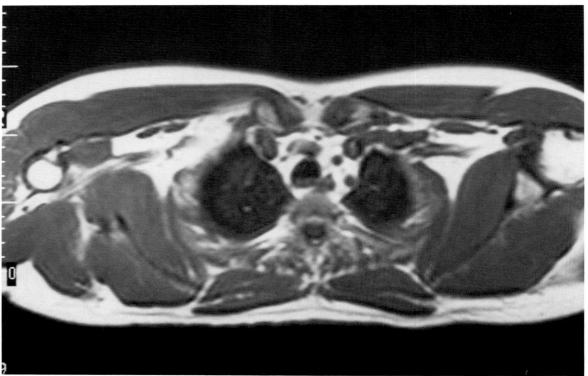

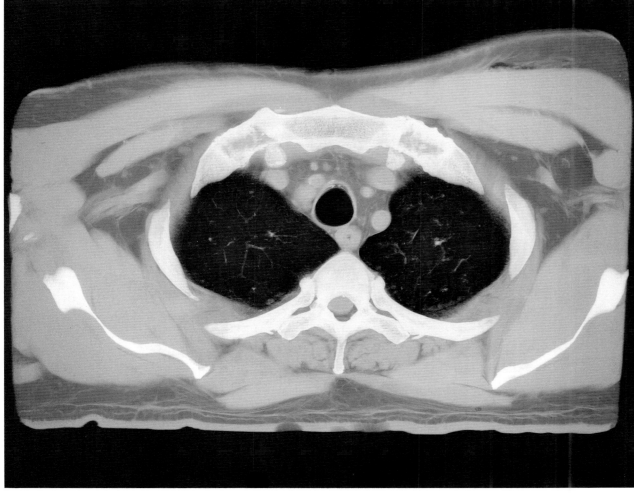

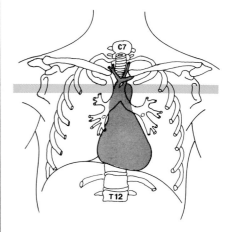

**Specimen Radiograph**

**Computed Tomogram—Mediastinum Technique**

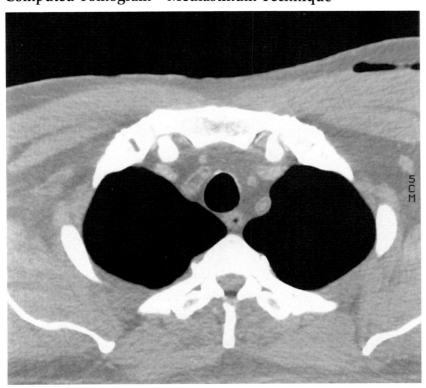

**Computed Tomogram—Lung Technique**

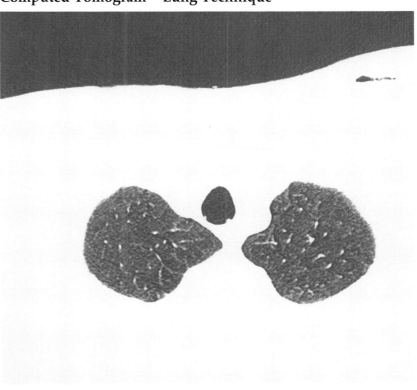

# Axial

## Section 7

## Anatomic Key

### Arteries
A 32 axillary, left
A 33 axillary, right
A 34 brachiocephalic (innominate) trunk
A 37 carotid, common, left
A 47 internal thoracic (mammary), left
A 48 internal thoracic (mammary), right
A 49 subclavian, left

### Esophagus
E 1 esophagus

### Fat
f 1 axillary
f 4 mediastinal

### Lobes
L 2 RUL, apical segment
L 3 RUL, posterior segment
L 15 LUL, apicoposterior segment

### Muscles
M 6 biceps brachii, short head
M 7 coracobrachialis
M 8 deltoid
M 14 iliocostalis, thoracis
M 15 infraspinatus
M 16 intercostal
M 27 longissimus, thoracis
M 31 pectoralis, major
M 32 pectoralis, minor
M 33 platysma
M 35 rhomboid, major
M 44 serratus, anterior
M 54 sternothyroid
M 57 subscapularis
M 60 teres minor
M 61 transversospinalis
M 63 trapezius

### Neural Structures
n 1 brachial plexus
n 5 spinal cord
n 6 spinal dura
n 10 vagus nerve

### Nodes
N 9 paratracheal

### Pleural Structures
P 2 costal (parietal)
P 5 mediastinal (parietal)
P 6 visceral

### Skeletal Structures
S 4 rib, body, anterior (2nd)
S 6 rib, body, posterior (3rd)
S 7 rib, costal cartilage (1st)
S 8 rib, head and neck (4th)
S 10 scapula (lateral margin)
S 18 sternum, manubrium
S 21 vertebra, body (T3)
S 25 vertebra, lamina (T3)
S 26 vertebra, pedicle (T3)
S 27 vertebra, spinous process (T3)

### Trachea
T 1 trachea

### Veins
V 6 axillary, right
V 8 brachiocephalic (innominate), left
V 9 brachiocephalic (innominate), right
V 15 intercostal, highest (superior), right
V 35 thyroid, inferior, right

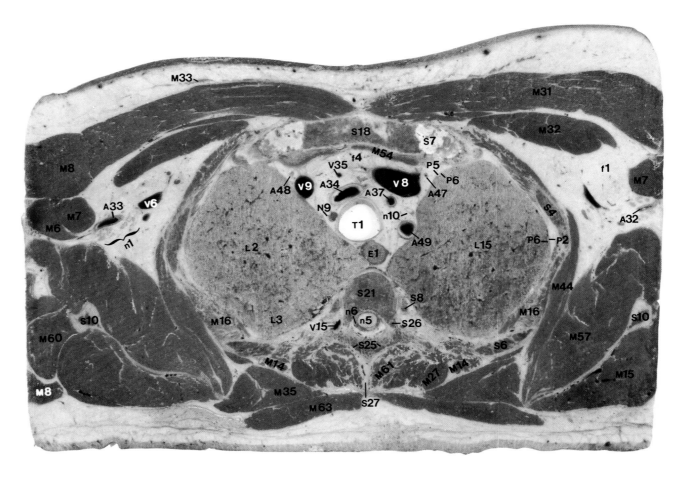

**Anatomic Specimen**

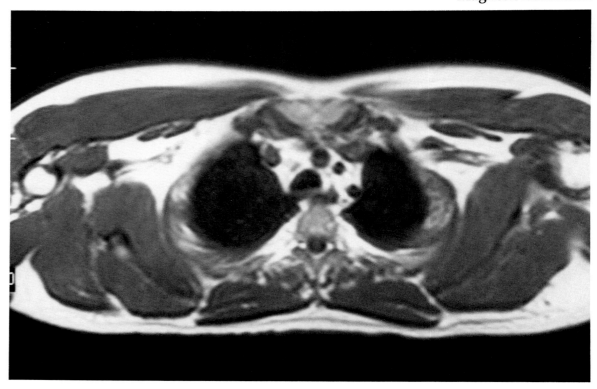

**Magnetic Resonance**

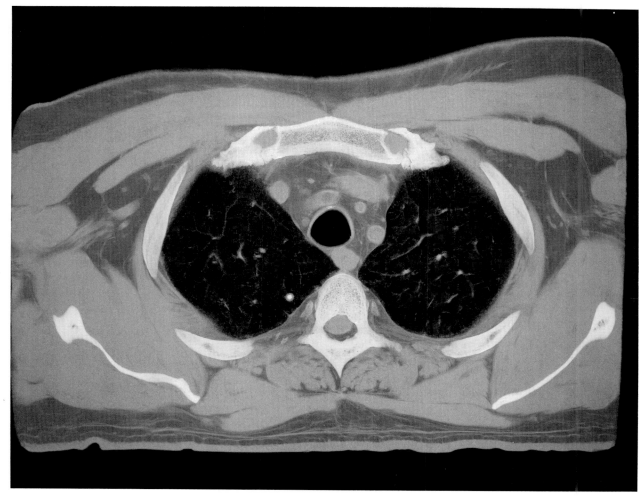

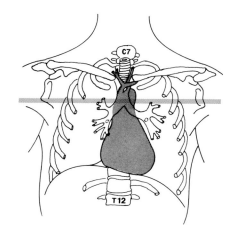

**Specimen Radiograph**

**Computed Tomogram—Mediastinum Technique**

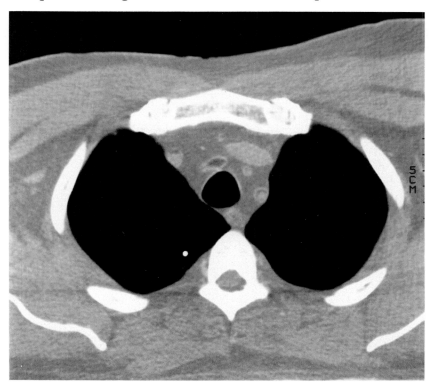

**Computed Tomogram—Lung Technique**

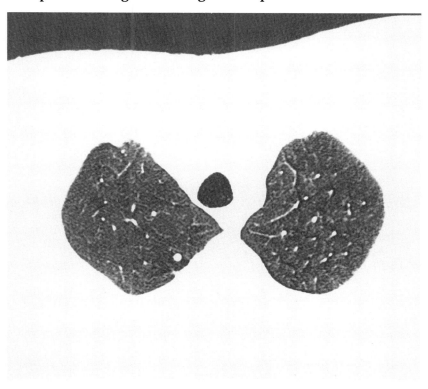

# Axial

## Section 8

## Anatomic Key

### Arteries
A **33** axillary, right
A **34** brachiocephalic (innominate)
   trunk
A **37** carotid, common, left
A **45** intercostal, posterior, left
A **46** intercostal, posterior, right
A **47** internal thoracic (mammary),
   left
A **48** internal thoracic (mammary),
   right
A **49** subclavian, left

### Esophagus
E **1** esophagus

### Fat
f **1** axillary
f **4** mediastinal
f **7** subcutaneous

### Lobes
L **2** RUL, apical segment
L **3** RUL, posterior segment
L **15** LUL, apicoposterior segment

### Muscles
M **7** coracobrachialis
M **8** deltoid
M **14** iliocostalis, thoracis
M **15** infraspinatus
M **16** intercostal
M **27** longissimus, thoracis
M **31** pectoralis, major
M **32** pectoralis, minor
M **35** rhomboid, major
M **44** serratus, anterior
M **49** spinalis, thoracis
M **57** subscapularis
M **60** teres minor
M **61** transversospinalis
M **63** trapezius

### Neural Structures
n **1** brachial plexus
n **4** recurrent laryngeal nerve

### Nodes
N **6** mediastinal, anterior
N **8** parasternal (internal thoracic)
N **9** paratracheal

### Pleural Structures
P **2** costal (parietal)
P **5** mediastinal (parietal)
P **6** visceral

### Skeletal Structures
S **4** rib, body, anterior (2nd)
S **6** rib, body, posterior (3rd)
S **8** rib, head and neck (4th)
S **9** rib, tubercle (4th)
S **10** scapula (medial margin)
S **18** sternum, manubrium
S **21** vertebra, body (T3)
S **22** vertebra, costal facet (T3)
S **23** vertebra, inferior articular
   process (T3)
S **27** vertebra, spinous process (T3)
S **28** vertebra, superior articular
   process (T4)
S **29** vertebra, transverse process
   (T4)

### Trachea
T **1** trachea

### Veins
V **6** axillary, right
V **8** brachiocephalic (innominate),
   left
V **9** brachiocephalic (innominate),
   right
V **14** intercostal, highest (superior),
   left
V **15** intercostal, highest (superior),
   right
V **20** internal thoracic (mammary),
   left
V **21** internal thoracic (mammary),
   right

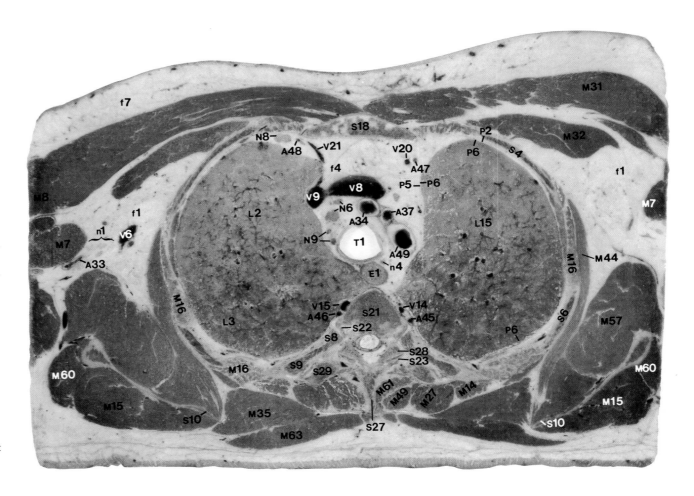

**Anatomic Specimen**

**Magnetic Resonance**

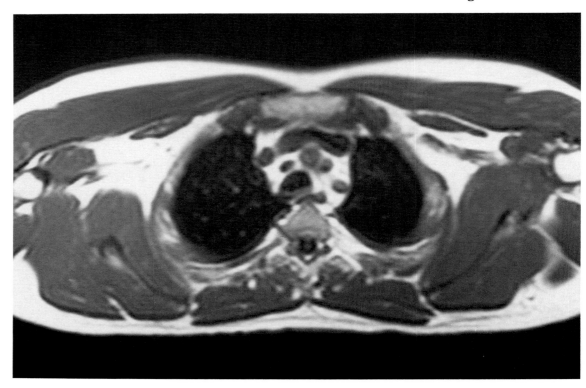

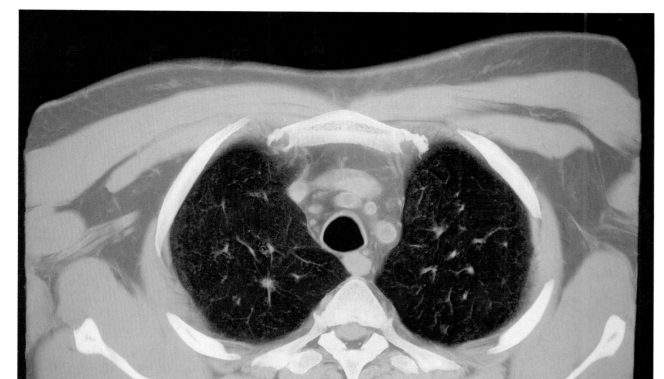

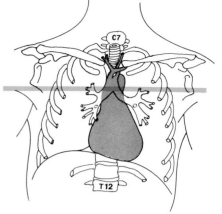

**Specimen Radiograph**

**Computed Tomogram—Mediastinum Technique**

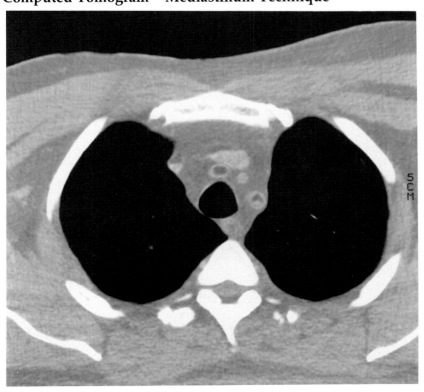

**Computed Tomogram—Lung Technique**

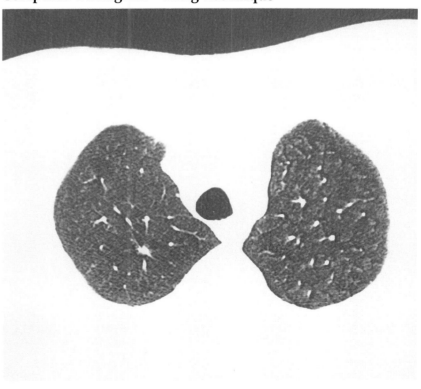

# Axial

## Section 9

## Anatomic Key

### Arteries
A 29 aortic arch
A 33 axillary, right
A 34 brachiocephalic (innominate) trunk
A 37 carotid, common, left
A 45 intercostal, posterior, left
A 47 internal thoracic (mammary), left
A 48 internal thoracic (mammary), right

### Esophagus
E 1 esophagus

### Fat
f 1 axillary
f 4 mediastinal

### Lobes
L 2 RUL, apical segment
L 3 RUL, posterior segment
L 4 RUL, anterior segment
L 15 LUL, apicoposterior segment
L 16 LUL, anterior segment

### Muscles
M 6 biceps brachii, short head
M 7 coracobrachialis
M 8 deltoid
M 14 iliocostalis, thoracis
M 15 infraspinatus
M 16 intercostal
M 27 longissimus, thoracis
M 31 pectoralis, major
M 32 pectoralis, minor
M 35 rhomboid, major
M 43 semispinalis, thoracis
M 44 serratus, anterior
M 49 spinalis, thoracis
M 57 subscapularis
M 60 teres minor
M 61 transversospinalis
M 63 trapezius

### Nodes
N 8 parasternal (internal thoracic)
N 9 paratracheal

### Pleural Structures
P 2 costal (parietal)
P 5 mediastinal (parietal)
P 6 visceral

### Skeletal Structures
S 4 rib, body, anterior (2nd)
S 5 rib, body, lateral (3rd)
S 7 rib, costal cartilage (2nd)
S 8 rib, head and neck (4th)
S 9 rib, tubercle (4th)
S 10 scapula (lateral margin)
S 18 sternum, manubrium
S 20 intervertebral disc (T3–T4)
S 21 vertebra, body (T4)
S 22 vertebra, costal facet (T4)
S 25 vertebra, lamina (T4)
S 26 vertebra, pedicle (T4)
S 27 vertebra, spinous process (T4)
S 29 vertebra, transverse process (T4)

### Trachea
T 1 trachea

### Veins
v 6 axillary, right
v 8 brachiocephalic (innominate), left
v 14 intercostal, highest (superior), left
v 15 intercostal, highest (superior), right
v 18 intercostal, posterior, left
v 20 internal thoracic (mammary), left
v 39 vena cava, superior
v 40 venous plexus, mediastinal

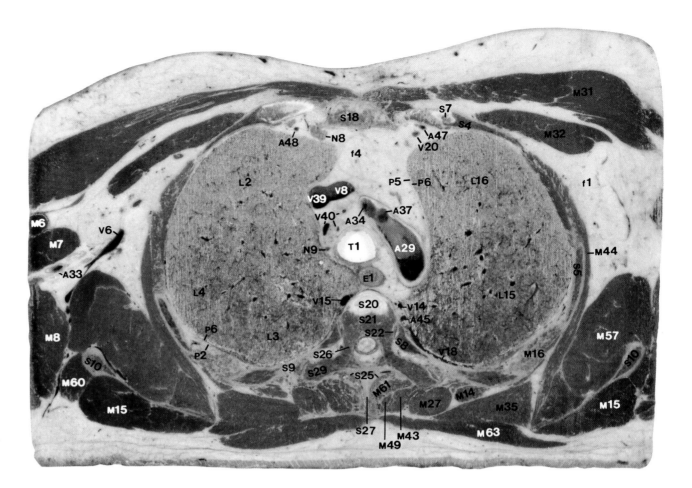

**Anatomic Specimen**

**Magnetic Resonance**

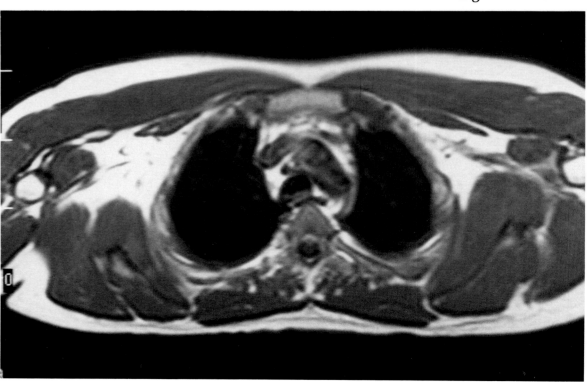

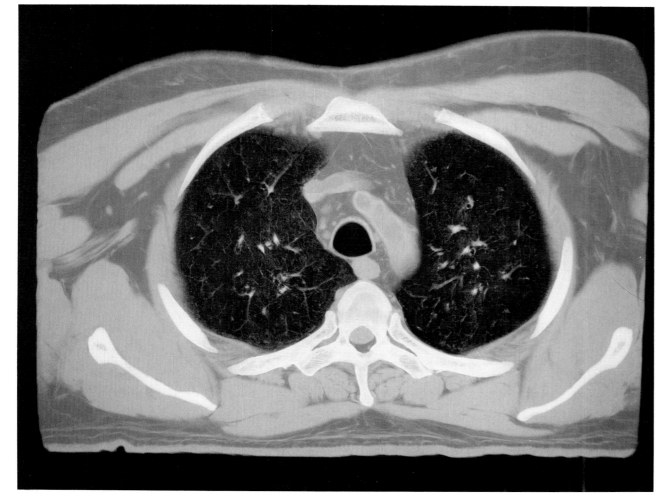

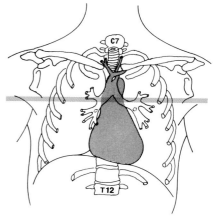

**Specimen Radiograph**

**Computed Tomogram—Mediastinum Technique**

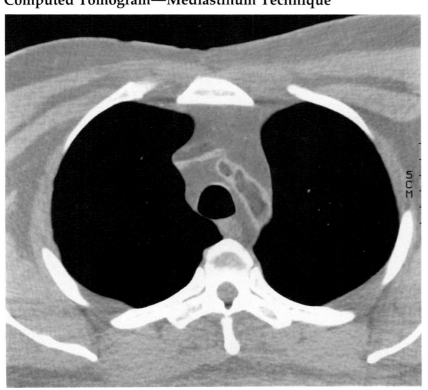

**Computed Tomogram—Lung Technique**

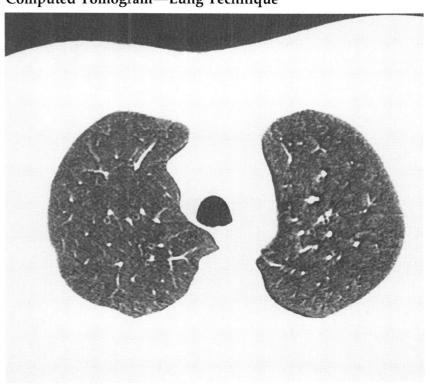

# Axial

## Section 10

### Anatomic Key

**Arteries**
A **29** aortic arch
A **47** internal thoracic (mammary), left
A **48** internal thoracic (mammary), right

**Bronchi**
B **3** RUL, apical segment
B **4** RUL, posterior segment
B **5** RUL, anterior segment
B **18** LUL, apicoposterior segment
B **19** LUL, anterior segment

**Esophagus**
E **1** esophagus

**Fat**
f **1** axillary
f **4** mediastinal

**Lobes**
L **2** RUL, apical segment
L **3** RUL, posterior segment
L **4** RUL, anterior segment
L **15** LUL, apicoposterior segment
L **16** LUL, anterior segment

**Muscles**
M **6** biceps brachii, short head
M **7** coracobrachialis
M **14** iliocostalis, thoracis
M **15** infraspinatus
M **16** intercostal
M **27** longissimus, thoracis
M **31** pectoralis, major
M **32** pectoralis, minor
M **35** rhomboid, major
M **43** semispinalis, thoracis
M **44** serratus, anterior
M **49** spinalis, thoracis
M **57** subscapularis
M **59** teres major
M **60** teres minor
M **61** transversospinalis
M **63** trapezius

**Neural Structures**
n **5** spinal cord
n **6** spinal dura

**Nodes**
N **6** mediastinal, anterior

**Pleural Structures**
P **2** costal (parietal)
P **5** mediastinal (parietal)
P **6** visceral

**Skeletal Structures**
S **5** rib, body, lateral (3rd)
S **6** rib, body, posterior (4th)
S **7** rib, costal cartilage (2nd)
S **10** scapula (medial margin)
S **18** sternum, manubrium
S **21** vertebra, body (T4)
S **25** vertebra, lamina (T4)
S **26** vertebra, pedicle (T4)
S **27** vertebra, spinous process (T4)

**Trachea**
T **1** trachea

**Veins**
V **12** hemiazygos
V **15** intercostal, highest (superior), right
V **18** intercostal, posterior, left
V **20** internal thoracic (mammary), left
V **21** internal thoracic (mammary), right
V **39** vena cava, superior
V **40** venous plexus, mediastinal

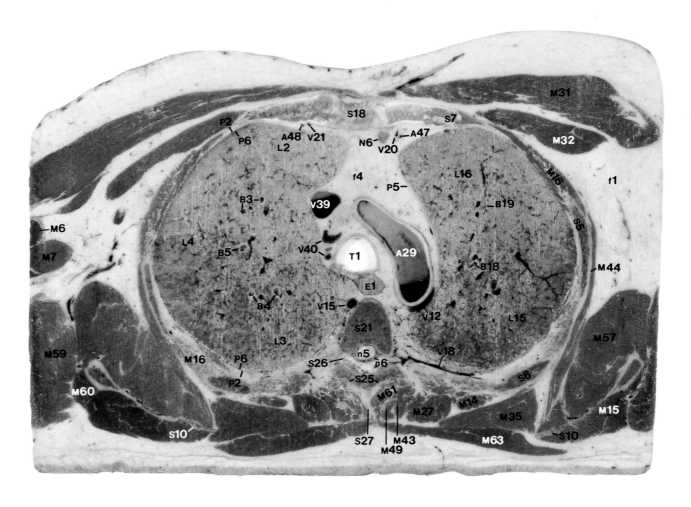

**Anatomic Specimen**

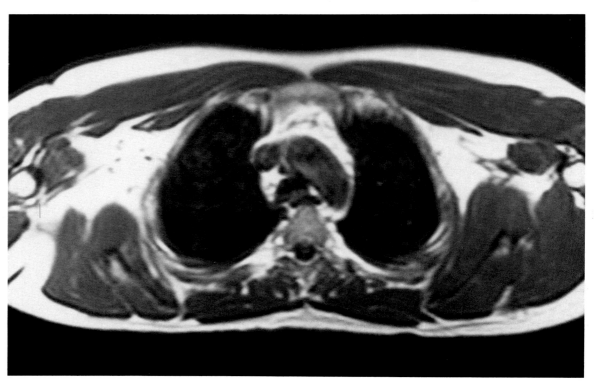

**Magnetic Resonance**

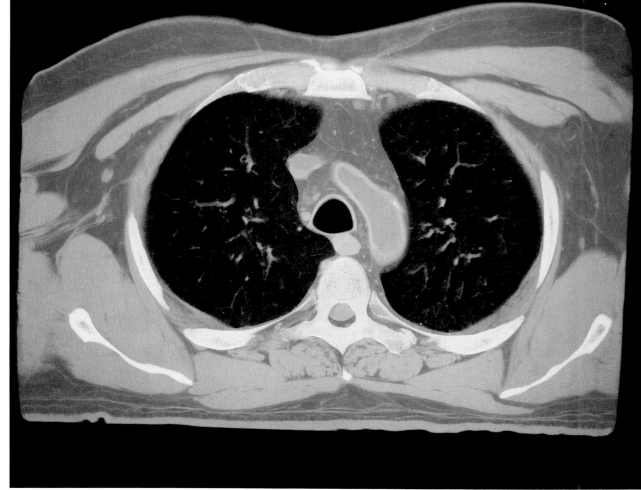

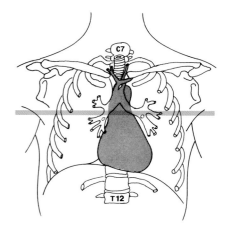

**Specimen Radiograph**

**Computed Tomogram—Mediastinum Technique**

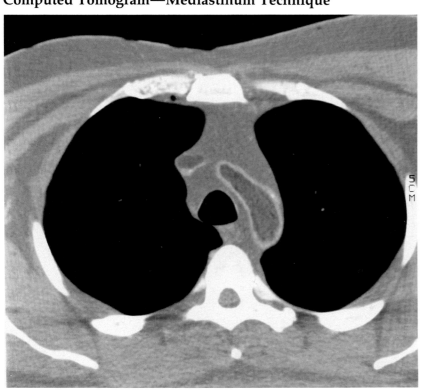

**Computed Tomogram—Lung Technique**

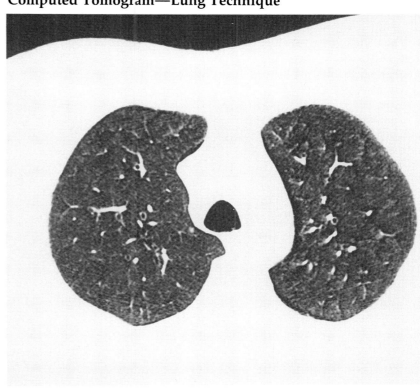

# Axial

## Section 11

### Anatomic Key

**Arteries**
A 29 aortic arch
A 47 internal thoracic (mammary), left
A 48 internal thoracic (mammary), right

**Bronchi**
B 3 RUL, apical segment
B 4 RUL, posterior segment
B 5 RUL, anterior segment
B 19 LUL, anterior segment

**Esophagus**
E 1 esophagus

**Fat**
f 1 axillary
f 4 mediastinal

**Lobes**
L 2 RUL, apical segment
L 3 RUL, posterior segment
L 4 RUL, anterior segment
L 15 LUL, apicoposterior segment
L 16 LUL, anterior segment

**Muscles**
M 6 biceps brachii, short head
M 7 coracobrachialis
M 14 iliocostalis, thoracis
M 15 infraspinatus
M 16 intercostal
M 27 longissimus, thoracis
M 31 pectoralis, major
M 32 pectoralis, minor
M 35 rhomboid, major
M 43 semispinalis, thoracis
M 44 serratus, anterior
M 49 spinalis, thoracis
M 57 subscapularis
M 59 teres major
M 60 teres minor
M 61 transversospinalis
M 63 trapezius

**Neural Structures**
n 5 spinal cord

**Nodes**
N 6 mediastinal, anterior

**Pleural Structures**
P 2 costal (parietal)
P 5 mediastinal (parietal)
P 6 visceral

**Skeletal Structures**
S 5 rib, body, lateral (3rd)
S 6 rib, body, posterior (4th)
S 7 rib, costal cartilage (2nd)
S 10 scapula (lateral margin)
S 17 sternum, body
S 21 vertebra, body (T4)
S 23 vertebra, inferior articular process (T4)
S 27 vertebra, spinous process (T4)
S 28 vertebra, superior articular process (T5)

**Trachea**
T 1 trachea

**Veins**
V 12 hemiazygos
V 15 intercostal, highest (superior), right
V 18 intercostal, posterior, left
V 20 internal thoracic (mammary), left
V 21 internal thoracic (mammary), right
V 39 vena cava, superior
V 40 venous plexus, mediastinal

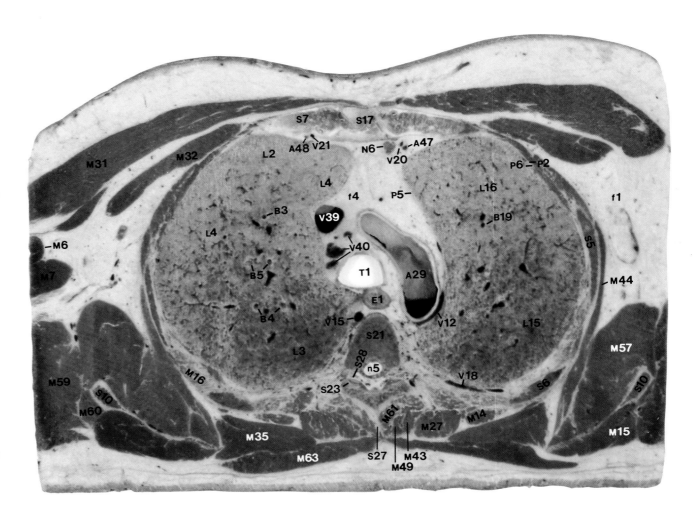

**Anatomic Specimen**

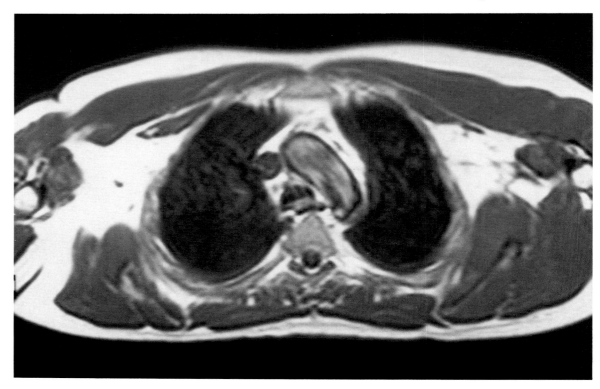

**Magnetic Resonance**

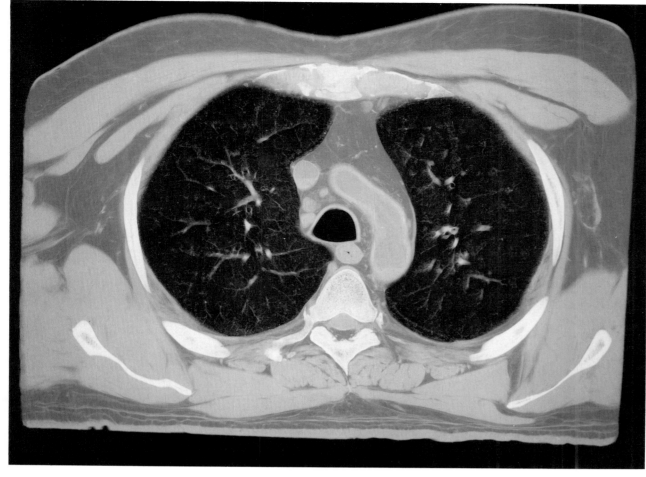
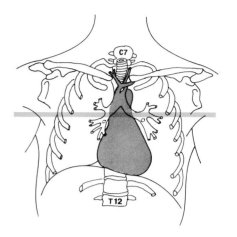

**Specimen Radiograph**

**Computed Tomogram—Mediastinum Technique**

**Computed Tomogram—Lung Technique**

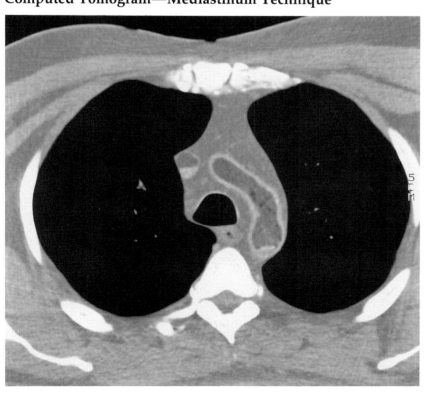
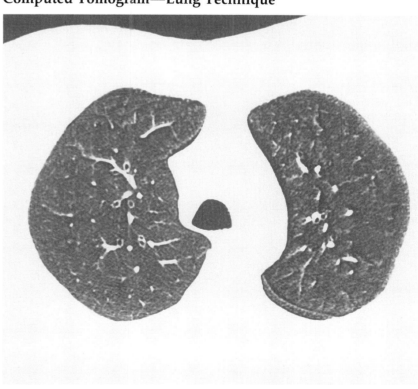

# Axial

## Section 12

## Anatomic Key

### Arteries
A **2** main pulmonary, right
A **4** RUL, apical branch
A **5** RUL, posterior branch
A **6** RUL, anterior branch
A **18** LUL, apicoposterior branch
A **19** LUL, anterior branch
A **30** aorta, ascending
A **31** aorta, descending
A **46** intercostal, posterior, right
A **47** internal thoracic (mammary), left
A **48** internal thoracic (mammary), right

### Bronchi
B **3** RUL, apical segment
B **4** RUL, posterior segment
B **5** RUL, anterior segment
B **18** LUL, apicoposterior segment
B **19** LUL, anterior segment

### Duct
D **1** thoracic

### Esophagus
E **1** esophagus

### Fat
f **1** axillary
f **4** mediastinal

### Fissures
F **1** left major (oblique)

### Lobes
L **2** RUL, apical segment
L **3** RUL, posterior segment
L **4** RUL, anterior segment
L **15** LUL, apicoposterior segment
L **16** LUL, anterior segment
L **21** LLL, superior segment

### Muscles
M **14** iliocostalis, thoracis
M **15** infraspinatus
M **16** intercostal
M **27** longissimus, thoracis
M **31** pectoralis, major
M **32** pectoralis, minor
M **35** rhomboid, major
M **43** semispinalis, thoracis
M **44** serratus, anterior
M **49** spinalis, thoracis
M **57** subscapularis
M **59** teres major
M **61** transversospinalis
M **63** trapezius

### Neural Structures
n **5** spinal cord
n **6** spinal dura

### Pleural Structures
P **2** costal (parietal)
P **5** mediastinal (parietal)
P **6** visceral

### Skeletal Structures
S **4** rib, body, anterior (3rd)
S **6** rib, body, posterior (4th)
S **8** rib, head and neck (5th)
S **9** rib, tubercle (5th)
S **10** scapula (lateral margin)
S **17** sternum, body
S **20** intervertebral disc (T4–T5)
S **26** vertebra, pedicle (T5)
S **27** vertebra, spinous process (T5)
S **29** vertebra, transverse process (T5)

### Trachea
T **1** trachea

### Veins
V **7** azygos (arch)
V **15** intercostal, highest (superior), right
V **20** internal thoracic (mammary), left
V **21** internal thoracic (mammary), right
V **39** vena cava, superior

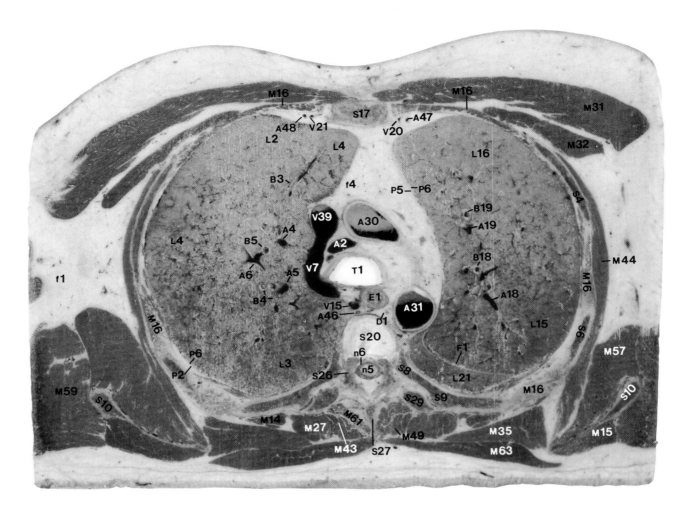

**Anatomic Specimen**

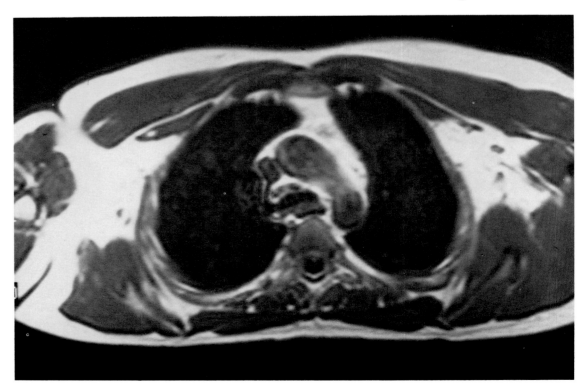

**Magnetic Resonance**

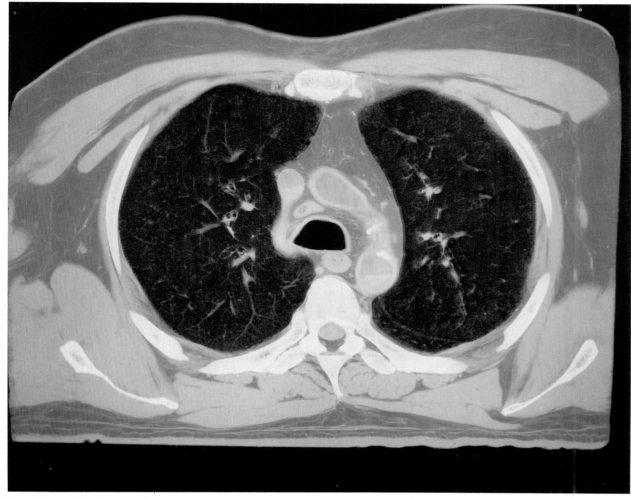

**Specimen Radiograph**

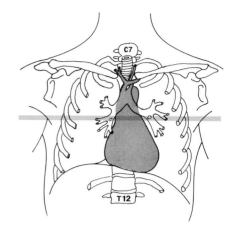

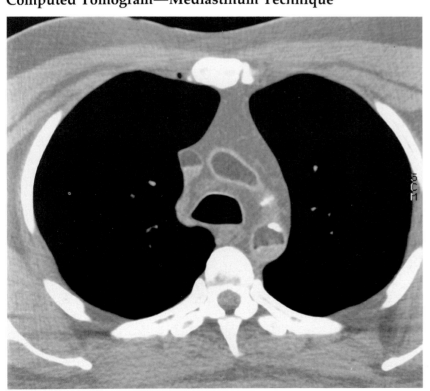

**Computed Tomogram—Mediastinum Technique**

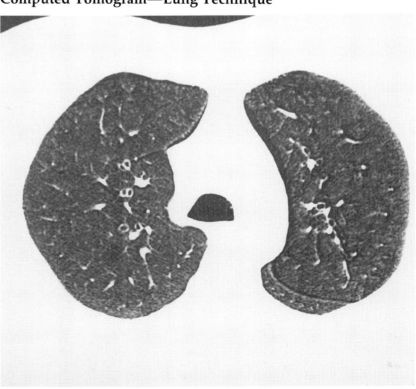

**Computed Tomogram—Lung Technique**

# Axial

## Section 13

## Anatomic Key

### Arteries
A  2  main pulmonary, right
A  5  RUL, posterior branch
A  6  RUL, anterior branch
A 18  LUL, apicoposterior branch
A 19  LUL, anterior branch
A 30  aorta, ascending
A 31  aorta, descending
A 46  intercostal, posterior, right
A 47  internal thoracic (mammary), left
A 48  internal thoracic (mammary), right

### Bronchi
B  4  RUL, posterior segment
B  5  RUL, anterior segment
B 18  LUL, apicoposterior segment
B 19  LUL, anterior segment

### Esophagus
E  1  esophagus

### Fat
f  1  axillary
f  4  mediastinal

### Fissures
F  1  left major (oblique)

### Lobes
L  2  RUL, apical segment
L  3  RUL, posterior segment
L  4  RUL, anterior segment
L 15  LUL, apicoposterior segment
L 16  LUL, anterior segment
L 21  LLL, superior segment

### Muscles
M 14  iliocostalis, thoracis
M 15  infraspinatus
M 16  intercostal
M 27  longissimus, thoracis
M 31  pectoralis, major
M 32  pectoralis, minor
M 35  rhomboid, major
M 44  serratus, anterior
M 49  spinalis, thoracis
M 57  subscapularis
M 59  teres major
M 61  transversospinalis
M 63  trapezius

### Neural Structures
n  5  spinal cord
n  6  spinal dura

### Nodes
N  6  mediastinal, anterior
N 12  tracheobronchial, superior

### Pleural Structures
P  2  costal (parietal)
P  5  mediastinal (parietal)
P  6  visceral

### Skeletal Structures
S  4  rib, body, anterior (3rd)
S  5  rib, body, lateral (4th)
S  8  rib, head and neck (5th)
S  9  rib, tubercle (5th)
S 10  scapula (medial margin)
S 17  sternum, body
S 21  vertebra, body (T5)
S 22  vertebra, costal facet (T5)
S 25  vertebra, lamina (T5)
S 26  vertebra, pedicle (T5)
S 27  vertebra, spinous process (T5)
S 29  vertebra, transverse process (T5)

### Trachea
T  2  carina (bifurcation of trachea)

### Veins
V  7  azygos (arch)
V 20  internal thoracic (mammary), left
V 21  internal thoracic (mammary), right
V 28  lateral thoracic, left
V 39  vena cava, superior
V 40  venous plexus, mediastinal

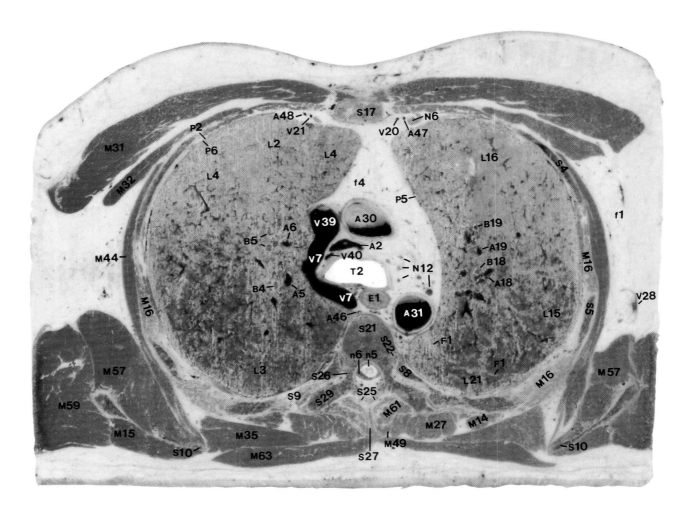

**Anatomic Specimen**

**Magnetic Resonance**

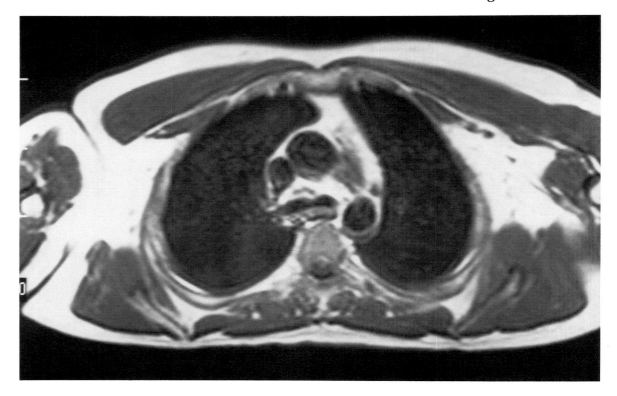

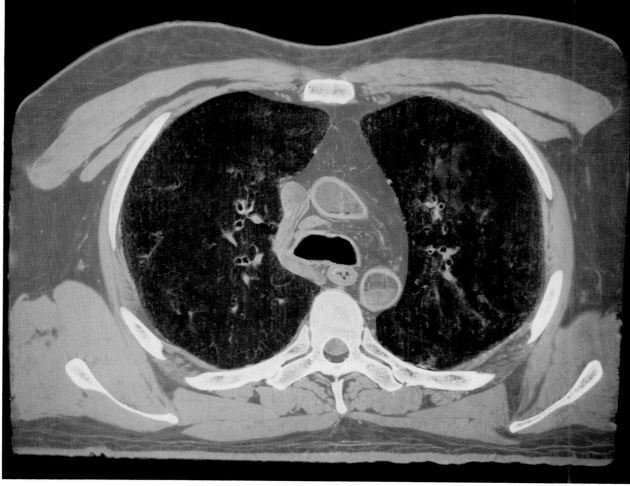

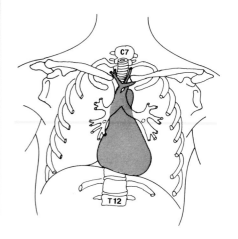

**Specimen Radiograph**

**Computed Tomogram—Mediastinum Technique**

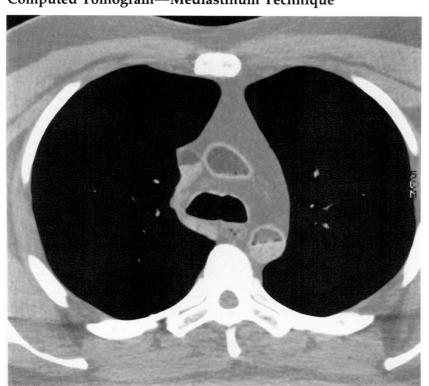

**Computed Tomogram—Lung Technique**

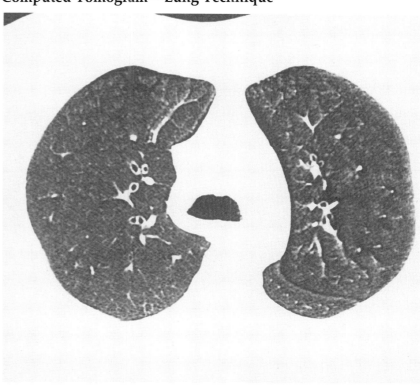

# Axial

## Section 14

### Anatomic Key

**Arteries**
A  2  main pulmonary, right
A  5  RUL, posterior branch
A  6  RUL, anterior branch
A 11  RLL, superior branch
A 18  LUL, apicoposterior branch
A 19  LUL, anterior branch
A 30  aorta, ascending
A 31  aorta, descending
A 46  intercostal, posterior, right
A 47  internal thoracic (mammary), left
A 48  internal thoracic (mammary), right

**Bronchi**
B  1  main, right
B  4  RUL, posterior segment
B  5  RUL, anterior segment
B 16  main, left
B 18  LUL, apicoposterior segment
B 19  LUL, anterior segment

**Duct**
D  1  thoracic

**Esophagus**
E  1  esophagus

**Fat**
f  1  axillary
f  4  mediastinal
f  5  pericardial

**Fissures**
F  1  left major (oblique)

**Heart**
H 13  pericardium, parietal serous

**Lobes**
L  3  RUL, posterior segment
L  4  RUL, anterior segment
L  9  RLL, superior segment
L 15  LUL, apicoposterior segment
L 16  LUL, anterior segment
L 21  LLL, superior segment

**Muscles**
M 14  iliocostalis, thoracis
M 15  infraspinatus
M 16  intercostal
M 27  longissimus, thoracis
M 31  pectoralis, major
M 32  pectoralis, minor
M 35  rhomboid, major
M 43  semispinalis, thoracis
M 44  serratus, anterior
M 49  spinalis, thoracis
M 57  subscapularis
M 59  teres major
M 61  transversospinalis
M 63  trapezius

**Neural Structures**
n  5  spinal cord
n  6  spinal dura

**Nodes**
N  8  parasternal (internal thoracic)
N 12  tracheobronchial, superior

**Pleural Structures**
P  2  costal (parietal)
P  5  mediastinal (parietal)
P  6  visceral

**Skeletal Structures**
S  4  rib, body, anterior (3rd)
S  5  rib, body, lateral (4th)
S  6  rib, body, posterior (5th)
S 10  scapula
S 17  sternum, body
S 21  vertebra, body (T5)
S 25  vertebra, lamina (T5)
S 26  vertebra, pedicle (T5)
S 27  vertebra, spinous process (T5)
S 29  vertebra, transverse process (T5)

**Veins**
v  0  pulmonary, superior, RUL
v  7  azygos
v 18  intercostal, posterior, left
v 19  intercostal, posterior, right
v 20  internal thoracic (mammary), left
v 21  internal thoracic (mammary), right
v 39  vena cava, superior

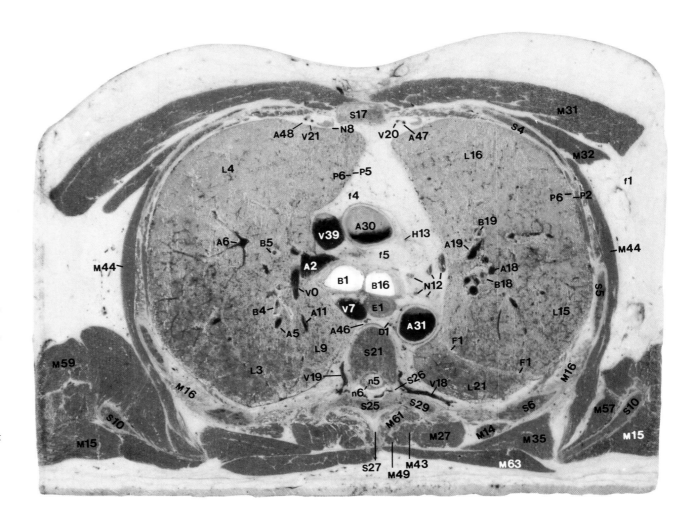

**Anatomic Specimen**

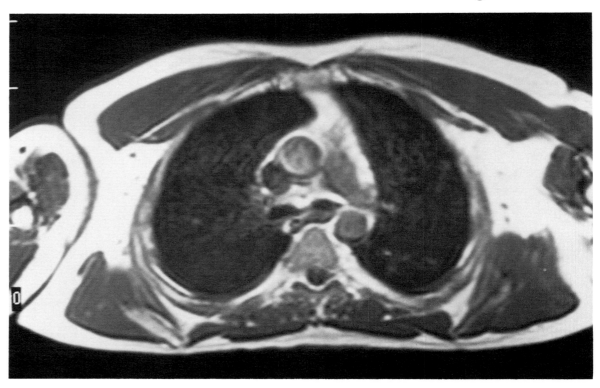

**Magnetic Resonance**

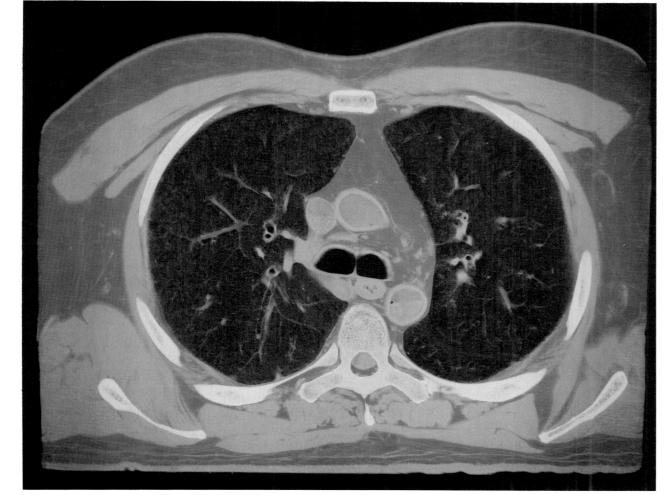

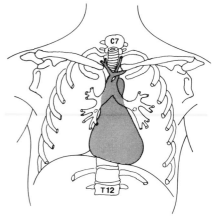

**Specimen Radiograph**

**Computed Tomogram—Mediastinum Technique**

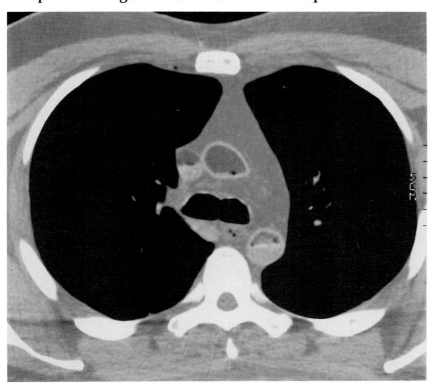

**Computed Tomogram—Lung Technique**

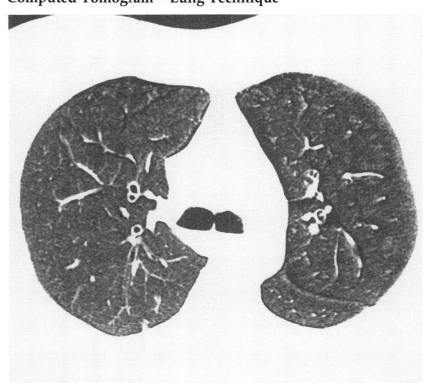

# Axial

## Section 15

## Anatomic Key

### Arteries
A **1** pulmonary trunk
A **2** main pulmonary, right
A **16** main pulmonary, left
A **18** LUL, apicoposterior branch
A **19** LUL, anterior branch
A **30** aorta, ascending
A **31** aorta, descending
A **45** intercostal, posterior, left
A **47** internal thoracic (mammary), left
A **48** internal thoracic (mammary), right

### Bronchi
B **1** main, right
B **2** right upper lobe (RUL)
B **4** RUL, posterior segment
B **5** RUL, anterior segment
B **16** main, left
B **18** LUL, apicoposterior segment
B **19** LUL, anterior segment

### Esophagus
E **1** esophagus

### Fascia
F **6** perivisceral

### Fat
f **1** axillary
f **4** mediastinal
f **5** pericardial
f **7** subcutaneous

### Fissures
F **1** left major (oblique)

### Heart
H **13** pericardium, parietal serous

### Lobes
L **3** RUL, posterior segment
L **4** RUL, anterior segment
L **9** RLL, superior segment
L **15** LUL, apicoposterior segment
L **16** LUL, anterior segment
L **21** LLL, superior segment

### Muscles
M **14** iliocostalis, thoracis
M **15** infraspinatus
M **16** intercostal
M **27** longissimus, thoracis
M **31** pectoralis, major
M **32** pectoralis, minor
M **35** rhomboid, major
M **43** semispinalis, thoracis
M **44** serratus, anterior
M **49** spinalis, thoracis
M **57** subscapularis
M **59** teres major
M **60** teres minor
M **61** transversospinalis
M **63** trapezius

### Neural Structures
n **5** spinal cord

### Nodes
N **2** bronchopulmonary (hilar)
N **8** parasternal (internal thoracic)

### Pleural Structures
P **2** costal (parietal)
P **5** mediastinal (parietal)
P **6** visceral

### Skeletal Structures
S **4** rib, body, anterior (3rd)
S **5** rib, body, lateral (4th)
S **6** rib, body, posterior (5th)
S **7** rib, costal cartilage (3rd)
S **10** scapula
S **17** sternum, body
S **21** vertebra, body (T5)
S **23** vertebra, inferior articular process (T5)
S **24** vertebra, intervertebral foramen (T5–T6)
S **25** vertebra, lamina (T5)
S **27** vertebra, spinous process (T5)
S **28** vertebra, superior articular process (T6)

### Veins
v **0** pulmonary, superior, RUL
v **7** azygos
v **18** intercostal, posterior, left
v **19** intercostal, posterior, right
v **20** internal thoracic (mammary), left
v **21** internal thoracic (mammary), right
v **39** vena cava, superior
v **40** venous plexus, mediastinal

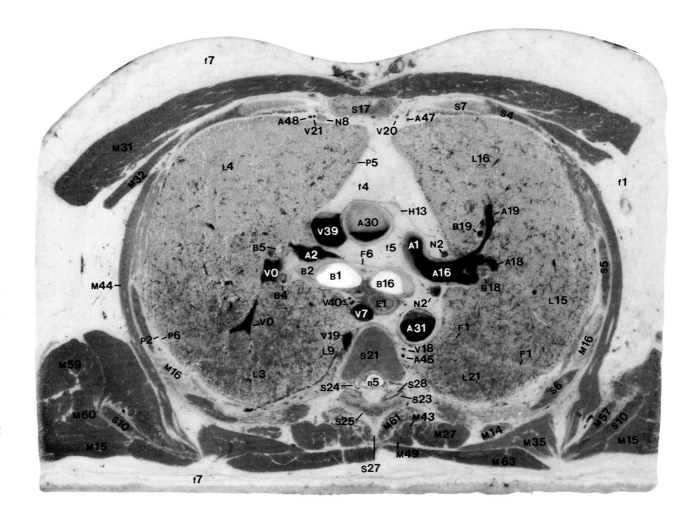

**Anatomic Specimen**

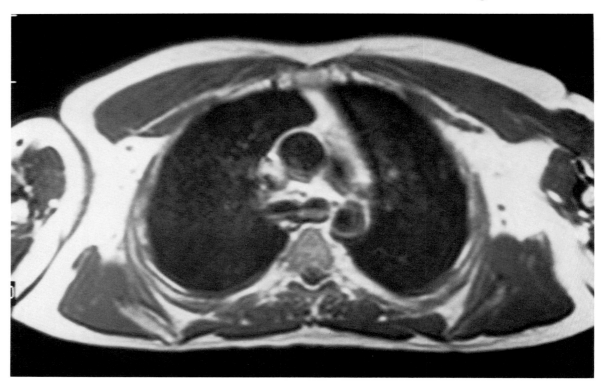

**Magnetic Resonance**

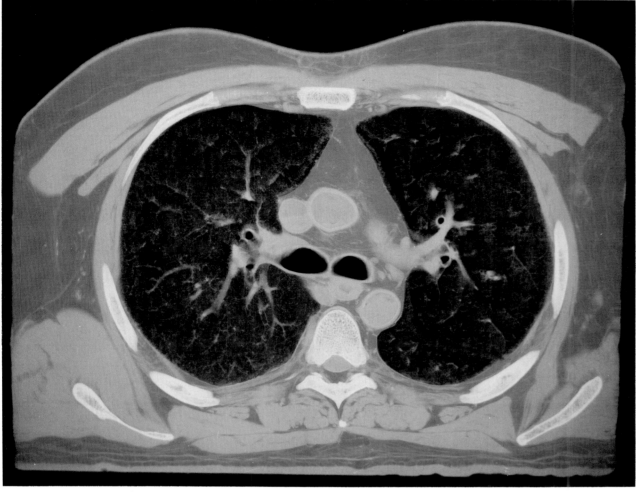

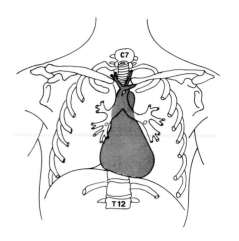

**Specimen Radiograph**

**Computed Tomogram—Mediastinum Technique**

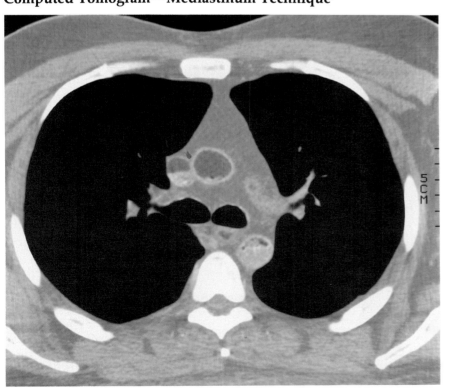

**Computed Tomogram—Lung Technique**

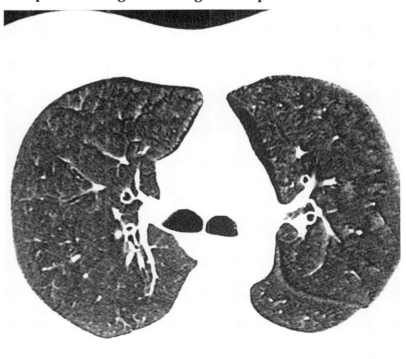

# Axial
## Section 16

## Anatomic Key

### Arteries
A **1** pulmonary trunk
A **2** main pulmonary, right
A **3** right upper lobe (RUL)
A **16** main pulmonary, left
A **18** LUL, apicoposterior branch
A **30** aorta, ascending
A **31** aorta, descending
A **48** internal thoracic (mammary), right

### Bronchi
B **1** main, right
B **2** right upper lobe (RUL)
B **4** RUL, posterior segment
B **5** RUL, anterior segment
B **16** main, left
B **18** LUL, apicoposterior segment
B **19** LUL, anterior segment

### Duct
D **1** thoracic

### Esophagus
E **1** esophagus

### Fascia
F **6** perivisceral

### Fat
f **1** axillary
f **4** mediastinal
f **5** pericardial

### Fissures
F **1** left major (oblique)

### Glands
G **3** thymus

### Heart
H **13** pericardium, parietal serous

### Lobes
L **3** RUL, posterior segment
L **4** RUL, anterior segment
L **9** RLL, superior segment
L **15** LUL, apicoposterior segment
L **16** LUL, anterior segment
L **21** LLL, superior segment

### Muscles
M **14** iliocostalis, thoracis
M **15** infraspinatus
M **16** intercostal
M **21** latissimus dorsi
M **27** longissimus thoracis
M **31** pectoralis, major
M **32** pectoralis, minor
M **35** rhomboid, major
M **44** serratus, anterior
M **57** subscapularis
M **59** teres major
M **60** teres minor
M **61** transversospinalis
M **63** trapezius

### Neural Structures
n **5** spinal cord
n **6** spinal dura

### Nodes
N **2** bronchopulmonary (hilar)

### Pleural Structures
P **2** costal (parietal)
P **5** mediastinal (parietal)
P **6** visceral

### Skeletal Structures
s **4** rib, body, anterior (3rd)
s **5** rib, body, lateral (4th)
s **6** rib, body, posterior (5th)
s **7** rib, costal cartilage (3rd)
s **8** rib, head and neck (6th)
s **10** scapula
s **17** sternum, body
s **20** intervertebral disc (T5–T6)
s **21** vertebra, body (T5)
s **22** vertebra, costal facet (T5)
s **23** vertebra, inferior articular process (T5)
s **27** vertebra, spinous process (T5)
s **28** vertebra, superior articular process (T6)

### Veins
v **0** pulmonary, superior, RUL
v **2** pulmonary, superior, LUL
v **7** azygos
v **19** intercostal, posterior, right
v **21** internal thoracic (mammary), right
v **29** lateral thoracic, right
v **39** vena cava, superior

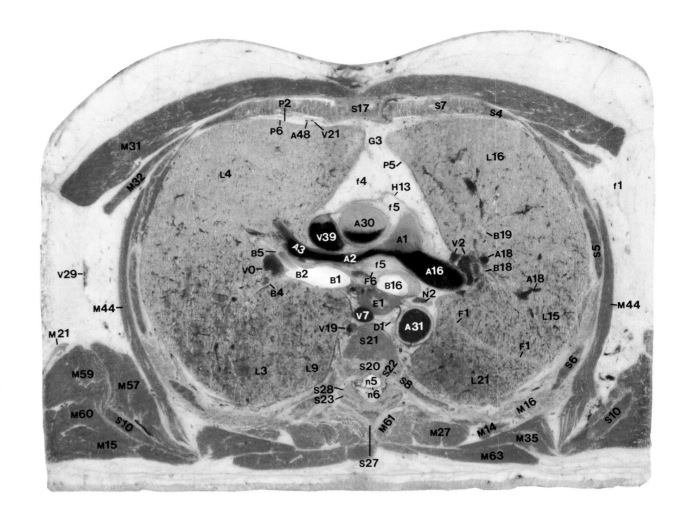

**Anatomic Specimen**

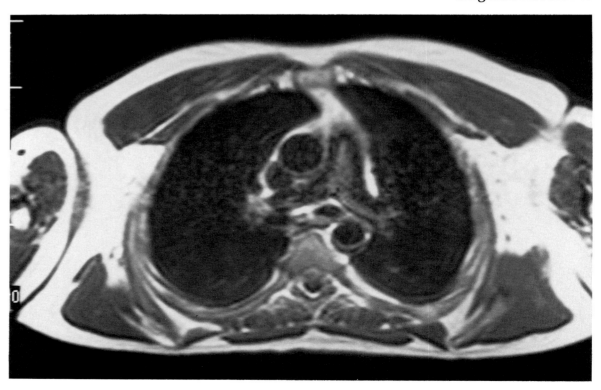

**Magnetic Resonance**

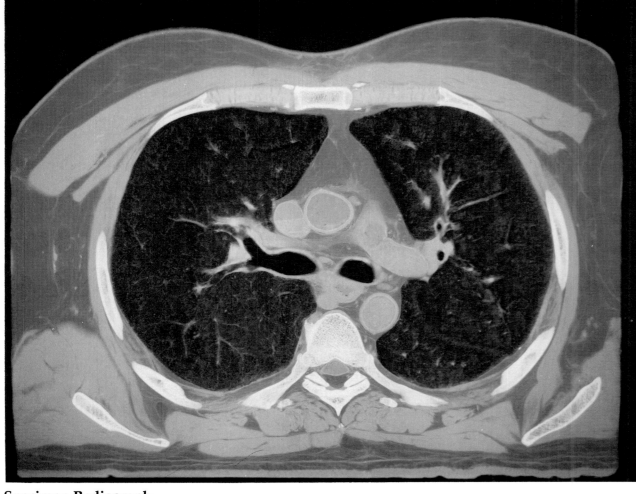

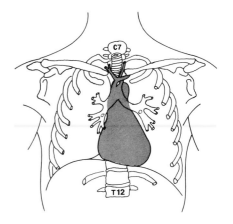

**Specimen Radiograph**

**Computed Tomogram—Mediastinum Technique**

**Computed Tomogram—Lung Technique**

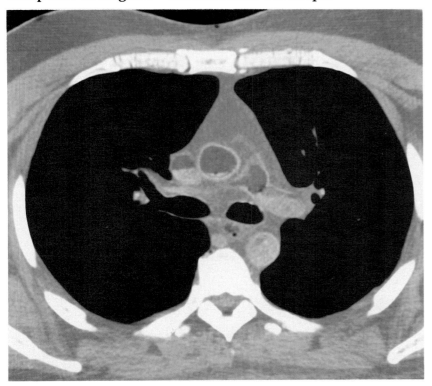

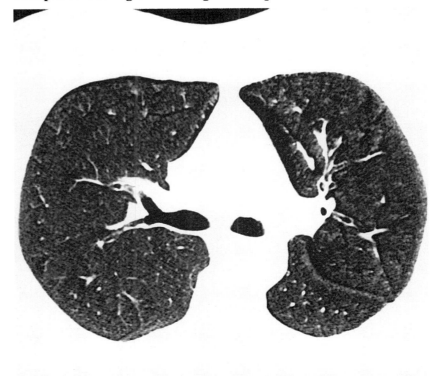

# Axial

## Section 17

### Anatomic Key

**Arteries**

A  1  pulmonary trunk
A  2  main pulmonary, right
A  5  RUL, posterior branch
A  6  RUL, anterior branch
A 16  main pulmonary, left
A 18  LUL, apicoposterior branch
A 19  LUL, anterior branch
A 30  aorta, ascending
A 31  aorta, descending
A 45  intercostal, posterior, left

**Bronchi**

B  4  RUL, posterior segment
B  5  RUL, anterior segment
B  6  intermediate, right
B 16  main, left
B 18  LUL, apicoposterior segment
B 19  LUL, anterior segment

**Esophagus**

E  1  esophagus

**Fascia**

F  6  perivisceral
F  7  prevertebral

**Fat**

f  4  mediastinal
f  5  pericardial
f  7  subcutaneous

**Fissures**

F  1  left major (oblique)
F  2  right major (oblique)

**Heart**

H  4  auricle, right
H 13  pericardium, parietal serous

**Lobes**

L  3  RUL, posterior segment
L  4  RUL, anterior segment
L  9  RLL, superior segment
L 15  LUL, apicoposterior segment
L 16  LUL, anterior segment
L 21  LLL, superior segment

**Muscles**

M 14  iliocostalis, thoracis
M 15  infraspinatus
M 16  intercostal
M 21  latissimus dorsi
M 27  longissimus, thoracis
M 31  pectoralis, major
M 32  pectoralis, minor
M 35  rhomboid, major
M 43  semispinalis, thoracis
M 44  serratus, anterior
M 49  spinalis, thoracis
M 57  subscapularis
M 59  teres major
M 61  transversospinalis
M 63  trapezius

**Neural Structures**

n  5  spinal cord

**Pleural Structures**

P  2  costal (parietal)
P  5  mediastinal (parietal)
P  6  visceral

**Skeletal Structures**

S  4  rib, body, anterior (4th)
S  5  rib, body, lateral (5th)
S  6  rib, body, posterior (6th)
S  7  rib, costal cartilage (3rd)
S  8  rib, head and neck (6th)
S  9  rib, tubercle (6th)
S 10  scapula
S 17  sternum, body
S 20  intervertebral disc (T5–T6)
S 21  vertebra, body (T6)
S 22  vertebra, costal facet (T6)

S 26  vertebra, pedicle (T6)
S 27  vertebra, spinous process (T5)
S 29  vertebra, transverse process (T6)

**Veins**

V  0  pulmonary, superior, RUL
V  2  pulmonary, superior, LUL
V  7  azygos
V 18  intercostal, posterior, left
V 29  lateral thoracic, right
V 39  vena cava, superior

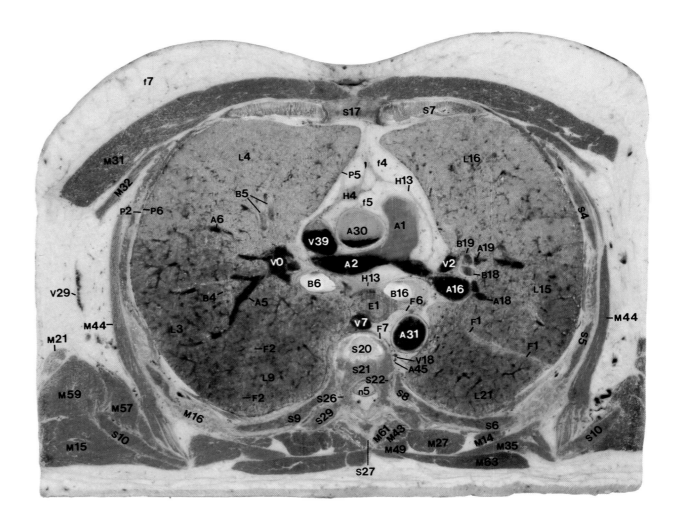

**Anatomic Specimen**

**Magnetic Resonance**

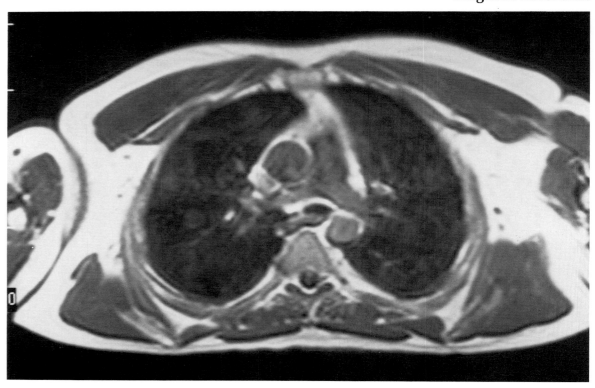

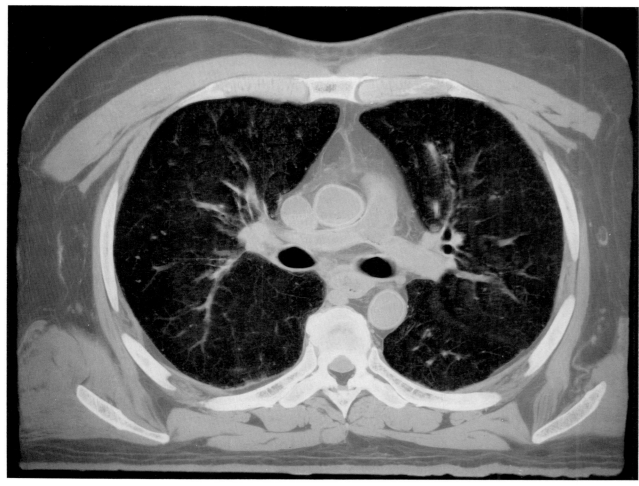

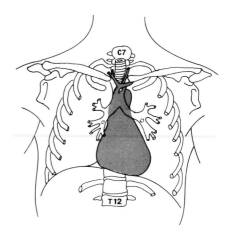

**Specimen Radiograph**

**Computed Tomogram—Mediastinum Technique**

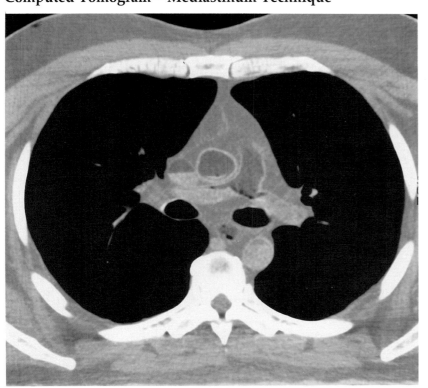

**Computed Tomogram—Lung Technique**

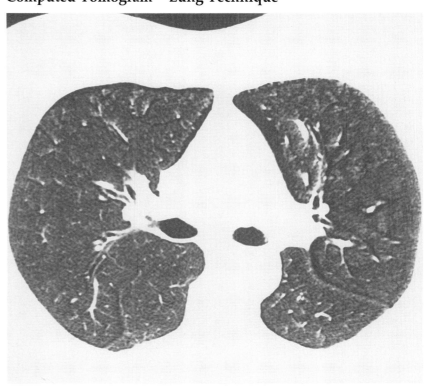

# Axial

## Section 18

### Anatomic Key

#### Arteries

A 1 pulmonary trunk
A 2 main pulmonary, right
A 2* main pulmonary, right (interlobar part)
A 6 RUL, anterior branch
A 17 left upper lobe (LUL)
A 23 left lower lobe (LLL)
A 30 aorta, ascending
A 31 aorta, descending
A 48 internal thoracic (mammary), right

#### Bronchi

B 4 RUL, posterior segment
B 5 RUL, anterior segment
B 6 intermediate, right
B 16 main, left
B 17 left upper lobe (LUL)

#### Duct

D 1 thoracic

#### Esophagus

E 1 esophagus

#### Fascia

F 6 perivisceral

#### Fat

f 3 extrapleural
f 4 mediastinal
f 5 pericardial

#### Fissures

F 1 left major (oblique)
F 2 right major (oblique)

#### Heart

H 3 auricle, left
H 4 auricle, right
H 13 pericardium, parietal serous
H 22 ventricle, right

#### Lobes

L 3 RUL, posterior segment
L 4 RUL, anterior segment
L 9 RLL, superior segment
L 16 LUL, anterior segment
L 18 LUL, superior lingular segment
L 19 LUL, inferior lingular segment
L 21 LLL, superior segment

#### Muscles

M 14 iliocostalis, thoracis
M 15 infraspinatus
M 16 intercostal
M 21 latissimus dorsi
M 27 longissimus, thoracis
M 31 pectoralis, major
M 32 pectoralis, minor
M 35 rhomboid, major
M 43 semispinalis, thoracis
M 44 serratus, anterior
M 49 spinalis, thoracis
M 57 subscapularis
M 59 teres major
M 61 transversospinalis
M 63 trapezius

#### Neural Structures

n 5 spinal cord
n 6 spinal dura

#### Pleural Structures

P 1 anterior junction line
P 2 costal (parietal)
P 6 visceral

#### Skeletal Structures

S 4 rib, body, anterior (4th)
S 5 rib, body, lateral (5th)
S 6 rib, body, posterior (6th)
S 7 rib, costal cartilage (3rd)
S 8 rib, head and neck (6th)
S 9 rib, tubercle (6th)
S 10 scapula
S 17 sternum, body
S 21 vertebra, body (T6)
S 22 vertebra, costal facet (T6)
S 26 vertebra, pedicle (T6)
S 27 vertebra, spinous process (T5)
S 29 vertebra, transverse process (T6)

#### Veins

V 0 pulmonary, superior, RUL
V 1 pulmonary, superior, RML
V 2 pulmonary, superior, LUL
V 3 pulmonary, superior, LUL, lingula
V 7 azygos
V 18 intercostal, posterior, left
V 19 intercostal, posterior, right
V 21 internal thoracic (mammary), right
V 29 lateral thoracic, right
V 39 vena cava, superior
V 41 venous plexus, vertebral (internal)

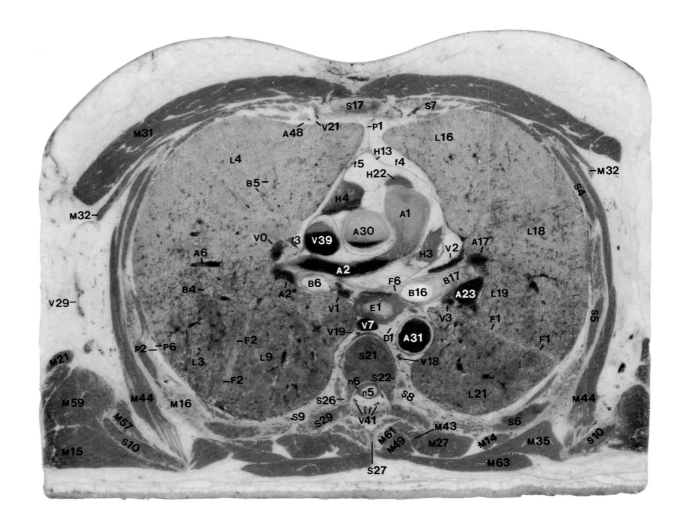

**Anatomic Specimen**

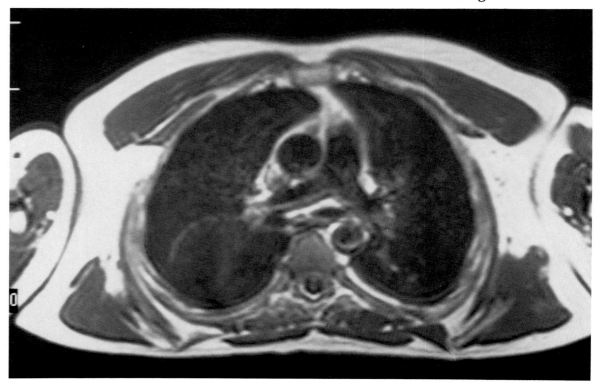

**Magnetic Resonance**

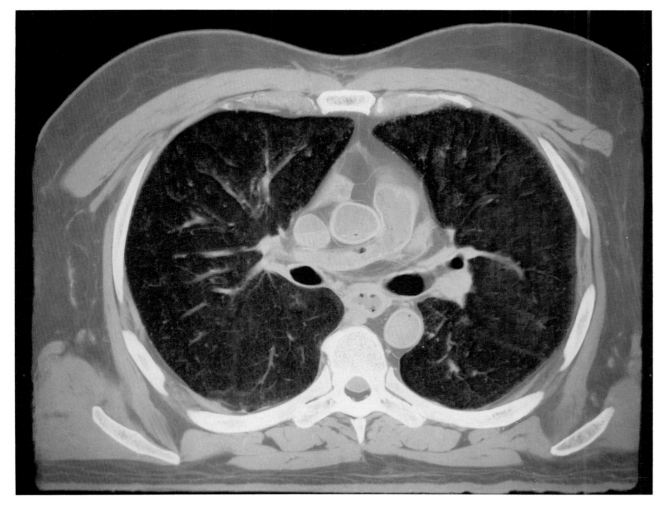

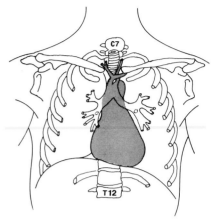

**Specimen Radiograph**

**Computed Tomogram—Mediastinum Technique**

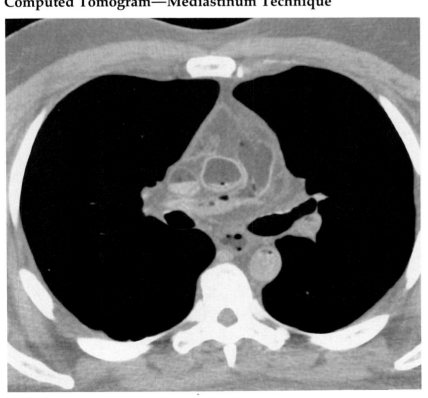

**Computed Tomogram—Lung Technique**

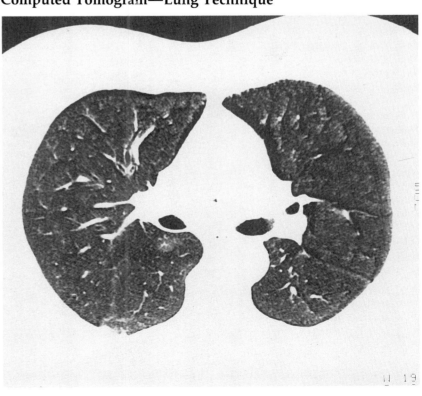

# Axial
## Section 19

## Anatomic Key

### Arteries
A  2  main pulmonary, right
A  2* main pulmonary, right
      (interlobar part)
A 20  LUL, lingula
A 23  left lower lobe (LLL)
A 24  LLL, superior branch
A 30  aorta, ascending
A 31  aorta, descending
A 47  internal thoracic (mammary),
      left
A 48  internal thoracic (mammary),
      right

### Bronchi
B  5  RUL, anterior segment
B  6  intermediate, right
B 16  main, left
B 17  left upper lobe (LUL)
B 24  LLL, superior segment

### Esophagus
E  1  esophagus

### Fat
f  3  extrapleural
f  4  mediastinal
f  5  pericardial

### Fissures
F  1  left major (oblique)
F  2  right major (oblique)
F  3  right minor (horizontal)

### Heart
H  3  auricle, left
H  4  auricle, right
H 13  pericardium, parietal serous
H 18  valve, pulmonary
H 22  ventricle, right

### Lobes
L  4  RUL, anterior segment
L  5  right middle lobe (RML)
L  6  RML, lateral segment
L  9  RLL, superior segment
L 18  LUL, superior lingular segment
L 19  LUL, inferior lingular segment
L 21  LLL, superior segment

### Muscles
M 14  iliocostalis, thoracis
M 15  infraspinatus
M 16  intercostal
M 21  latissimus dorsi
M 27  longissimus, thoracis
M 31  pectoralis, major
M 32  pectoralis, minor
M 35  rhomboid, major
M 43  semispinalis, thoracis
M 44  serratus, anterior
M 49  spinalis, thoracis
M 57  subscapularis
M 59  teres major
M 61  transversospinalis
M 62  transversus thoracis
M 63  trapezius

### Neural Structures
n  5  spinal cord
n  6  spinal dura

### Pleural Structures
P  1  anterior junction line
P  2  costal (parietal)
P  6  visceral

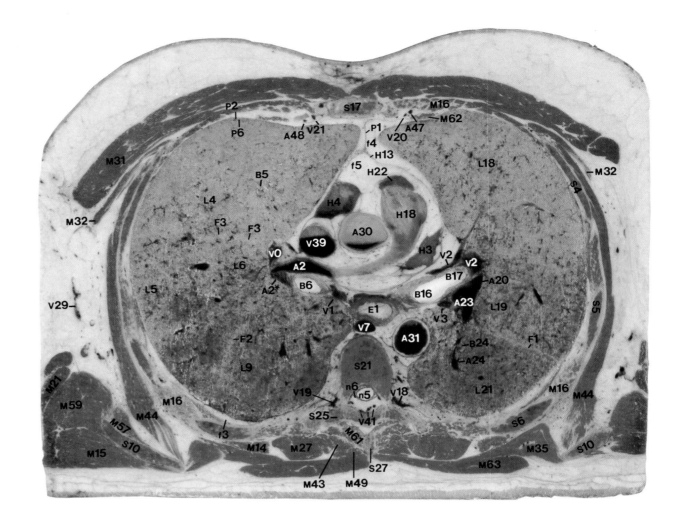

**Anatomic Specimen**

### Skeletal Structures
S  4  rib, body, anterior (4th)
S  5  rib, body, lateral (5th)
S  6  rib, body, posterior (6th)
S 10  scapula
S 17  sternum, body
S 21  vertebra, body (T6)
S 25  vertebra, lamina (T6)
S 27  vertebra, spinous process (T6)

### Veins
V  0  pulmonary, superior (RUL)
V  1  pulmonary, superior (RML)
V  2  pulmonary, superior (LUL)
V  3  pulmonary, superior (LUL,
      lingula)
V  7  azygos
V 18  intercostal, posterior, left
V 19  intercostal, posterior, right

V 20  internal thoracic (mammary),
      left
V 21  internal thoracic (mammary),
      right
V 29  lateral thoracic, right
V 39  vena cava, superior
V 41  venous plexus, vertebral
      (internal)

**Magnetic Resonance**

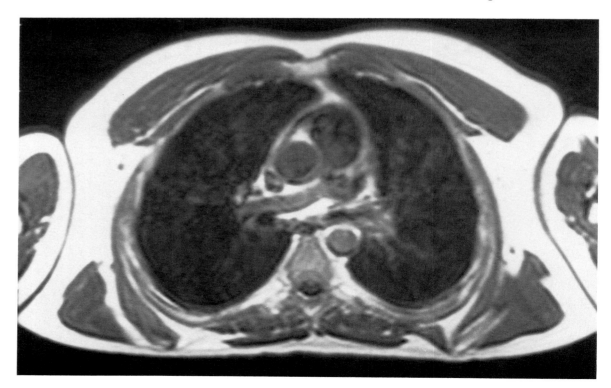

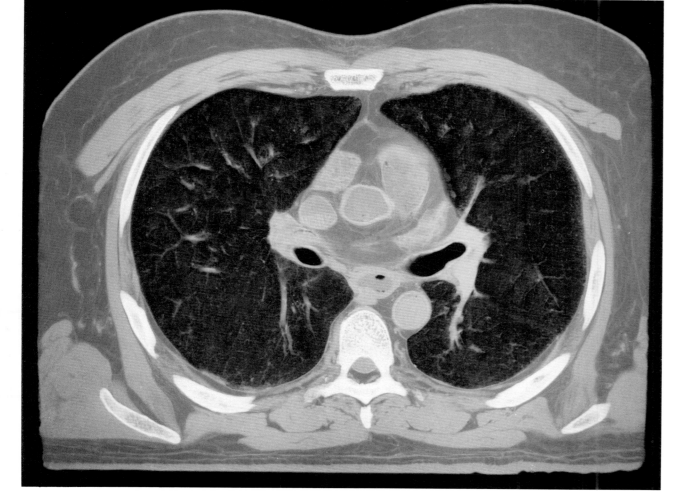

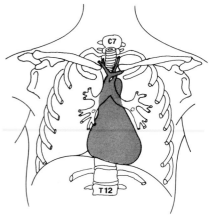

**Specimen Radiograph**

**Computed Tomogram—Mediastinum Technique**

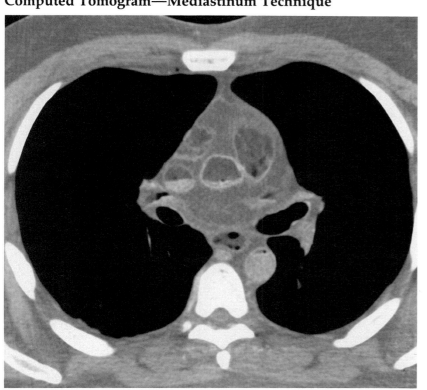

**Computed Tomogram—Lung Technique**

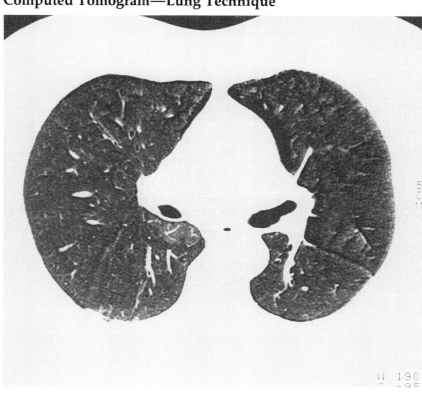

# Axial

## Section 20

### Anatomic Key

#### Arteries
A  2* main pulmonary, right
       (interlobar part)
A 11 RLL, superior branch
A 21 LUL, superior lingular branch
A 22 LUL, inferior lingular branch
A 23 left lower lobe (LLL)
A 24 LLL, superior branch
A 30 aorta, ascending
A 31 aorta, descending
A 45 intercostal, posterior, left

#### Bronchi
B  5 RUL, anterior segment
B  6 intermediate, right
B 21 LUL, superior lingular segment
B 22 LUL, inferior lingular segment
B 23 left lower lobe (LLL)
B 24 LLL, superior segment

#### Duct
D  1 thoracic

#### Esophagus
E  1 esophagus

#### Fascia
F  7 prevertebral

#### Fat
f  3 extrapleural
f  4 mediastinal
f  5 pericardial

#### Fissures
F  1 left major (oblique)
F  2 right major (oblique)

#### Heart
H  1 atrium, left
H  2 atrium, right
H  3 auricle, left
H  9 coronary artery, anterior
       descending
H 10 coronary artery, left
H 13 pericardium, parietal serous
H 14 pericardium, visceral serous
H 22 ventricle, right

#### Lobes
L  4 RUL, anterior segment
L  6 RML, lateral segment
L  9 RLL, superior segment
L 18 LUL, superior lingular segment
L 19 LUL, inferior lingular segment
L 21 LLL, superior segment

#### Muscles
M 14 iliocostalis, thoracis
M 16 intercostal
M 21 latissimus dorsi
M 27 longissimus, thoracis
M 31 pectoralis, major
M 32 pectoralis, minor
M 35 rhomboid, major
M 43 semispinalis, thoracis
M 44 serratus, anterior
M 49 spinalis, thoracis
M 59 teres major
M 61 transversospinalis
M 62 transversus thoracis
M 63 trapezius

#### Neural Structures
n  5 spinal cord
n  6 spinal dura

#### Nodes
N  8 parasternal (internal thoracic)

#### Pleural Structures
P  1 anterior junction line
P  2 costal (parietal)
P  6 visceral

#### Skeletal Structures
S  4 rib, body, anterior (4th)
S  5 rib, body, lateral (5th)
S  6 rib, body, posterior (6th)
S  7 rib, costal cartilage (4th)
S  8 rib, head and neck (7th)
S 17 sternum, body
S 21 vertebra, body (T6)
S 22 vertebra, costal facet (T6)
S 23 vertebra, inferior articular
       process (T6)
S 27 vertebra, spinous process (T6)
S 28 vertebra, superior articular
       process (T7)

#### Veins
V  0 pulmonary, superior, RUL
V  1 pulmonary, superior, RML
V  2 pulmonary, superior, LUL
V  3 pulmonary, superior, LUL,
       lingula
V  4 pulmonary, inferior, RLL
V  7 azygos
V 18 intercostal, posterior, left
V 19 intercostal, posterior, right
V 39 vena cava, superior

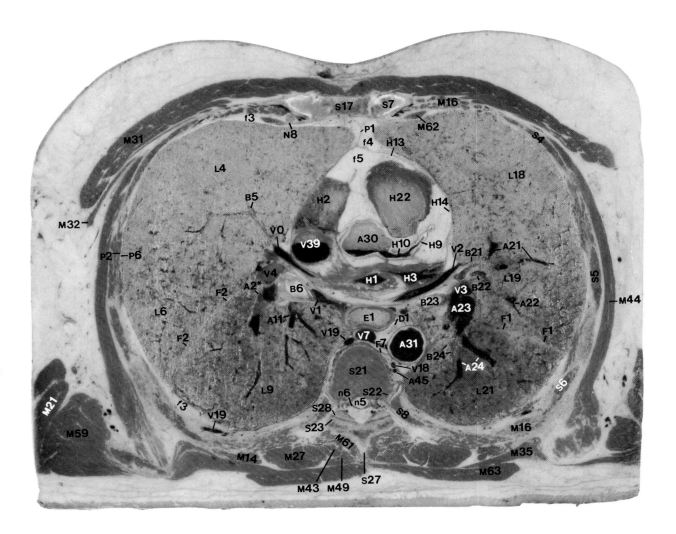

**Anatomic Specimen**

**Magnetic Resonance**

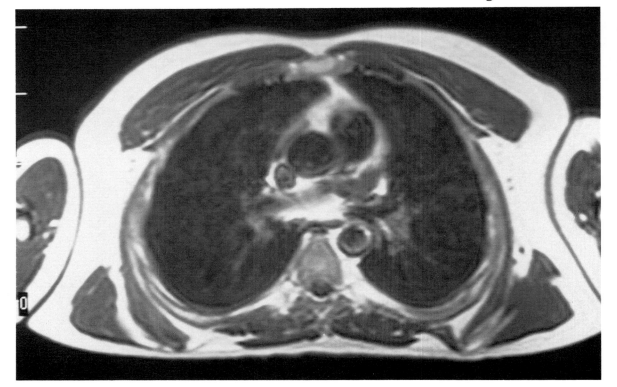

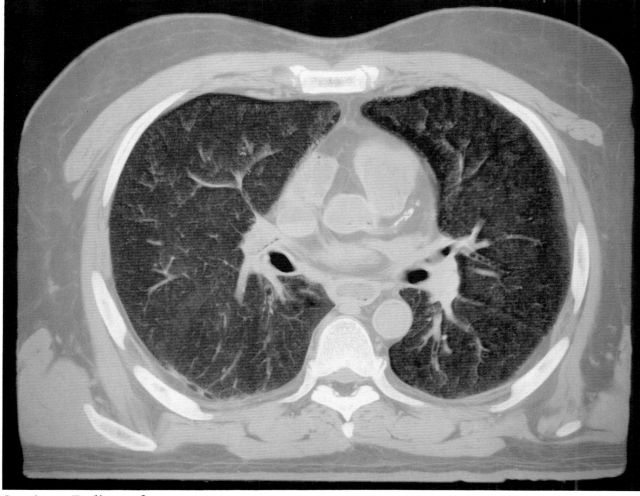

**Specimen Radiograph**

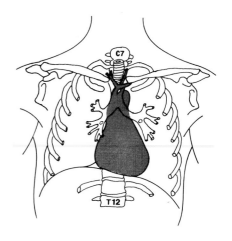

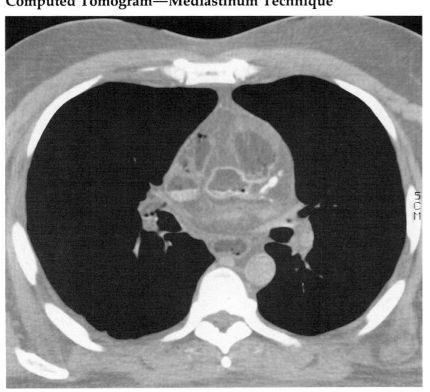

**Computed Tomogram—Mediastinum Technique**

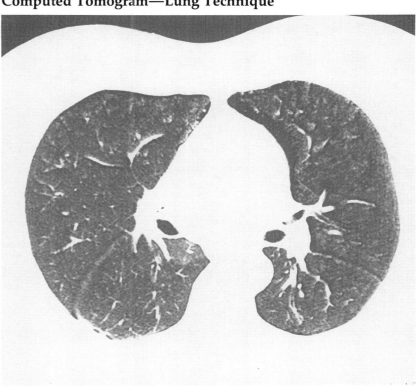

**Computed Tomogram—Lung Technique**

# Axial

## Section 21

## Anatomic Key

### Arteries
A 2* main pulmonary, right (interlobar part)
A 11 RLL, superior branch
A 21 LUL, superior lingular branch
A 22 LUL, inferior lingular branch
A 23 left lower lobe (LLL)
A 30 aorta, ascending
A 31 aorta, descending
A 45 intercostal, posterior, left
A 46 intercostal, posterior, right
A 47 internal thoracic (mammary), left

### Bronchi
B 6 intermediate, right
B 7 right middle lobe (RML)
B 8 RML, lateral segment
B 9 RML, medial segment
B 11 RLL, superior segment
B 20 LUL, lingula
B 21 LUL, superior lingular segment
B 22 LUL, inferior lingular segment
B 23 left lower lobe (LLL)
B 24 LLL, superior segment

### Duct
D 1 thoracic

### Esophagus
E 1 esophagus

### Fat
f 3 extrapleural
f 5 pericardial

### Fissures
F 1 left major (oblique)
F 2 right major (oblique)

### Heart
H 1 atrium, left
H 2 atrium, right
H 3 auricle, left
H 8 coronary artery, circumflex
H 9 coronary artery, anterior descending
H 10 coronary artery, left
H 13 pericardium, parietal serous
H 14 pericardium, visceral serous
H 15 septum, interatrial
H 22 ventricle, right

### Lobes
L 4 RUL, anterior segment
L 6 RML, lateral segment
L 7 RML, medial segment
L 9 RLL, superior segment
L 18 LUL, superior lingular segment
L 19 LUL, inferior lingular segment
L 21 LLL, superior segment

### Muscles
M 14 iliocostalis, thoracis
M 16 intercostal
M 21 latissimus dorsi
M 27 longissimus, thoracis
M 31 pectoralis, major
M 32 pectoralis, minor
M 35 rhomboid, major
M 43 semispinalis, thoracis
M 44 serratus, anterior
M 49 spinalis, thoracis
M 59 teres major
M 61 transversospinalis
M 62 transversus thoracis
M 63 trapezius

### Neural Structures
n 5 spinal cord
n 6 spinal dura

### Nodes
N 8 parasternal (internal thoracic)

### Pleural Structures
P 1 anterior junction line
P 2 costal (parietal)
P 6 visceral

### Skeletal Structures
S 4 rib, body, anterior (4th)
S 5 rib, body, lateral (5th)
S 6 rib, body, posterior (6th)
S 7 rib, costal cartilage (4th)
S 8 rib, head and neck (7th)
S 17 sternum, body
S 21 vertebra, body (T6)
S 22 vertebra, costal facet (T6)
S 23 vertebra, inferior articular process (T6)
S 27 vertebra, spinous process (T6)
S 28 vertebra, superior articular process (T7)

### Veins
v 0 pulmonary, superior, RUL
v 1 pulmonary, superior, RML
v 2 pulmonary, superior, LUL
v 3 pulmonary, superior, LUL, lingula
v 4 pulmonary, inferior, RLL
v 7 azygos
v 18 intercostal, posterior, left
v 19 intercostal, posterior, right
v 20 internal thoracic (mammary), left
v 28 lateral thoracic, left
v 39 vena cava, superior

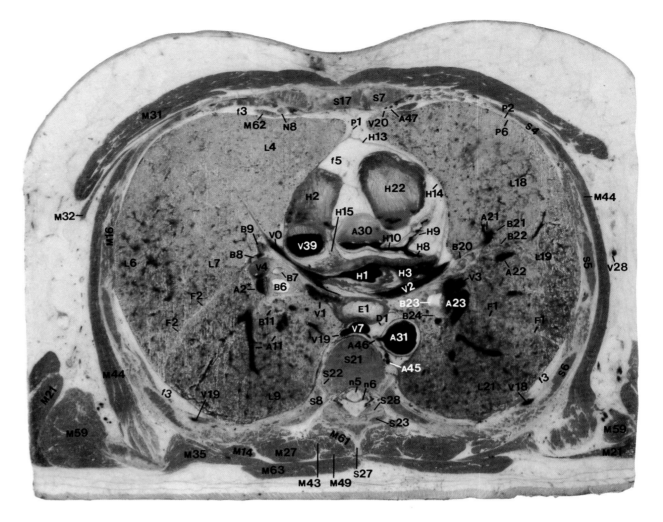

**Anatomic Specimen**

**Magnetic Resonance**

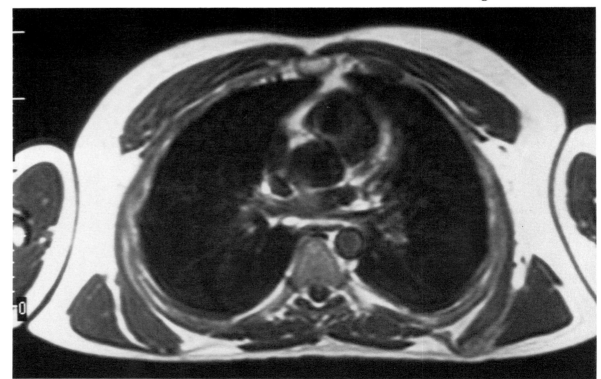

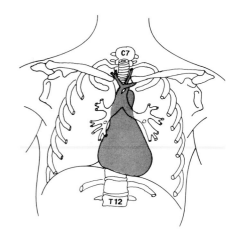

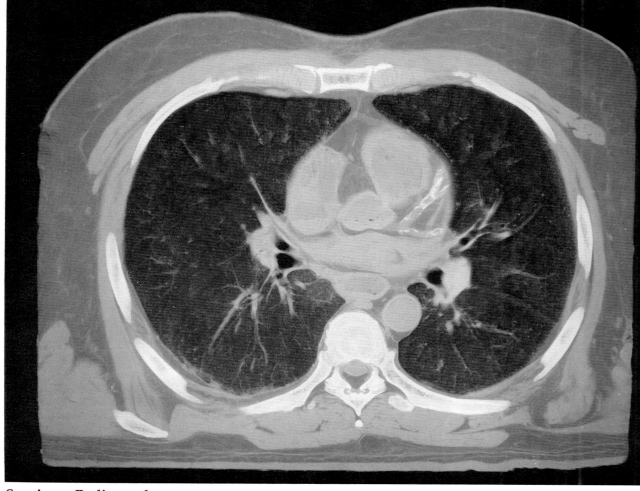

**Specimen Radiograph**

**Computed Tomogram—Mediastinum Technique**

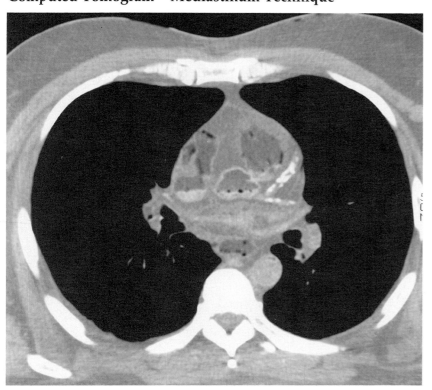

**Computed Tomogram—Lung Technique**

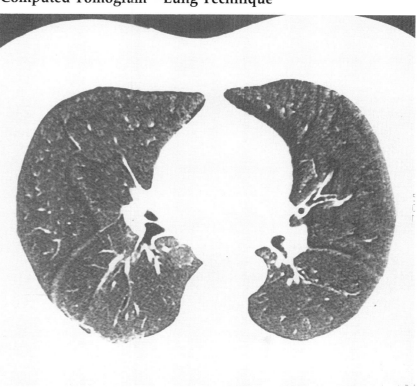

# Axial

## Section 22

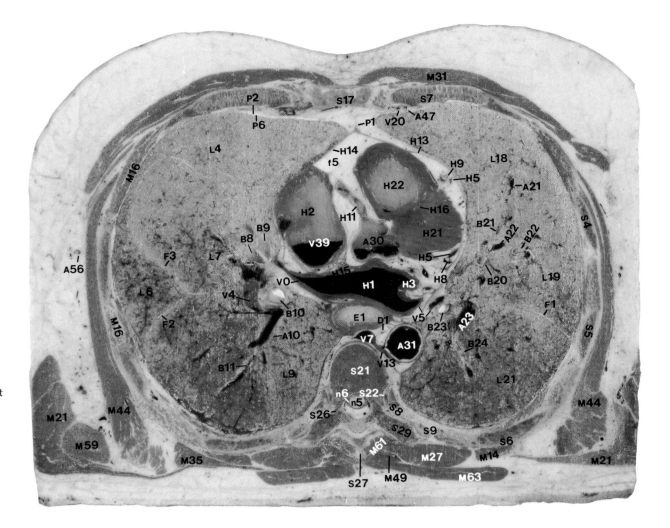

**Anatomic Specimen**

## Anatomic Key

### Arteries
A 10 right lower lobe (RLL)
A 21 LUL, superior lingular branch
A 22 LUL, inferior lingular branch
A 23 left lower lobe (LLL)
A 30 aorta, ascending
A 31 aorta, descending
A 47 internal thoracic (mammary), left
A 56 thoracic, lateral, right

### Bronchi
B 8 RML, lateral segment
B 9 RML, medial segment
B 10 right lower lobe (RLL)
B 11 RLL, superior segment
B 20 LUL, lingula
B 21 LUL, superior lingular segment
B 22 LUL, inferior lingular segment
B 23 left lower lobe (LLL)
B 24 LLL, superior segment

### Duct
D 1 thoracic

### Esophagus
E 1 esophagus

### Fat
f 5 pericardial

### Fissures
F 1 left major (oblique)
F 2 right major (oblique)
F 3 right minor (horizontal)

### Heart
H 1 atrium, left
H 2 atrium, right
H 3 auricle, left
H 5 cardiac vein, great
H 8 coronary artery, circumflex
H 9 coronary artery, anterior descending
H 11 coronary artery, right
H 13 pericardium, parietal serous
H 14 pericardium, visceral serous
H 15 septum, interatrial
H 16 septum, interventricular
H 21 ventricle, left
H 22 ventricle, right

### Lobes
L 4 RUL, anterior segment
L 6 RML, lateral segment
L 7 RML, medial segment
L 9 RLL, superior segment
L 18 LUL, superior lingular segment
L 19 LUL, inferior lingular segment
L 21 LLL, superior segment

### Muscles
M 14 iliocostalis, thoracis
M 16 intercostal
M 21 latissimus dorsi
M 27 longissimus, thoracis
M 31 pectoralis, major
M 35 rhomboid, major
M 44 serratus, anterior
M 49 spinalis, thoracis
M 59 teres major
M 61 transversospinalis
M 63 trapezius

### Neural Structures
n 5 spinal cord
n 6 spinal dura

### Pleural Structures
P 1 anterior junction line
P 2 costal (parietal)
P 6 visceral

### Skeletal Structures
S 4 rib, body, anterior (5th)
S 5 rib, body, lateral (6th)
S 6 rib, body, posterior (7th)
S 7 rib, costal cartilage (4th)
S 8 rib, head and neck (7th)
S 9 rib, tubercle (7th)
S 17 sternum, body
S 21 vertebra, body (T7)
S 22 vertebra, costal facet (T7)
S 26 vertebra, pedicle (T7)
S 27 vertebra, spinous process (T6)
S 29 vertebra, transverse process (T7)

### Veins
v 0 pulmonary, superior, RUL
v 4 pulmonary, inferior, RLL
v 5 pulmonary, inferior, LLL
v 7 azygos
v 13 hemiazygos, accessory
v 20 internal thoracic (mammary), left
v 39 vena cava, superior

**Magnetic Resonance**

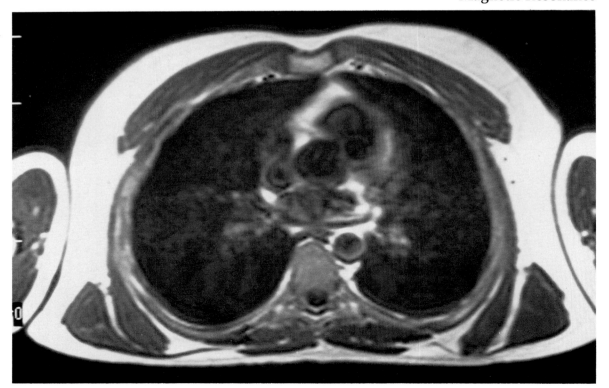

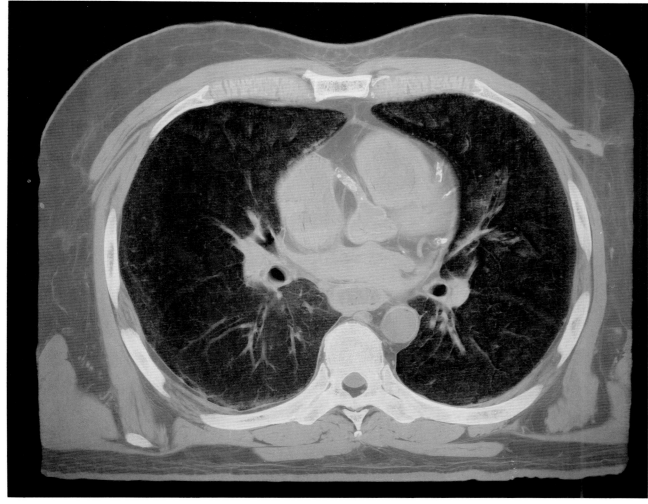

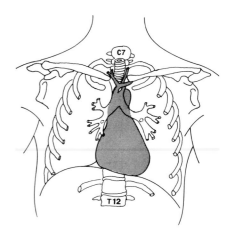

**Specimen Radiograph**

**Computed Tomogram—Mediastinum Technique**

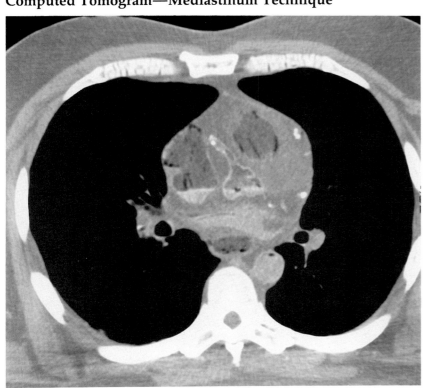

**Computed Tomogram—Lung Technique**

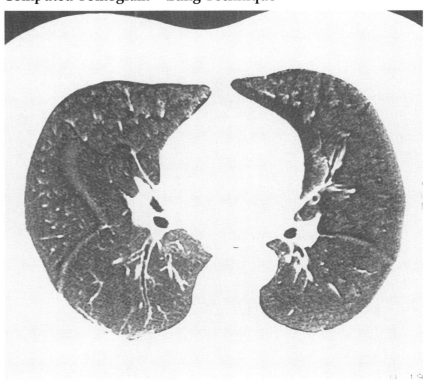

# Axial

## Section 23

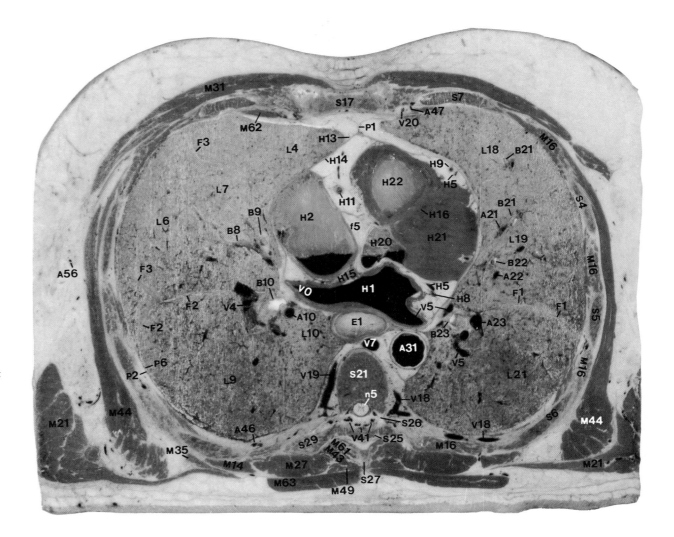

**Anatomic Specimen**

## Anatomic Key

### Arteries
A 10 right lower lobe (RLL)
A 21 LUL, superior lingular branch
A 22 LUL, inferior lingular branch
A 23 left lower lobe (LLL)
A 31 aorta, descending
A 46 intercostal, posterior, right
A 47 internal thoracic (mammary), left
A 56 thoracic, lateral, right

### Bronchi
B 8 RML, lateral segment
B 9 RML, medial segment
B 10 right lower lobe (RLL)
B 21 LUL, superior lingular segment
B 22 LUL, inferior lingular segment
B 23 left lower lobe (LLL)

### Esophagus
E 1 esophagus

### Fat
f 5 pericardial

### Fissures
F 1 left major (oblique)
F 2 right major (oblique)
F 3 right minor (horizontal)

### Heart
H 1 atrium, left
H 2 atrium, right
H 5 cardiac vein, great
H 8 coronary artery, circumflex
H 9 coronary artery, anterior descending
H 11 coronary artery, right
H 13 pericardium, parietal serous
H 14 pericardium, visceral serous
H 15 septum, interatrial
H 16 septum, interventricular
H 20 valve, aortic
H 21 ventricle, left
H 22 ventricle, right

### Lobes
L 4 RUL, anterior segment
L 6 RML, lateral segment
L 7 RML, medial segment
L 9 RLL, superior segment
L 10 RLL, medial basal segment
L 18 LUL, superior lingular segment
L 19 LUL, inferior lingular segment
L 21 LLL, superior segment

### Muscles
M 14 iliocostalis, thoracis
M 16 intercostal
M 21 latissimus dorsi
M 27 longissimus, thoracis
M 31 pectoralis, major
M 35 rhomboid, major
M 43 semispinalis, thoracis
M 44 serratus, anterior
M 49 spinalis, thoracis
M 61 transversospinalis
M 62 transversus thoracis
M 63 trapezius

### Neural Structures
n 5 spinal cord

### Pleural Structures
P 1 anterior junction line
P 2 costal (parietal)
P 6 visceral

### Skeletal Structures
s 4 rib, body, anterior (5th)
s 5 rib, body, lateral (6th)
s 6 rib, body, posterior (7th)
s 7 rib, costal cartilage (4th)
s 17 sternum, body
s 21 vertebra, body (T7)
s 25 vertebra, lamina (T7)
s 26 vertebra, pedicle (T7)
s 27 vertebra, spinous process (T6)
s 29 vertebra, transverse process (T7)

### Veins
v 0 pulmonary, superior, RUL
v 4 pulmonary, inferior, RLL
v 5 pulmonary, inferior, LLL
v 7 azygos
v 18 intercostal, posterior, left
v 19 intercostal, posterior, right
v 20 internal thoracic (mammary), left
v 41 venous plexus, vertebral (internal)

**Magnetic Resonance**

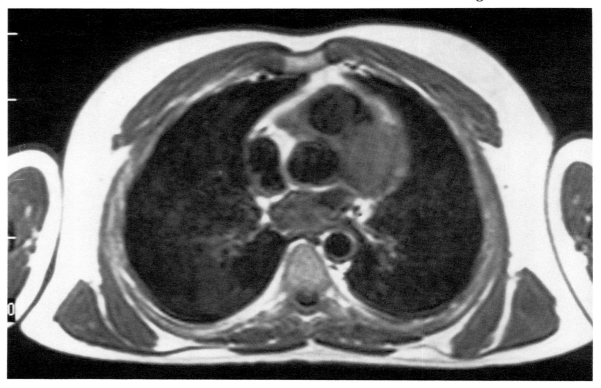

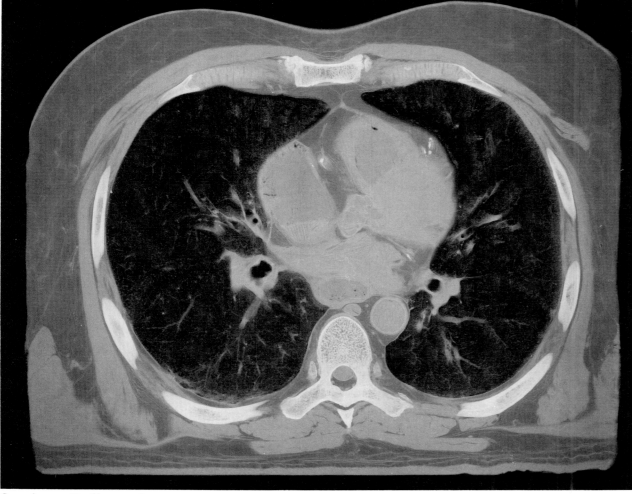

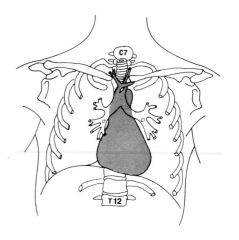

**Specimen Radiograph**

**Computed Tomogram—Mediastinum Technique**

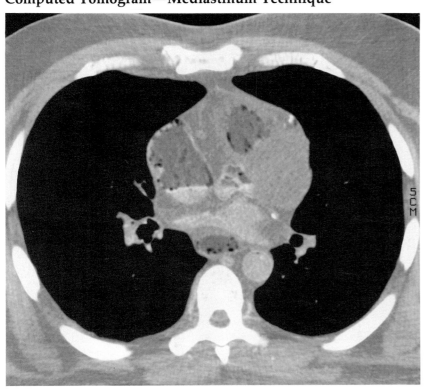

**Computed Tomogram—Lung Technique**

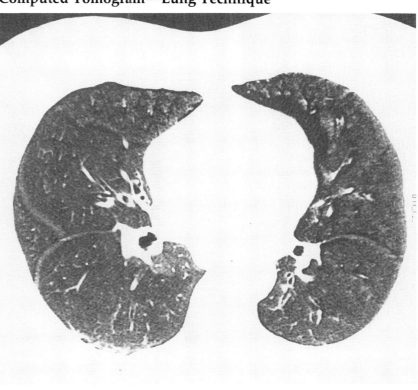

## Anatomic Key

### Arteries
A 12 RLL, medial basal branch
A 14 RLL, lateral basal branch
A 21 LUL, superior lingular branch
A 23 left lower lobe (LLL)
A 31 aorta, descending
A 47 internal thoracic (mammary), left
A 55 thoracic, lateral, left
A 56 thoracic, lateral, right

### Bronchi
B 8 RML, lateral segment
B 9 RML, medial segment
B 10 right lower lobe (RLL)
B 12 RLL, medial basal segment
B 13 RLL, anterior basal segment
B 14 RLL, lateral basal segment
B 15 RLL, posterior basal segment
B 21 LUL, superior lingular segment
B 22 LUL, inferior lingular segment
B 23 left lower lobe (LLL)
B 25 LLL, anteromedial basal segment

### Esophagus
E 1 esophagus

### Fat
f 3 extrapleural
f 5 pericardial

### Fissures
F 1 left major (oblique)
F 2 right major (oblique)
F 3 right minor (horizontal)

### Heart
H 1 atrium, left
H 2 atrium, right
H 5 cardiac vein, great
H 8 coronary artery, circumflex
H 9 coronary artery, anterior descending
H 11 coronary artery, right
H 13 pericardium, parietal serous
H 14 pericardium, visceral serous
H 15 septum, interatrial
H 16 septum, interventricular
H 20 valve, aortic
H 21 ventricle, left
H 22 ventricle, right

### Lobes
L 4 RUL, anterior segment
L 6 RML, lateral segment
L 7 RML, medial segment
L 10 RLL, medial basal segment
L 11 RLL, anterior basal segment
L 12 RLL, lateral basal segment
L 13 RLL, posterior basal segment
L 18 LUL, superior lingular segment
L 19 LUL, inferior lingular segment
L 21 LLL, superior segment
L 22 LLL, anteromedial basal segment

### Muscles
M 14 iliocostalis, thoracis
M 16 intercostal
M 21 latissimus dorsi
M 27 longissimus, thoracis
M 31 pectoralis, major
M 44 serratus, anterior
M 49 spinalis, thoracis
M 61 transversospinalis
M 63 trapezius

### Neural Structures
n 5 spinal cord
n 6 spinal dura

### Pleural Structures
P 1 anterior junction line
P 2 costal (parietal)
P 6 visceral

### Skeletal Structures
s 4 rib, body, anterior (5th)
s 5 rib, body, lateral (6th)
s 6 rib, body, posterior (7th)
s 7 rib, costal cartilage (4th)
s 8 rib, head and neck (8th)
s 17 sternum, body
s 21 vertebra, body (T7)
s 24 vertebra, intervertebral foramen (T7–T8)
s 25 vertebra, lamina
s 27 vertebra, spinous process (T7)

### Veins
v 4 pulmonary, inferior, RLL
v 5 pulmonary, inferior, LLL
v 7 azygos
v 18 intercostal, posterior, left
v 19 intercostal, posterior, right
v 20 internal thoracic (mammary), left
v 28 lateral thoracic, left

**Anatomic Specimen**

**Magnetic Resonance**

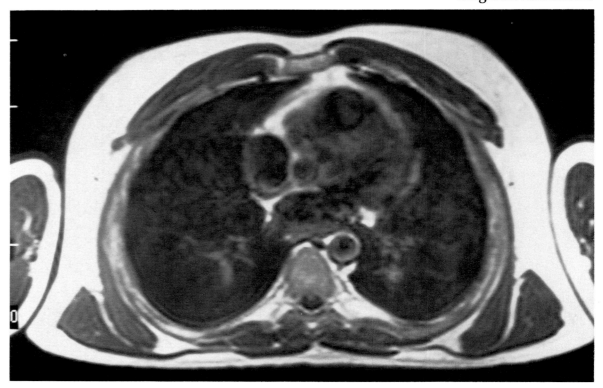

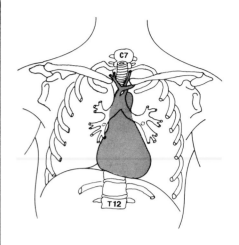

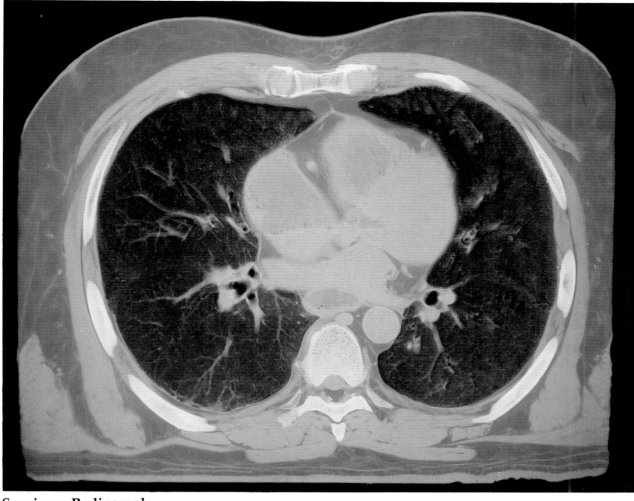

**Specimen Radiograph**

**Computed Tomogram—Mediastinum Technique**

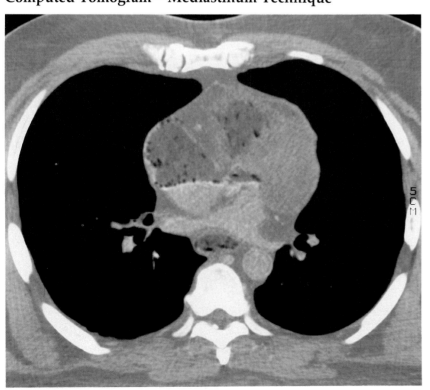

**Computed Tomogram—Lung Technique**

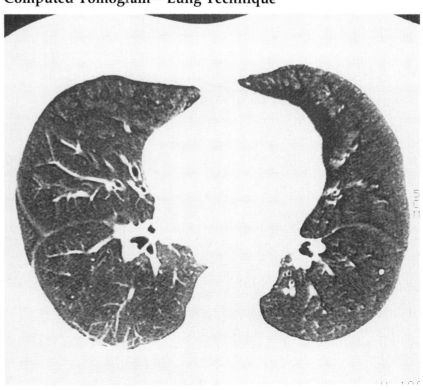

# Axial

## Section 25

### Anatomic Key

#### Arteries
A 12 RLL, medial basal branch
A 14 RLL, lateral basal branch
A 15 RLL, posterior basal branch
A 21 LUL, superior lingular branch
A 25 LLL, anteromedial basal branch
A 27 LLL, posterior basal branch
A 31 aorta, descending
A 55 thoracic, lateral, left
A 56 throacic, lateral, right

#### Bronchi
B 8 RML, lateral segment
B 9 RML, medial segment
B 12 RLL, medial basal segment
B 13 RLL, anterior basal segment
B 14 RLL, lateral basal segment
B 15 RLL, posterior basal segment
B 21 LUL, superior lingular segment
B 22 LUL, inferior lingular segment
B 25 LLL, anteromedial basal
     segment
B 26 LLL, lateral basal segment
B 27 LLL, posterior basal segment

#### Esophagus
E 1 esophagus

#### Fat
f 3 extrapleural
f 4 mediastinal
f 5 pericardial

#### Fissures
F 1 left major (oblique)
F 2 right major (oblique)

#### Heart
H 1 atrium, left
H 2 atrium, right
H 5 cardiac vein, great
H 8 coronary artery, circumflex
H 9 coronary artery, anterior
     descending
H 11 coronary artery, right
H 15 septum, interatrial
H 16 septum, interventricular
H 21 ventricle, left
H 22 ventricle, right

#### Lobes
L 4 RUL, anterior segment
L 6 RML, lateral segment
L 7 RML, medial segment
L 10 RLL, medial basal segment
L 11 RLL, anterior basal segment
L 12 RLL, lateral basal segment
L 13 RLL, posterior basal segment
L 18 LUL, superior lingular segment
L 19 LUL, inferior lingular segment
L 22 LLL, anteromedial basal
     segment
L 23 LLL, lateral basal segment
L 24 LLL, posterior basal segment

#### Muscles
M 14 iliocostalis, thoracis
M 16 intercostal
M 21 latissimus dorsi
M 27 longissimus, thoracis
M 31 pectoralis, major
M 44 serratus, anterior
M 49 spinalis, thoracis
M 61 transversospinalis
M 63 trapezius

#### Neural Structures
n 5 spinal cord

#### Pleural Structures
P 1 anterior junction line
P 2 costal (parietal)
P 6 visceral

#### Skeletal Structures
S 4 rib, body, anterior (4th)
S 5 rib, body, lateral (6th)
S 6 rib, body, posterior (7th)
S 7 rib, costal cartilage (4th)
S 8 rib, head and neck (8th)
S 17 sternum, body
S 20 intervertebral disc (T7–T8)
S 22 vertebra, costal facet (T8)
S 23 vertebra, inferior articular
     process (T7)
S 27 vertebra, spinous process (T7)
S 28 vertebra, superior articular
     process (T8)

#### Veins
V 4 pulmonary, inferior, RLL
V 5 pulmonary, inferior, LLL
V 7 azygos
V 18 intercostal, posterior, left
V 19 intercostal, posterior, right
V 28 lateral thoracic, left
V 41 venous plexus, vertebral
     (internal)

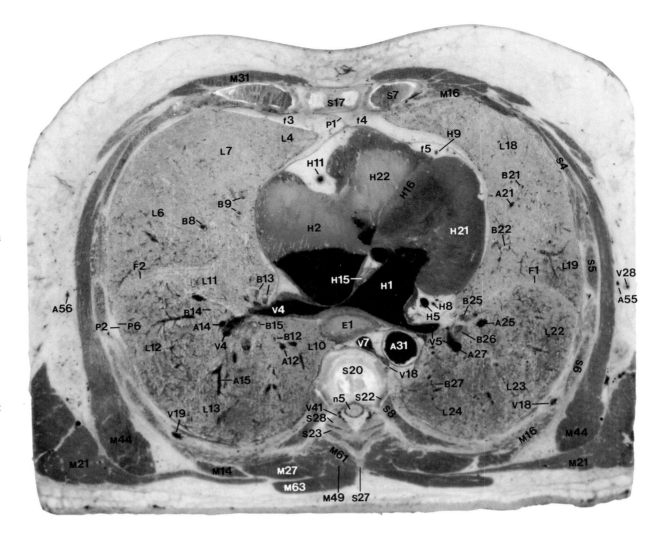

**Anatomic Specimen**

**Magnetic Resonance**

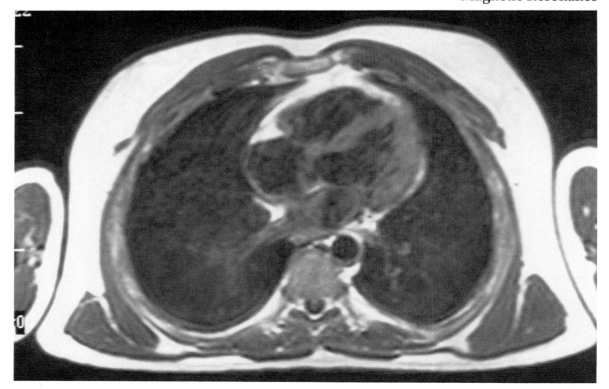

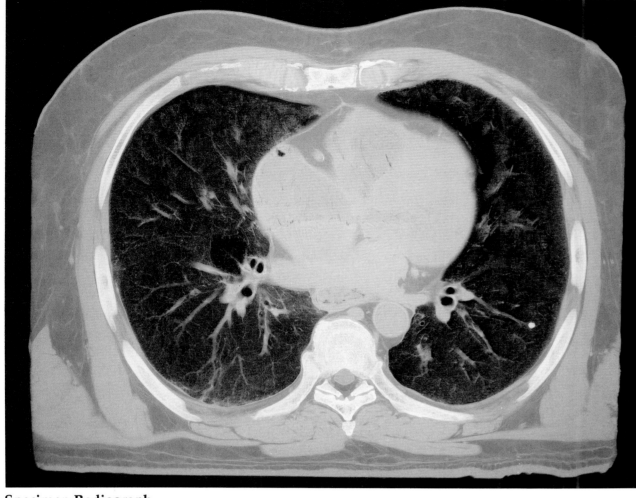

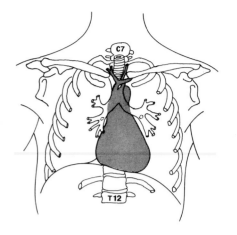

**Specimen Radiograph**

**Computed Tomogram—Mediastinum Technique**

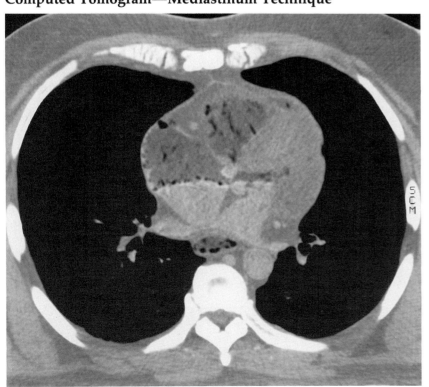

**Computed Tomogram—Lung Technique**

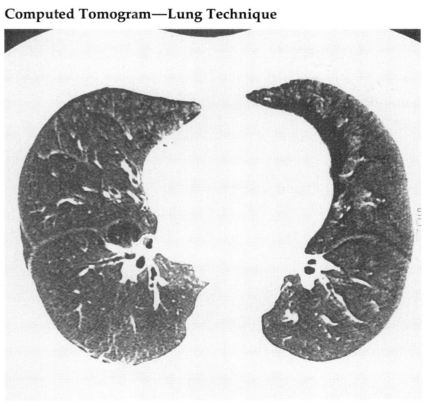

# Axial

## Section 26

### Anatomic Key

**Arteries**
A **13** RLL, anterior basal branch
A **14** RLL, lateral basal branch
A **15** RLL, posterior basal branch
A **27** LLL, posterior basal branch
A **31** aorta, descending
A **55** thoracic, lateral, left
A **56** thoracic, lateral, right

**Bronchi**
B **8** RML, lateral segment
B **9** RML, medial segment
B **12** RLL, medial basal segment
B **13** RLL, anterior basal segment
B **14** RLL, lateral basal segment
B **15** RLL, posterior basal segment
B **21** LUL, superior lingular segment
B **22** LUL, inferior lingular segment
B **25** LLL, anteromedial basal segment
B **26** LLL, lateral basal segment
B **27** LLL, posterior basal segment

**Esophagus**
E **1** esophagus

**Fascia**
F **6** perivisceral
F **7** prevertebral

**Fat**
f **3** extrapleural
f **4** mediastinal
f **5** pericardial
f **7** subcutaneous

**Fissures**
F **1** left major (oblique)
F **2** right major (oblique)

**Heart**
H **1** atrium, left
H **2** atrium, right
H **5** cardiac vein, great
H **8** coronary artery, circumflex
H **9** coronary artery, descending
H **11** coronary artery, right
H **13** pericardium, parietal serous
H **14** pericardium, visceral serous
H **15** septum, interatrial
H **16** septum, interventricular
H **17** valve, tricuspid
H **19** valve, mitral
H **21** ventricle, left
H **22** ventricle, right

**Lobes**
L **4** RUL, anterior segment
L **6** RML, lateral segment
L **7** RML, medial segment
L **10** RLL, medial basal segment
L **11** RLL, anterior basal segment
L **12** RLL, lateral basal segment
L **13** RLL, posterior basal segment
L **18** LUL, superior lingular segment
L **19** LUL, inferior lingular segment
L **22** LLL, anteromedial basal segment
L **23** LLL, lateral basal segment
L **24** LLL, posterior basal segment

**Muscles**
M **14** iliocostalis, thoracis
M **16** intercostal
M **21** latissimus dorsi
M **27** longissimus, thoracis
M **44** serratus, anterior
M **49** spinalis, thoracis
M **61** transversospinalis
M **62** transversus thoracis
M **63** trapezius

**Neural Structures**
n **5** spinal cord
n **6** spinal dura

**Pleural Structures**
P **1** anterior junction line
P **2** costal (parietal)
P **6** visceral

**Skeletal Structures**
S **4** rib, body, anterior (5th)
S **5** rib, body, lateral (6th)
S **6** rib, body, posterior (7th)
S **7** rib, costal cartilage (4th)
S **8** rib, head and neck (8th)
S **9** rib, tubercle (8th)
S **17** sternum, body
S **21** vertebra, body (T8)
S **22** vertebra, costal facet (T8)
S **27** vertebra, spinous process (T8)
S **29** vertebra, transverse process (T8)

**Veins**
V **4** pulmonary, inferior, RLL
V **5** pulmonary, inferior, LLL
V **7** azygos
V **18** intercostal, posterior, left
V **19** intercostal, posterior, right
V **28** lateral thoracic, left
V **29** lateral thoracic, right
V **41** venous plexus, vertebral (internal)

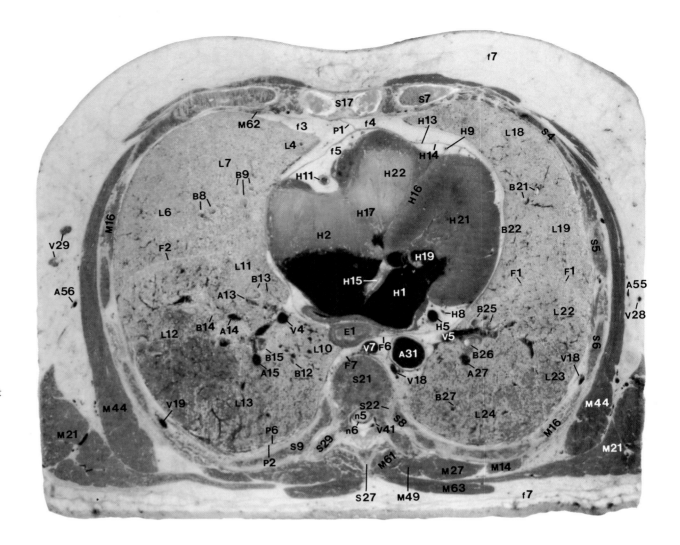

**Anatomic Specimen**

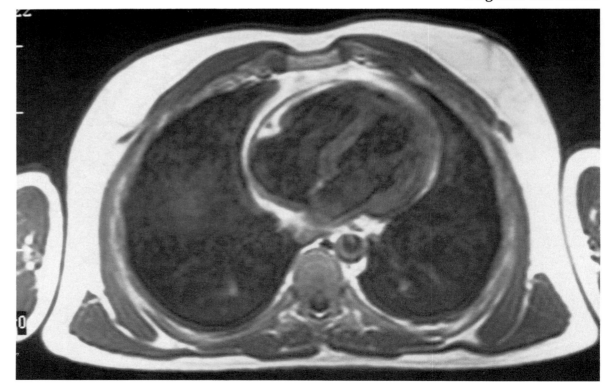

**Magnetic Resonance**

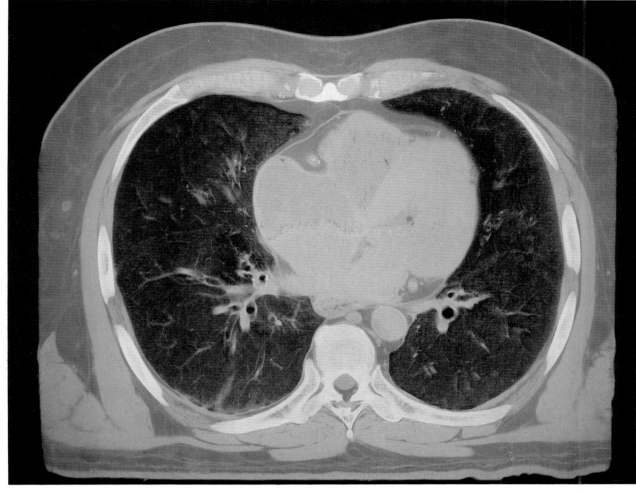

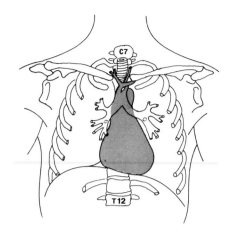

**Specimen Radiograph**

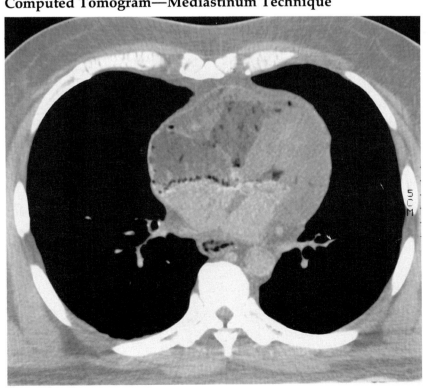

**Computed Tomogram—Mediastinum Technique**

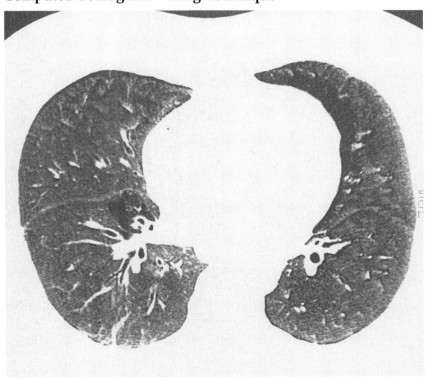

**Computed Tomogram—Lung Technique**

# Axial

## Section 27

### Anatomic Key

**Arteries**
A 14 RLL, lateral basal branch
A 15 RLL, posterior basal branch
A 27 LLL, posterior basal branch
A 31 aorta, descending
A 45 intercostal, posterior, left
A 47 internal thoracic (mammary), left
A 56 thoracic, lateral, right

**Bronchi**
B 8 RML, lateral segment
B 9 RML, medial segment
B 12 RLL, medial basal segment
B 13 RLL, anterior basal segment
B 14 RLL, lateral basal segment
B 15 RLL, posterior basal segment
B 25 LLL, anteromedial basal segment
B 26 LLL, lateral basal segment
B 27 LLL, posterior basal segment

**Duct**
D 1 thoracic

**Esophagus**
E 1 esophagus

**Fissures**
F 1 left major (oblique)
F 2 right major (oblique)

**Heart**
H 1 atrium, left
H 2 atrium, right
H 5 cardiac vein, great
H 8 coronary artery, circumflex
H 9 coronary artery, descending
H 11 coronary artery, right
H 13 pericardium, parietal serous
H 14 pericardium, visceral serous
H 15 septum, interatrial
H 16 septum, interventricular
H 19 valve, mitral
H 21 ventricle, left
H 22 ventricle, right

**Lobes**
L 6 RML, lateral segment
L 7 RML, medial segment
L 10 RLL, medial basal segment
L 11 RLL, anterior basal segment
L 12 RLL, lateral basal segment
L 13 RLL, posterior basal segment
L 18 LUL, superior lingular segment
L 19 LUL, inferior lingular segment
L 22 LLL, anteromedial basal segment
L 23 LLL, lateral basal segment
L 24 LLL, posterior basal segment

**Muscles**
M 14 iliocostalis, thoracis
M 21 latissimus dorsi
M 27 longissimus, thoracis
M 31 pectoralis, major
M 44 serratus, anterior
M 49 spinalis, thoracis
M 61 transversospinalis
M 62 transversus thoracis
M 63 trapezius

**Neural Structures**
n 5 spinal cord
n 6 spinal dura

**Pleural Structures**
P 1 anterior junction line
P 2 costal (parietal)
P 6 visceral

**Skeletal Structures**
S 4 rib, body, anterior (6th)
S 5 rib, body, lateral (7th)
S 6 rib, body, posterior (8th)
S 7 rib, costal cartilage (5th)
S 9 rib, tubercle (8th)
S 17 sternum, body
S 21 vertebra, body (T8)
S 26 vertebra, pedicle (T8)
S 27 vertebra, spinous process (T8)
S 29 vertebra, transverse process (T8)

**Veins**
V 4 pulmonary, inferior, RLL
V 5 pulmonary, inferior, LLL
V 7 azygos
V 18 intercostal, posterior, left
V 19 intercostal, posterior, right
V 20 internal thoracic (mammary), left
V 28 lateral thoracic, left
V 29 lateral thoracic, right
V 38 vena cava, inferior
V 41 venous plexus, vertebral (internal)

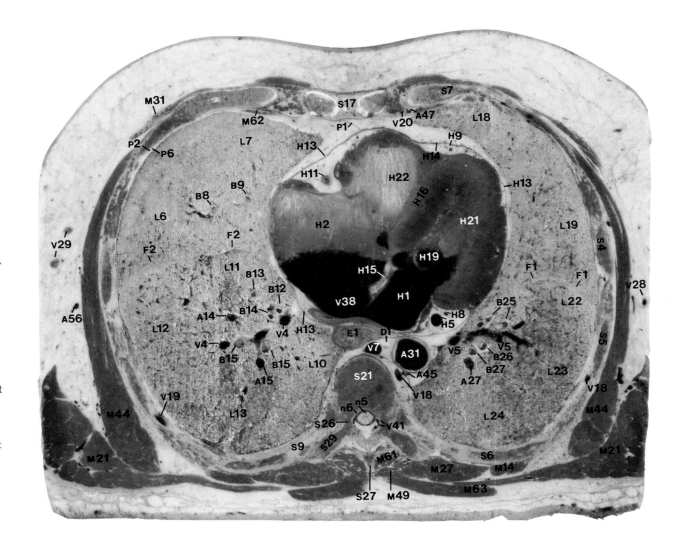

**Anatomic Specimen**

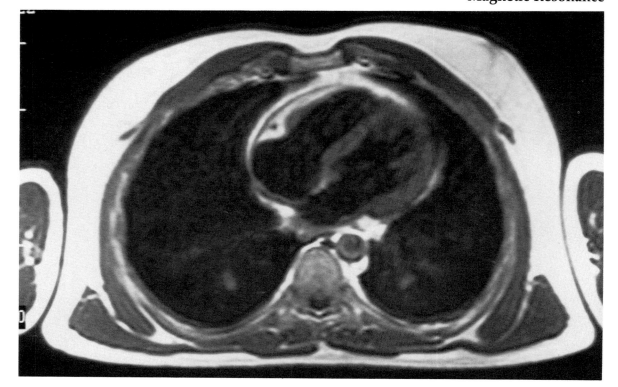

**Magnetic Resonance**

64

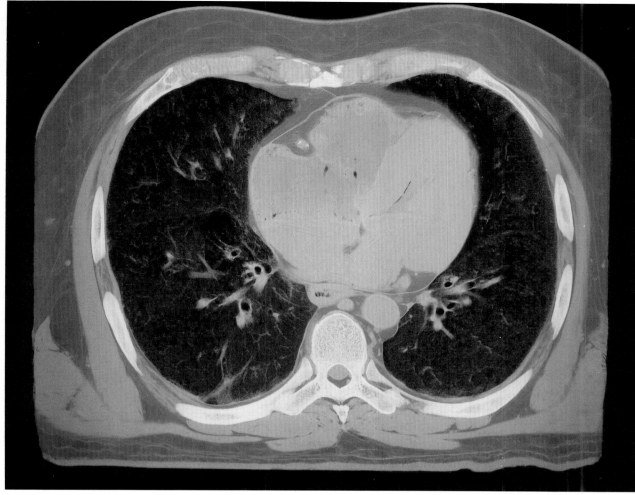

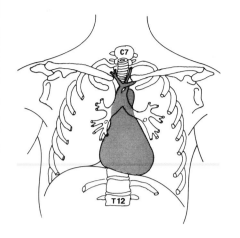

**Specimen Radiograph**

**Computed Tomogram—Mediastinum Technique**

**Computed Tomogram—Lung Technique**

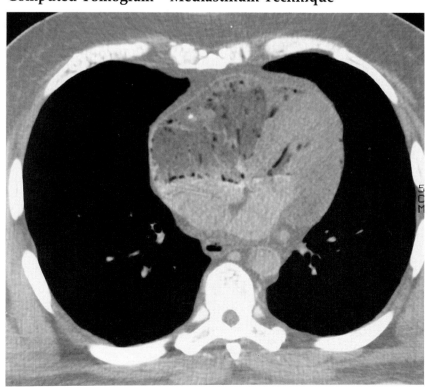

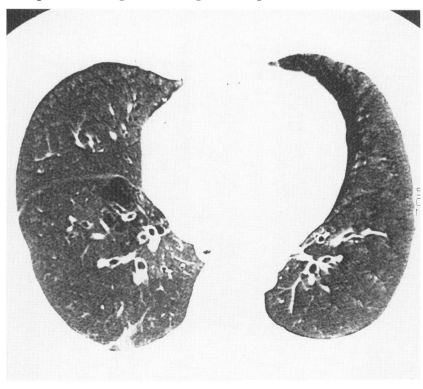

# Axial

## Section 28

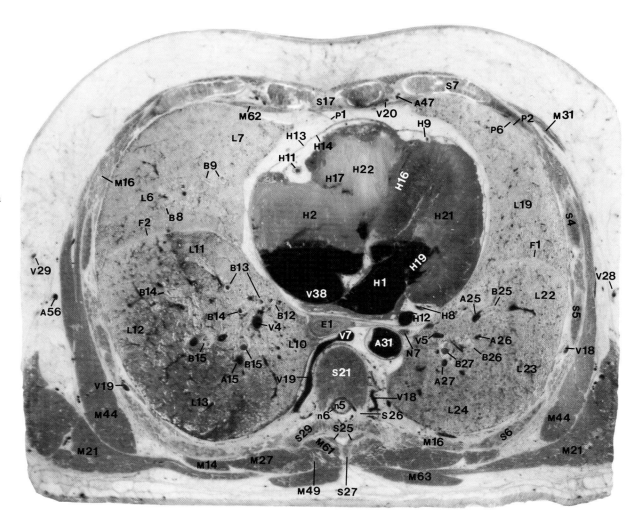

**Anatomic Specimen**

## Anatomic Key

### Arteries
A **15** RLL, posterior basal branch
A **25** LLL, anteromedial basal branch
A **26** LLL, lateral basal branch
A **27** LLL, posterior basal branch
A **31** aorta, descending
A **47** internal thoracic (mammary), left
A **56** thoracic, lateral, right

### Bronchi
B **8** RML, lateral segment
B **9** RML, medial segment
B **12** RLL, medial basal segment
B **13** RLL, anterior basal segment
B **14** RLL, lateral basal segment
B **15** RLL, posterior basal segment
B **25** LLL, anteromedial basal segment
B **26** LLL, lateral basal segment
B **27** LLL, posterior basal segment

### Esophagus
E **1** esophagus

### Fissures
F **1** left major (oblique)
F **2** right major (oblique)

### Heart
H **1** atrium, left
H **2** atrium, right
H **8** coronary artery, circumflex
H **9** coronary artery, anterior descending
H **11** coronary artery, right
H **12** coronary sinus
H **13** pericardium, parietal serous
H **14** pericardium, visceral serous
H **16** septum, interventricular
H **17** valve, tricuspid
H **19** valve, mitral
H **21** ventricle, left
H **22** ventricle, right

### Lobes
L **6** RML, lateral segment
L **7** RML, medial segment
L **10** RLL, medial basal segment
L **11** RLL, anterior basal segment
L **12** RLL, lateral basal segment
L **13** RLL, posterior basal segment
L **19** LUL, inferior lingular segment
L **22** LLL, anteromedial basal segment
L **23** LLL, lateral basal segment
L **24** LLL, posterior basal segment

### Muscles
M **14** iliocostalis, thoracis
M **16** intercostal
M **21** latissimus dorsi
M **27** longissimus, thoracis
M **31** pectoralis, major
M **44** serratus, anterior
M **49** spinalis, thoracis
M **61** transversospinalis
M **62** transversus thoracis
M **63** trapezius

### Neural Structures
n **5** spinal cord
n **6** spinal dura

### Nodes
N **7** mediastinal, posterior

### Pleural Structures
P **1** anterior junction line
P **2** costal (parietal)
P **6** visceral

### Skeletal Structures
s **4** rib, body, anterior (6th)
s **5** rib, body, lateral (7th)
s **6** rib, body, posterior (8th)
s **7** rib, costal cartilage (5th)
s **17** sternum, body
s **21** vertebra, body (T8)
s **25** vertebra, lamina (T8)
s **26** vertebra, pedicle (T8)
s **27** vertebra, spinous process (T8)
s **29** vertebra, transverse process (T8)

### Veins
v **4** pulmonary, inferior (RLL)
v **5** pulmonary, inferior (LLL)
v **7** azygos
v **18** intercostal, posterior, left
v **19** intercostal, posterior, right
v **20** internal thoracic (mammary), left
v **28** lateral thoracic, left
v **29** lateral thoracic, right
v **38** vena cava, inferior

**Magnetic Resonance**

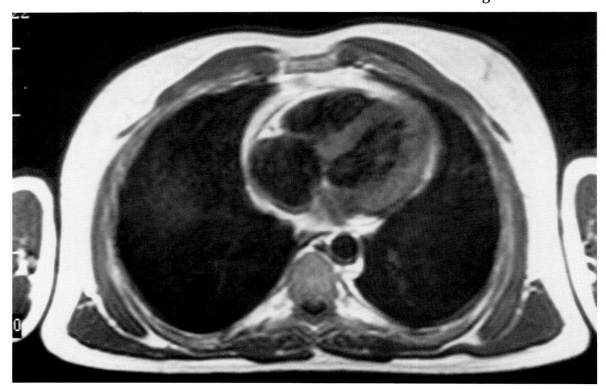

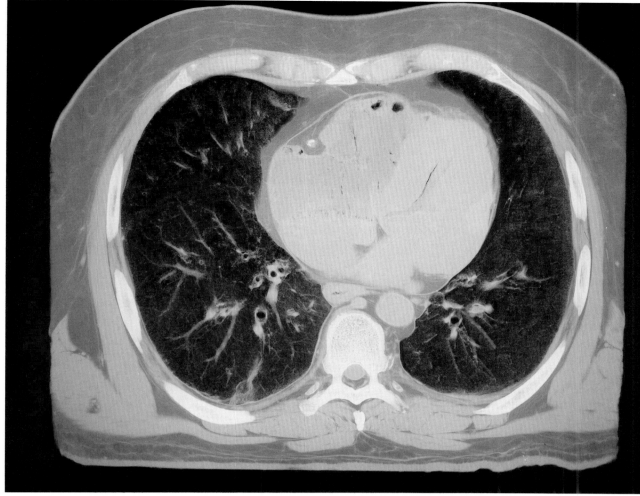

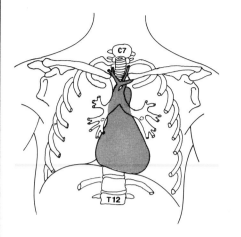

**Specimen Radiograph**

**Computed Tomogram—Mediastinum Technique**

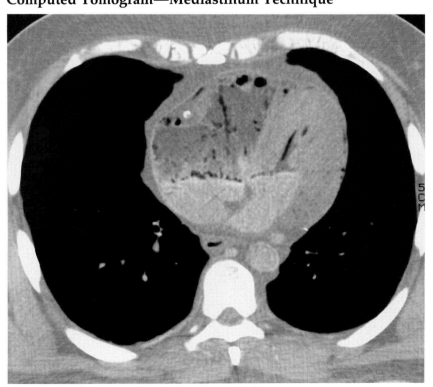

**Computed Tomogram—Lung Technique**

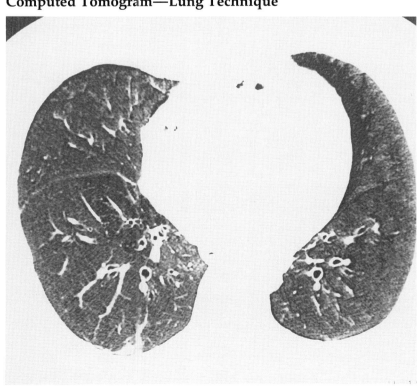

## Anatomic Key

### Arteries
A **14** RLL, lateral basal branch
A **15** RLL, posterior basal branch
A **25** LLL, anteromedial basal branch
A **26** LLL, lateral basal branch
A **27** LLL, posterior basal branch
A **31** aorta, descending
A **48** internal thoracic (mammary), right
A **55** thoracic, lateral, left
A **56** thoracic, lateral, right

### Bronchi
B **9** RML, medial segment
B **12** RLL, medial basal segment
B **13** RLL, anterior basal segment
B **14** RLL, lateral basal segment
B **15** RLL, posterior basal segment
B **25** LLL, anteromedial basal segment
B **26** LLL, lateral basal segment
B **27** LLL, posterior basal segment

### Esophagus
E **1** esophagus

### Fat
f **3** extrapleural
f **4** mediastinal
f **5** pericardial

### Fissures
F **1** left major (oblique)
F **2** right major (oblique)

### Heart
H **1** atrium, left
H **2** atrium, right
H **8** coronary artery, circumflex
H **11** coronary artery, right
H **12** coronary sinus
H **13** pericardium, parietal serous
H **14** pericardium, visceral serous
H **16** septum, interventricular
H **19** valve, mitral
H **21** ventricle, left
H **22** ventricle, right

### Lobes
L **6** RML, lateral segment
L **7** RML, medial segment
L **10** RLL, medial basal segment
L **11** RLL, anterior basal segment
L **12** RLL, lateral basal segment
L **13** RLL, posterior basal segment
L **19** LUL, inferior lingular segment
L **22** LLL, anteromedial basal segment
L **23** LLL, lateral basal segment
L **24** LLL, posterior basal segment

### Muscles
M **14** iliocostalis, thoracis
M **16** intercostal
M **21** latissimus dorsi
M **27** longissimus, thoracis
M **31** pectoralis, major
M **44** serratus, anterior
M **49** spinalis, thoracis
M **61** transversospinalis
M **62** transversus thoracis
M **63** trapezius

### Neural Structures
n **3** phrenic nerve
n **5** spinal cord

### Nodes
N **7** mediastinal, posterior

### Pleural Structures
P **1** anterior junction line
P **2** costal (parietal)
P **5** mediastinal (parietal)
P **6** visceral

### Skeletal Structures
S **3** rib (5th)
S **4** rib, body, anterior (6th)
S **5** rib, body, lateral (7th)
S **6** rib, body, posterior (8th)
S **7** rib, costal cartilage (6th)
S **8** rib, head and neck (9th)
S **19** sternum, xiphoid
S **21** vertebra, body (T8)
S **22** vertebra, costal facet (T8)
S **23** vertebra, inferior articular process (T8)
S **27** vertebra, spinous process (T8)
S **28** vertebra, superior articular process (T9)

### Veins
v **7** azygos
v **18** intercostal, posterior, left
v **19** intercostal, posterior, right
v **21** internal thoracic (mammary), right
v **38** vena cava, inferior
v **41** venous plexus, vertebral (internal)

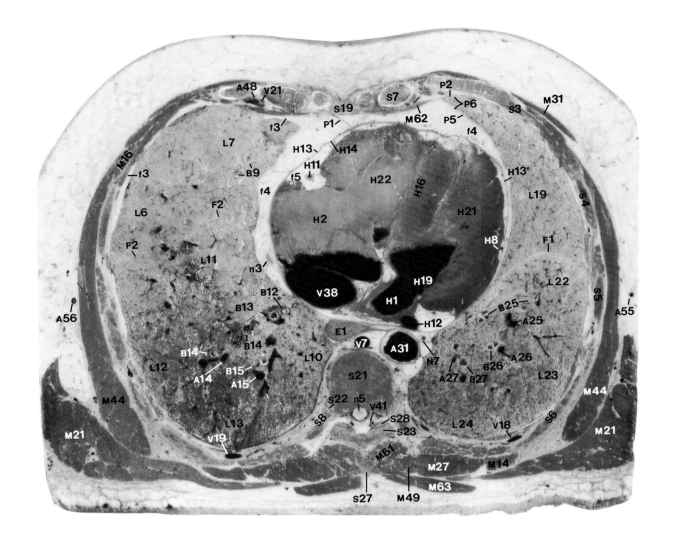

**Anatomic Specimen**

**Magnetic Resonance**

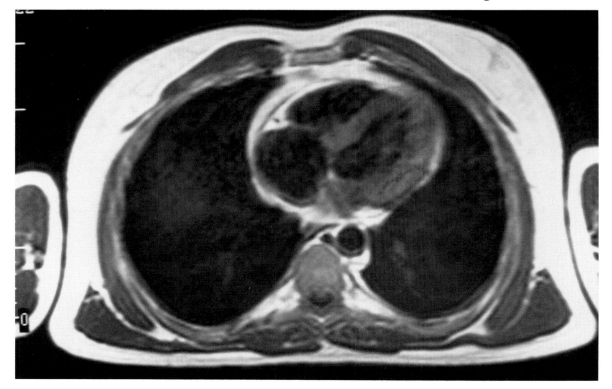

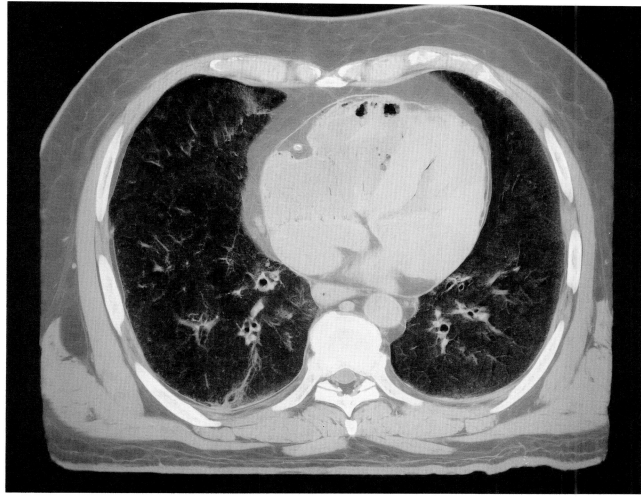

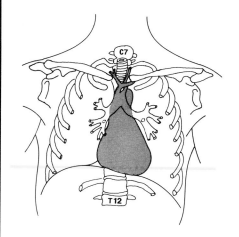

**Specimen Radiograph**

**Computed Tomogram—Mediastinum Technique**

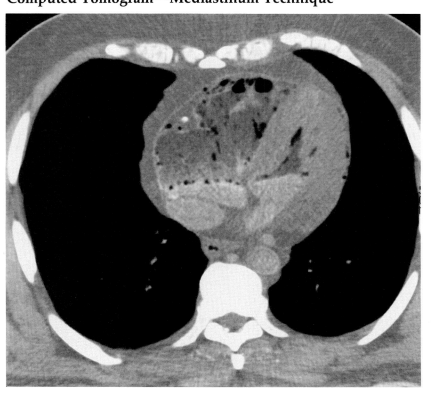

**Computed Tomogram—Lung Technique**

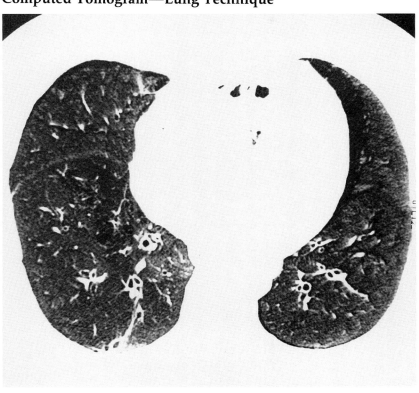

# Axial

## Section 30

## Anatomic Key

### Arteries
A 15 RLL, posterior basal branch
A 31 aorta, descending
A 55 thoracic, lateral, left
A 56 thoracic, lateral, right

### Bronchi
B 12 RLL, medial basal segment
B 13 RLL, anterior basal segment
B 14 RLL, lateral basal segment
B 15 RLL, posterior basal segment
B 25 LLL, anteromedial basal
    segment
B 26 LLL, lateral basal segment
B 27 LLL, posterior basal segment

### Esophagus
E 1 esophagus

### Fascia
F 6 perivisceral
F 7 prevertebral

### Fat
f 3 extrapleural
f 4 mediastinal
f 5 pericardial

### Fissures
F 1 left major (oblique)
F 2 right major (oblique)

### Heart
H 2 atrium, right
H 8 coronary artery, circumflex
H 11 coronary artery, right
H 12 coronary sinus
H 13 pericardium, parietal serous
H 14 pericardium, visceral serous
H 16 septum, interventricular
H 21 ventricle, left
H 22 ventricle, right

### Lobes
L 6 RML, lateral segment
L 7 RML, medial segment
L 10 RLL, medial basal segment
L 11 RLL, anterior basal segment
L 12 RLL, lateral basal segment
L 13 RLL, posterior basal segment
L 19 LUL, inferior lingular segment
L 22 LLL, anteromedial basal
    segment
L 23 LLL, lateral basal segment
L 24 LLL, posterior basal segment

### Muscles
M 14 iliocostalis, thoracis
M 16 intercostal
M 21 latissimus dorsi
M 27 longissimus, thoracis
M 31 pectoralis, major
M 44 serratus, anterior
M 49 spinalis, thoracis
M 61 transversospinalis
M 62 transversus thoracis
M 63 trapezius

### Neural Structures
n 3 phrenic nerve
n 5 spinal cord
n 6 spinal dura

### Skeletal Structures
S 4 rib, body, anterior (6th)
S 5 rib, body, lateral (7th)
S 6 rib, body, posterior (8th)
S 7 rib, costal cartilage (6th)
S 8 rib, head and neck (9th)
S 19 sternum, xiphoid
S 20 intervertebral disc (T8–T9)
S 23 vertebra, inferior articular
    process (T8)
S 27 vertebra, spinous process (T8)
S 28 vertebra, superior articular
    process (T9)

### Veins
v 7 azygos
v 18 intercostal, posterior, left
v 19 intercostal, posterior, right
v 29 lateral thoracic, right
v 38 vena cava, inferior
v 41 venous plexus, vertebral
    (internal)

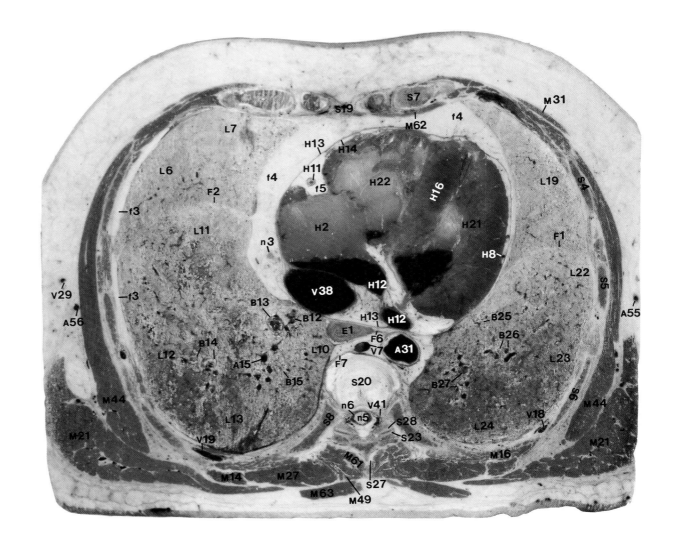

**Anatomic Specimen**

**Magnetic Resonance**

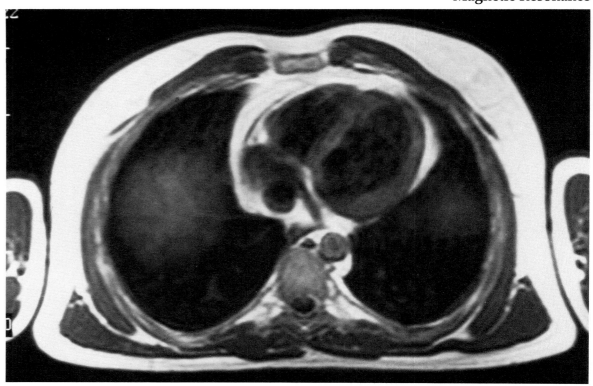

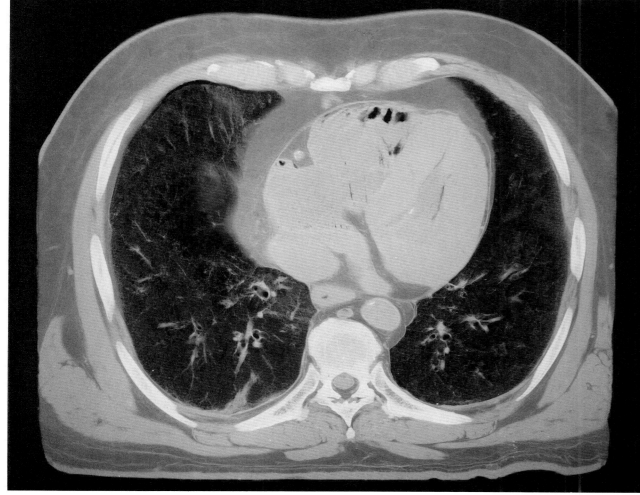

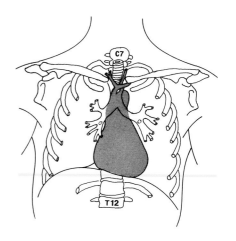

**Specimen Radiograph**

**Computed Tomogram—Mediastinum Technique**

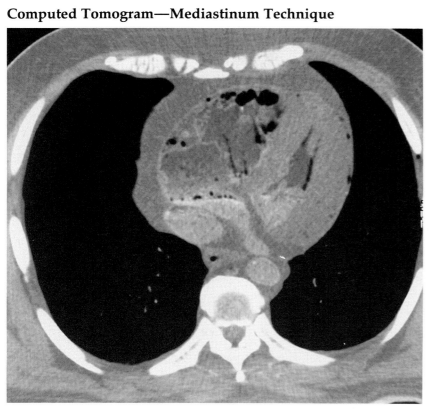

**Computed Tomogram—Lung Technique**

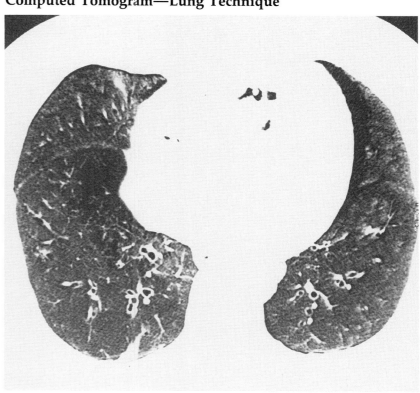

# Axial

## Section 31

## Anatomic Key

### Arteries
A 31 aorta, descending
A 47 internal thoracic (mammary), left
A 56 thoracic, lateral, right

### Bronchi
B 12 RLL, medial basal segment
B 14 RLL, lateral basal segment
B 15 RLL, posterior basal segment
B 26 LLL, lateral basal segment
B 27 LLL, posterior basal segment

### Esophagus
E 1 esophagus

### Fat
f 3 extrapleural
f 4 mediastinal
f 5 pericardial

### Fissures
F 1 left major (oblique)
F 2 right major (oblique)

### Heart
H 6 cardiac vein, small
H 9 coronary artery, posterior descending
H 11 coronary artery, right
H 12 coronary sinus
H 13 pericardium, parietal serous
H 14 pericardium, visceral serous
H 16 septum, interventricular
H 21 ventricle, left
H 22 ventricle, right

### Lobes
L 7 RML, medial segment
L 10 RLL, medial basal segment
L 11 RLL, anterior basal segment
L 12 RLL, lateral basal segment
L 13 RLL, posterior basal segment
L 19 LUL, inferior lingular segment
L 22 LLL, anteromedial basal segment
L 23 LLL, lateral basal segment
L 24 LLL, posterior basal segment

### Muscles
M 1 abdominal, external oblique
M 3 abdominis, rectus
M 9 diaphragm
M 14 iliocostalis, thoracis
M 16 intercostal
M 21 latissimus dorsi
M 27 longissimus, thoracis
M 29 multifidus
M 37 rotator
M 43 semispinalis thoracis
M 44 serratus, anterior
M 49 spinalis, thoracis
M 62 transversus thoracis
M 63 trapezius

### Neural Structures
n 5 spinal cord
n 9 sympathetic nerve chain

### Organs
O 3 liver

### Pleural Structures
P 2 costal (parietal)
P 3 costodiaphragmatic recess
P 6 visceral

### Skeletal Structures
S 4 rib, body, anterior (6th)
S 5 rib, body, lateral (7th)
S 6 rib, body posterior (8th)
S 7 rib, costal cartilage (6th)
S 9 rib, tubercle (9th)
S 19 sternum, xiphoid
S 21 vertebra, body (T9)
S 25 vertebra, lamina (T9)
S 26 vertebra, pedicle (T9)
S 27 vertebra, spinous process (T9)
S 29 vertebra, transverse process (T9)

### Veins
v 7 azygos
v 12 hemiazygos
v 18 intercostal, posterior, left
v 19 intercostal, posterior, right
v 20 internal thoracic (mammary), left
v 38 vena cava, inferior
v 41 venous plexus, vertebral (internal)

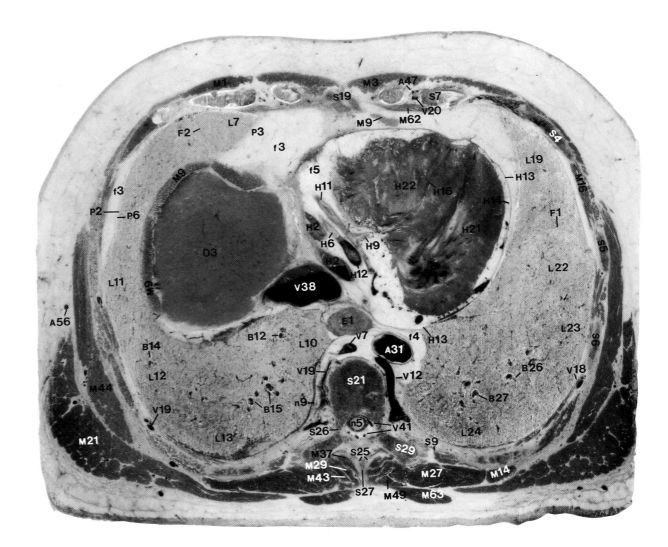

**Anatomic Specimen**

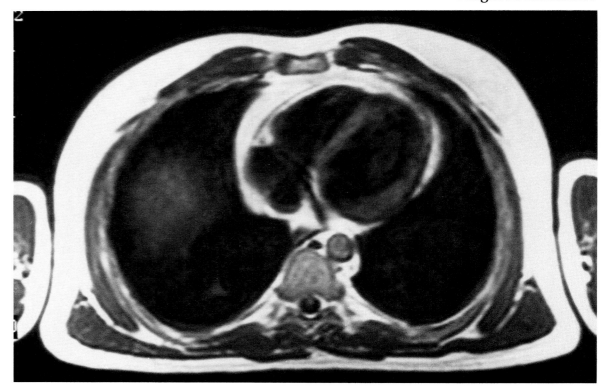

**Magnetic Resonance**

72

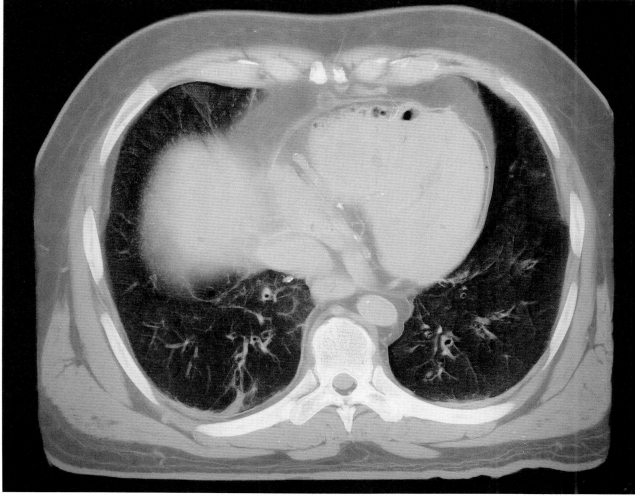

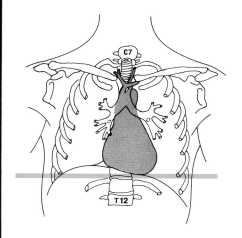

**Specimen Radiograph**

**Computed Tomogram—Mediastinum Technique**

**Computed Tomogram—Lung Technique**

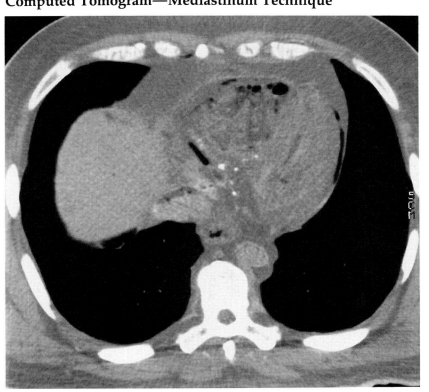

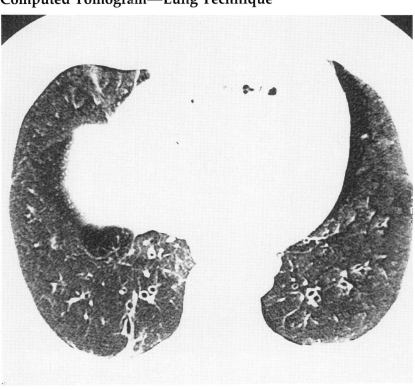

# Axial
## Section 32

### Anatomic Key

**Arteries**
A 31 aorta, ascending
A 45 intercostal, posterior, left
A 46 intercostal, posterior, right
A 56 thoracic, lateral, right

**Bronchi**
B 15 RLL, posterior basal segment
B 26 LLL, lateral basal segment
B 27 LLL, posterior basal segment

**Esophagus**
E 1 esophagus

**Fat**
f 3 extrapleural
f 4 mediastinal
f 5 pericardial

**Fissures**
F 2 right major (oblique)

**Heart**
H 13 pericardium, parietal serous
H 14 pericardium, visceral serous
H 21 ventricle, left
H 22 ventricle, right

**Lobes**
L 10 RLL, medial basal segment
L 11 RLL, anterior basal segment
L 12 RLL, lateral basal segment
L 13 RLL, posterior basal segment
L 19 LUL, inferior lingular segment
L 22 LLL, anteromedial basal segment
L 23 LLL, lateral basal segment
L 24 LLL, posterior basal segment

**Muscles**
M 1 abdominal, external oblique
M 3 abdominis, rectus
M 4 abdominis, transversus
M 9 diaphragm
M 14 iliocostalis, thoracis
M 21 latissimus dorsi
M 27 longissimus, thoracis
M 44 serratus, anterior
M 49 spinalis, thoracis
M 61 transversospinalis
M 63 trapezius

**Neural Structures**
n 5 spinal cord
n 6 spinal dura

**Organs**
O 3 liver

**Pleural Structures**
P 2 costal (parietal)
P 3 costodiaphragmatic recess
P 6 visceral

**Skeletal Structures**
S 4 rib, body, anterior (7th)
S 5 rib, body, lateral (8th)
S 6 rib, body, posterior (9th)
S 7 rib, costal cartilage (6th)
S 8 rib, head and neck (10th)
S 20 intervertebral disc (T9–T10)
S 23 vertebra, inferior articular process (T9)
S 27 vertebra, spinous process (T9)
S 28 vertebra, superior articular process (T10)

**Veins**
v 7 azygos
v 18 intercostal, posterior, left
v 19 intercostal, posterior, right
v 38 vena cava, inferior

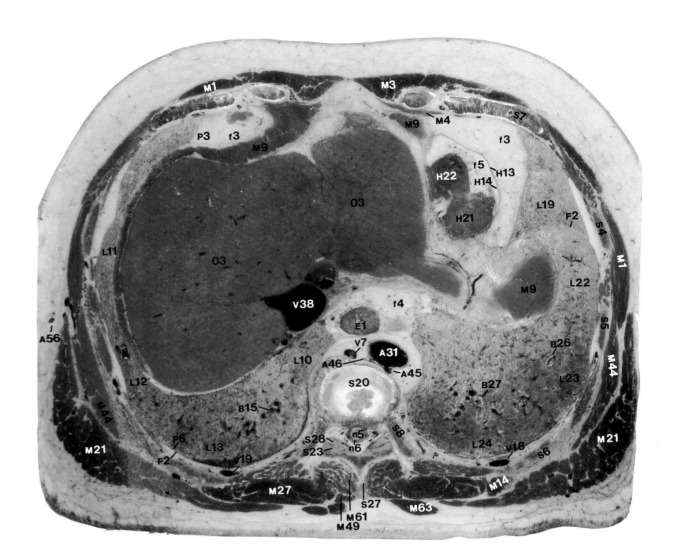

**Anatomic Specimen**

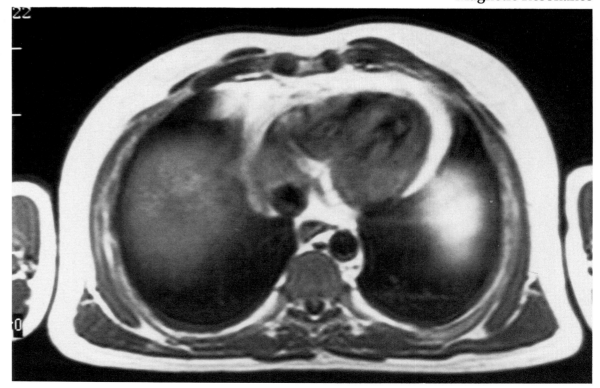

**Magnetic Resonance**

74

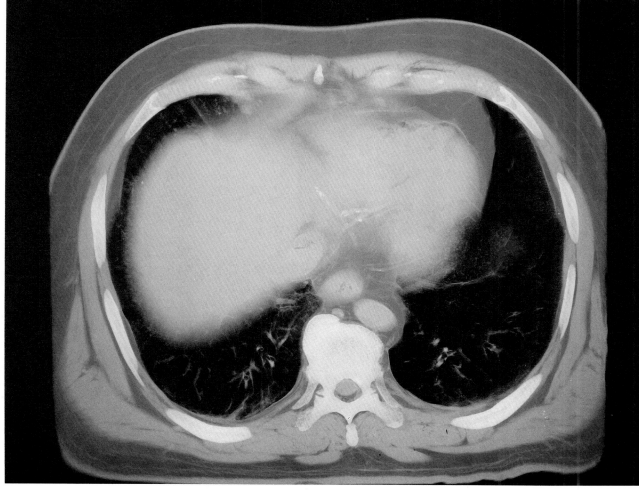

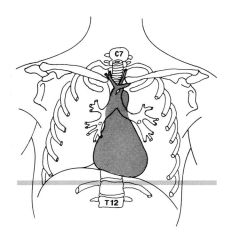

**Specimen Radiograph**

**Computed Tomogram—Mediastinum Technique**

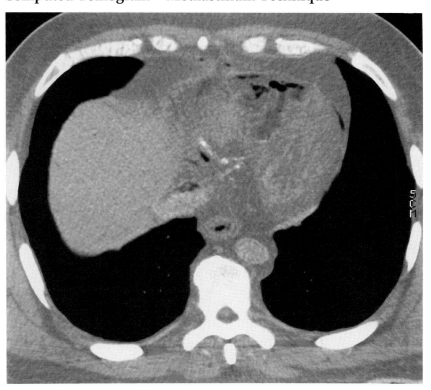

**Computed Tomogram—Lung Technique**

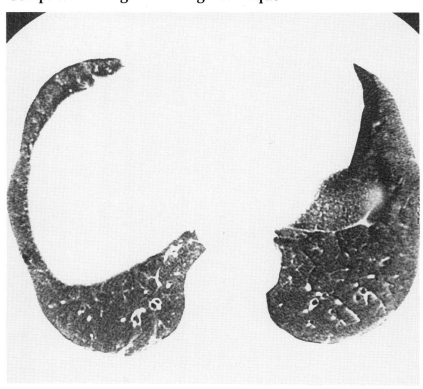

# Axial

## Section 33

### Anatomic Key

**Arteries**
A 31  aorta, descending

**Bronchi**
B 15  RLL, posterior basal segment
B 26  LLL, lateral basal segment
B 27  LLL, posterior basal segment

**Esophagus**
E  1  esophagus

**Fat**
f  3  extrapleural
f  4  mediastinal

**Lobes**
L 10  RLL, medial basal segment
L 11  RLL, anterior basal segment
L 13  RLL, posterior basal segment
L 22  LLL, anteromedial basal
       segment
L 23  LLL, lateral basal segment
L 24  LLL, posterior basal segment

**Muscles**
M  1  abdominal, external oblique
M  3  abdominis, rectus
M  4  abdominis, transversus
M  9  diaphragm
M 11  diaphragm, crus, left
M 12  diaphragm, crus, right
M 14  iliocostalis, thoracis
M 16  intercostal
M 21  latissimus dorsi
M 27  longissimus, thoracis
M 44  serratus, anterior
M 49  spinalis, thoracis
M 61  transversospinalis
M 63  trapezius

**Neural Structures**
n  5  spinal cord
n  6  spinal dura
n  8  splanchnic nerve

**Organs**
O  3  liver

**Pleural Structures**
P  3  costodiaphragmatic recess

**Skeletal Structures**
S  4  rib, body, anterior (7th)
S  5  rib, body, lateral (8th)
S  6  rib, body, posterior (9th)
S  7  rib, costal cartilage (6th)
S  8  rib, head and neck (10th)
S  9  rib, tubercle (10th)
S 21  vertebra, body (T10)
S 22  vertebra, costal facet (T10)
S 25  vertebra, lamina (T10)
S 26  vertebra, pedicle (T10)
S 27  vertebra, spinous process (T10)

**Veins**
V  7  azygos
V 12  hemiazygos
V 18  intercostal, posterior, left
V 19  intercostal, posterior, right
V 38  vena cava, inferior

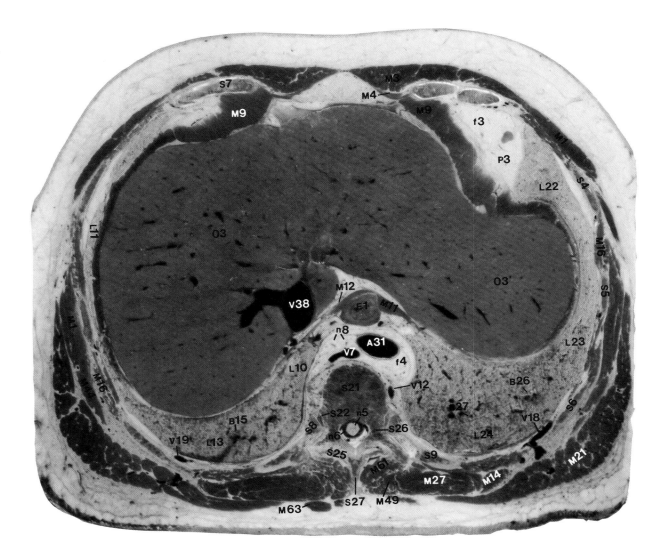

**Anatomic Specimen**

**Magnetic Resonance**

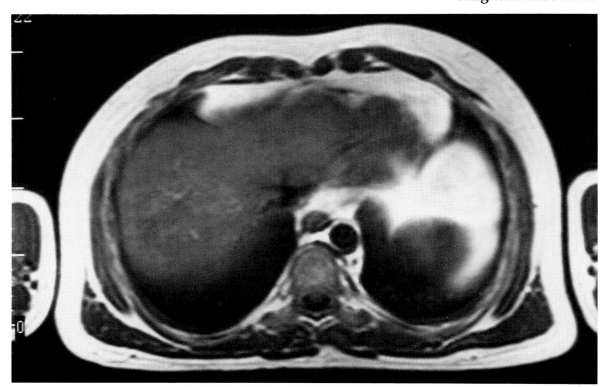

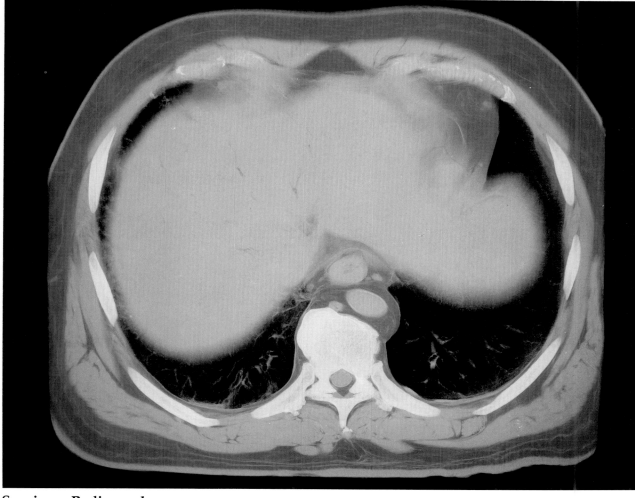

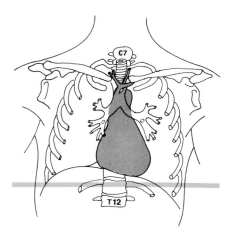

**Specimen Radiograph**

**Computed Tomogram—Mediastinum Technique**

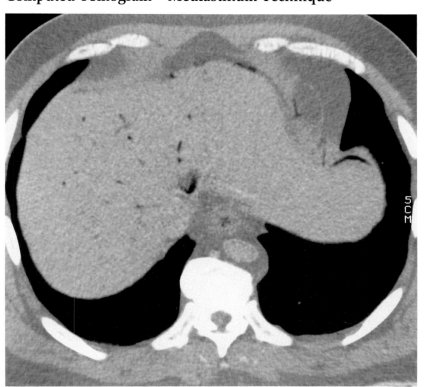

**Computed Tomogram—Lung Technique**

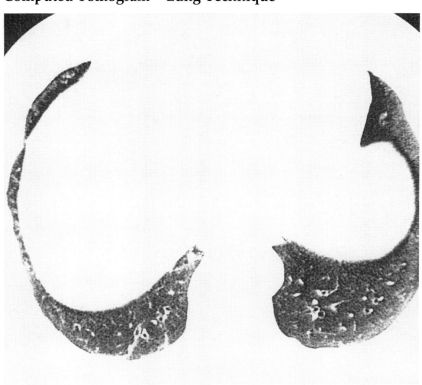

# Axial
## Section 34

### Anatomic Key

**Arteries**
A **31** aorta, descending

**Bronchi**
B **27** LLL, posterior basal segment

**Esophagus**
E **1** esophagus

**Fat**
f **3** extrapleural
f **4** mediastinal

**Lobes**
L **13** RLL, posterior basal segment
L **22** LLL, anteromedial basal segment
L **23** LLL, lateral basal segment
L **24** LLL, posterior basal segment

**Muscles**
M **1** abdominal, external oblique
M **3** abdominis, rectus
M **4** abdominis, transversus
M **9** diaphragm
M **11** diaphragm, crus, left
M **12** diaphragm, crus, right
M **14** iliocostalis, thoracis
M **16** intercostal
M **21** latissimus dorsi
M **27** longissimus, thoracis
M **29** multifidus
M **37** rotator
M **45** serratus, posterior inferior
M **49** spinalis, thoracis
M **61** transversospinalis

**Neural Structures**
n **5** spinal cord
n **6** spinal dura

**Organs**
O **3** liver
O **5** spleen
O **6** stomach

**Pleural Structures**
P **3** costodiaphragmatic recess

**Skeletal Structures**
S **3** rib (9th)
S **4** rib, body, anterior (7th)
S **5** rib, body, lateral (8th)
S **6** rib, body, posterior (10th)
S **7** rib, costal cartilage (7th)
S **21** vertebra, body (T10)
S **25** vertebra, lamina (T10)
S **27** vertebra, spinous process (T10)
S **29** vertebra, transverse process (T10)

**Veins**
v **7** azygos
v **12** hemiazygos
v **18** intercostal, posterior, left
v **19** intercostal, posterior, right
v **38** vena cava, inferior
v **41** venous plexus, vertebral (internal)

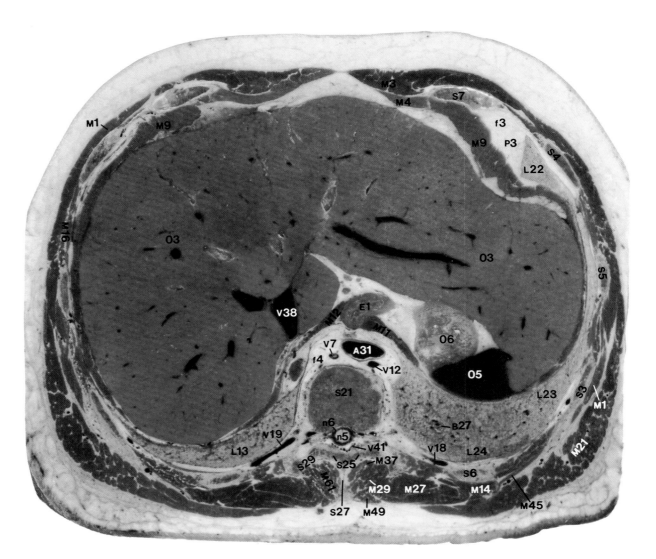

**Anatomic Specimen**

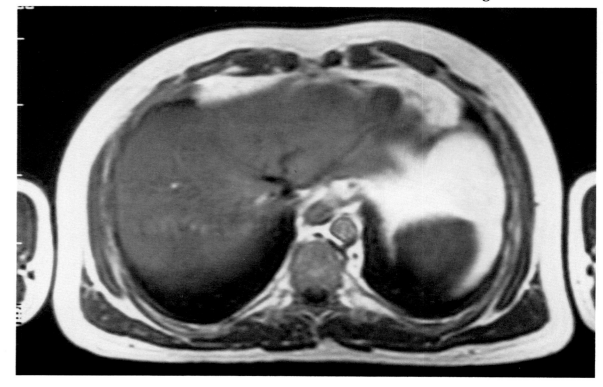

**Magnetic Resonance**

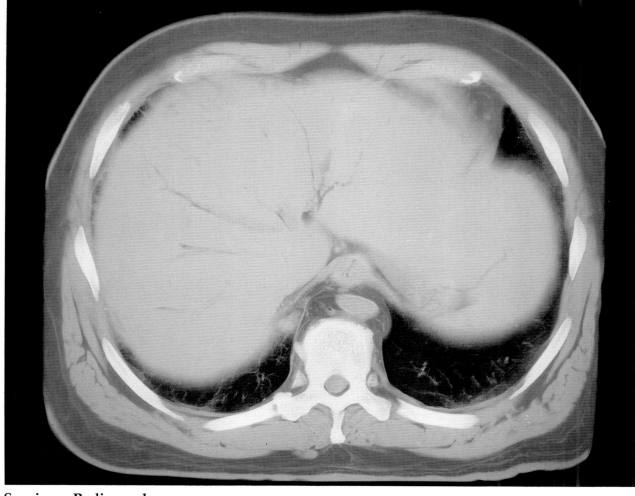

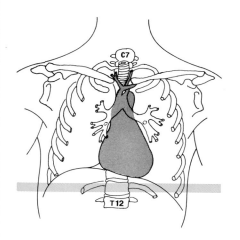

**Specimen Radiograph**

**Computed Tomogram—Mediastinum Technique**

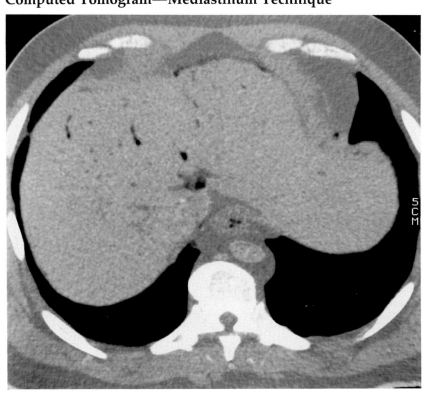

**Computed Tomogram—Lung Technique**

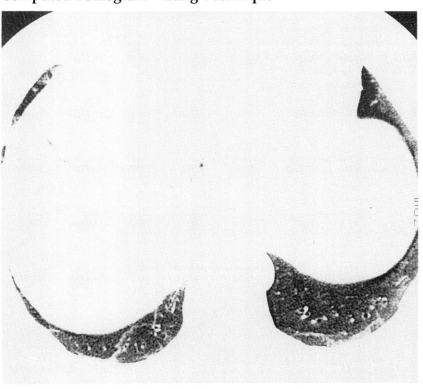

# Axial

## Section 35

### Anatomic Key

**Arteries**
A 28 aorta
A 31 aorta, descending

**Fascia**
F 5 infradiaphragmatic
(transversalis)
F 10 supradiaphragmatic

**Fat**
f 4 mediastinal
f 7 subcutaneous

**Lobes**
L 8 right lower lobe (RLL)
L 13 RLL, posterior basal segment
L 20 left lower lobe (LLL)
L 22 LLL, anteromedial basal
segment
L 24 LLL, posterior basal segment

**Muscles**
M 1 abdominal, external oblique
M 3 abdominis, rectus
M 4 abdominis, transversus
M 9 diaphragm
M 11 diaphragm, crus, left
M 12 diaphragm, crus, right
M 14 iliocostalis, thoracis
M 16 intercostal
M 21 latissimus dorsi
M 27 longissimus, thoracis
M 45 serratus, posterior inferior
M 49 spinalis, thoracis
M 61 transversospinalis

**Organs**
O 3 liver
O 5 spleen
O 6 stomach

**Pleural Structures**
P 2 costal (parietal)
P 6 visceral

**Skeletal Structures**
S 4 rib, body, anterior (8th)
S 5 rib, body, lateral (9th)
S 6 rib, body, posterior (10th)
S 7 rib, costal cartilage (7th)
S 8 rib, head and neck (11th)
S 21 vertebra, body (T11)
S 23 vertebra, inferior articular
process (T11)
S 27 vertebra, spinous process (T11)
S 28 vertebra, superior articular
process (T12)

**Veins**
V 7 azygos
V 18 intercostal, posterior, left
V 19 intercostal, posterior, right
V 38 vena cava, inferior

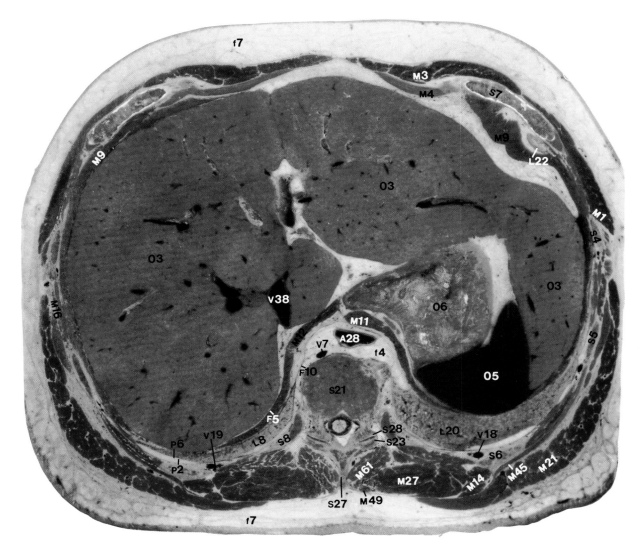

**Anatomic Specimen**

**Magnetic Resonance**

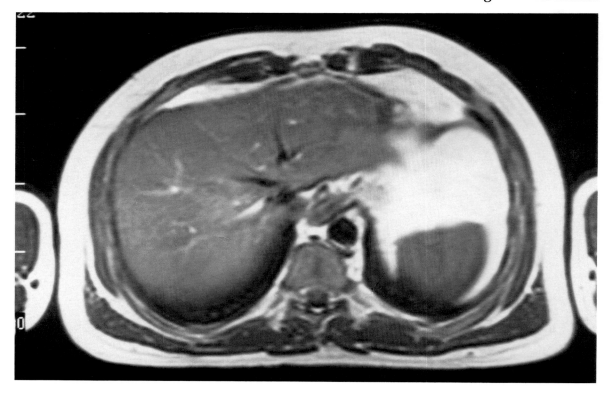

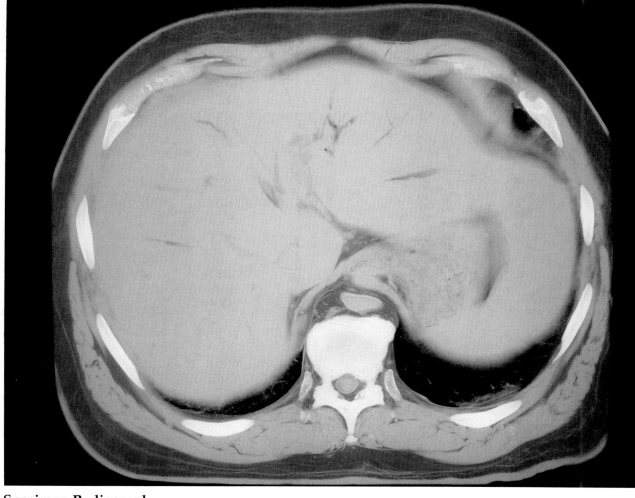

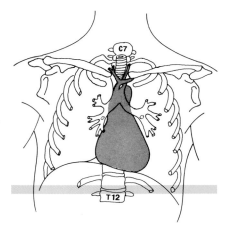

**Specimen Radiograph**

**Computed Tomogram—Mediastinum Technique**

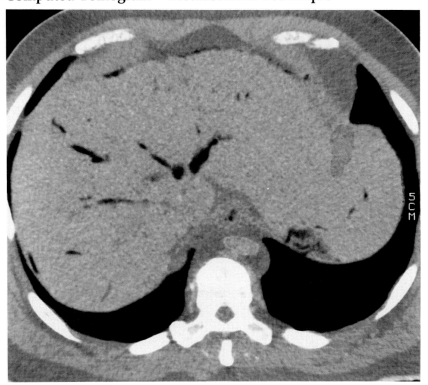

**Computed Tomogram—Lung Technique**

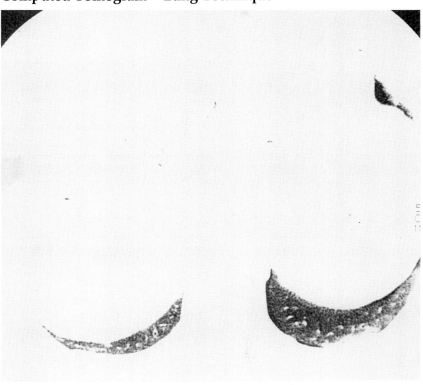

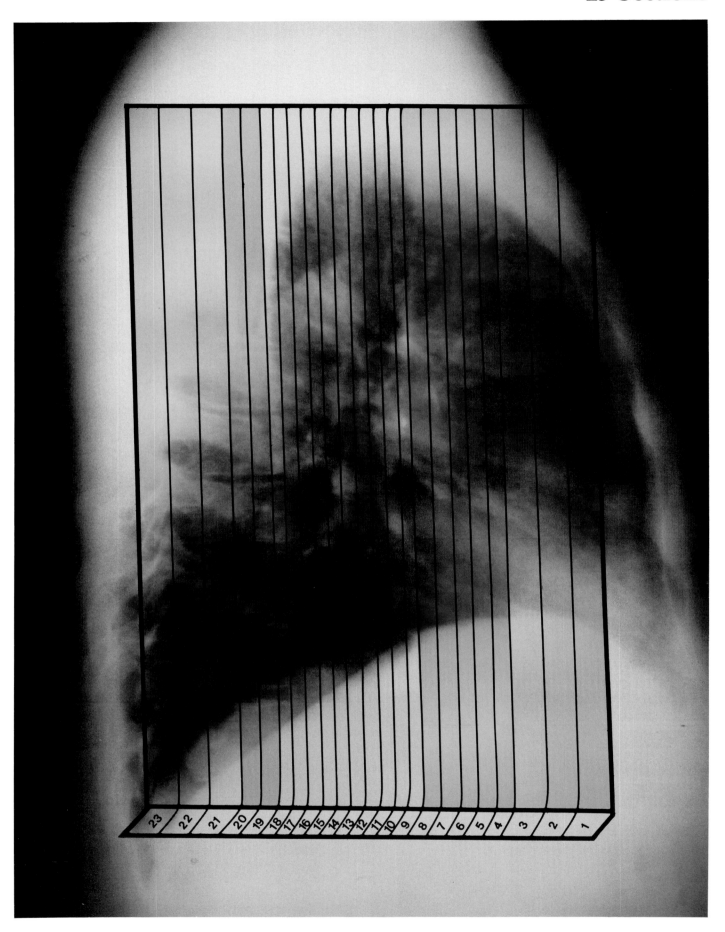

# Chapter 3

# Coronal Plane

## The Anatomic Specimen

The cadaver for the coronal sections was that of a 59-year-old white male who died while on life support following an acute myocardial occlusion. The donation was received from the Alabama Regional Organ and Tissue Center. Both corneas were taken for the Eye Bank, and the kidneys were harvested after consultation with us. An endotracheal tube that was already in position was double-clamped at full inspiration. Special care was taken to ligate all vessels during the immediate postmortem removal of the kidneys. Although an effort was made not to disturb other normal intraabdominal organs, some alteration of structures below the diaphragm can be noted.

Upon arrival at the University of South Alabama Medical Center, a routine radiograph confirmed that the chest was acceptable for the atlas and the cadaver was placed in the freezer. After 3 months, the head was removed for temporal bone study at the Helen Keller Institute, Florence, Alabama. The authors feel it worthy of mention here that this unusual cooperation between a family, an organ donation center, and an Anatomical Board resulted in such extensive utilization of a single human donation. The thorax remained in the freezer for 22 months prior to sectioning.

## Section Plan

The plan of section for the coronal series (facing page) is shown on a routine lateral radiograph of the chest as a colored overlay representing the precise layout of the positions of the gross sections. Each section is color-coded according to its thickness: **blue**, 10 mm; **pink**, 5 mm; **yellow**, 3 mm. The level and thickness of section is indicated throughout the chapter by the color bar overlying the miniature reference key at the top right of each double-page layout. The sectioning process began anteriorly just behind the sternum but anterior to the cardiac structures and continued posteriorly through both lungs.

Sections 1–3 are 10 mm thick and extend posteriorly to the anterior portion of the cardiac ventricles, the sternoclavicular junction, and the costal cartilage of the first rib. Sections 4–8 are 5 mm thick and continue in a posterior direction ending at the level of the ascending aorta. Sections 9–18 are approximately 3 mm thick and pass through the principal cardiac, mediastinal, and hilar structures. Sections 19 and 20 are again 5 mm thick. Section 20 is at the level of the midportion of the thoracic

vertebral bodies and the descending aorta. The final three posterior sections (Sections 21–23) are 10 mm thick and conclude at the level of the posterior elements of the thoracic spine, and the most posterior portion of both lungs.

As the saw blade passed through the site of the harvesting of the kidneys, several steel clips were encountered, which resulted in significant damage to the saw blade. The blade wandered badly throughout the remainder of the cut, damaging the tissue and causing the section to be thin and ragged, as noted in the specimen radiograph of Section 7. It was elected to use the section despite the tissue distortion.

Postmortem routine chest radiographs did not identify premortem pneumonic consolidation of the posterior apical segment of the left upper lobe, which became evident principally in the specimen radiographs. An area of increased density is noted in the inferior portion of the left upper lobe in Specimen Radiographs 15–19, developing into frank consolidation of the apex in Sections 20–22. This portion of the lung is excluded in Section 23. Since there was no distortion of normal anatomy in the left lung other than the color change seen in the gross anatomic specimens and the change in radiographic density noted in the specimen radiographs in the LUL, we elected to continue processing this specimen. Again, fluid settling into the dependent portions of the lung resulted in increase in density in the posterior planes of the anatomic specimens. This postmortem phenomenon, together with the sludging that occurred in the vascular cavities, has been described in greater detail in Chapter 1, Methods.

## Orientation of Illustrations

The coronal sections are viewed as seen from front to back such that structures of the right hemithorax are seen at the reader's left and structures of the left hemithorax are seen at the reader's right.

## Anatomic–Radiographic Correlation

The coronal projection of the chest is probably the most familiar image in all of radiology. The plain anteroposterior (AP) or posteroanterior (PA) film of the chest is equally familiar to nonradiologists. Whether common use of the coronal orientation is the result of fortuitous circumstance or careful scientific design is not important; the result is that most of the thoracic structures are best seen in this projection.

**Bronchial Structures**

The near vertical orientation of the bronchial tree lends itself well to sectional imaging in the coronal plane, as can be seen in the trispiral tomograms. Contrary to the axial plane, most of the major airways, with only a few exceptions, are imaged in longitudinal profile in the coronal plane. In addition, it is necessary to scan only a relatively thin zone (3–5 cm) in the midplane of the chest to see most of the bronchopulmonary segments on both sides. In this average male specimen (5 ft 10 in tall, 180 lb), a distance of 5 cm, extending from Section 5 through Section 18, is sufficient to image all of the lobar and the major segmental bronchi.

Since 1966, the authors have used pluridirectional tomography to view the tracheobronchial tree, taking 1-mm-thick coronal tomogram sections at 2-mm increments on 10″ × 12″ film (25). The clinical images are consistently comparable to the coronal trispiral tomogram images displayed here. A summary of the optimal bronchial display in coronal projection is seen in Table 3.1.

The entire extent of the trachea (T1) can be followed with contiguous longitudinal assessment from Sections 6 through 14. The longest segment of the trachea and its bifurcation into the right (B1) and left (B16) mainstem bronchi at the carina (T2) are seen in Sections 12–14. This display permits an excellent assessment of extrinsic impingement of the trachea and mainstem bronchi by mediastinal tumors, adenopathy, and subcarinal masses.

**Right Lung**: In addition to the right mainstem bronchus (B1), the apical segment to the RUL is seen in profile in Sections 12 and 13, and in Section 11. The RUL bronchus (B2) is best seen in Section 13. An excellent display of the intermediate bronchus (B6) is seen in Sections 12–14. The RLL bronchus (B10) with its medial basal (B12) and anterior basal (B13) divisions is best seen in Section 13. In Section 14, in addition to the right mainstem bronchus (B1) and the intermediate bronchus (B6), the RUL bronchus (B2) and its posterior segment (B4) are visualized. The posterior basal segment to the RLL (B15) is seen in Sections 15 and 16.

**Left Lung**: The left main bronchus (B16) is seen in Sections 12–14 and is most clearly imaged in Section 14. The LUL bronchi (B17) is seen in Sections 11–13, and branches of the apicoposterior division of the LUL (B18) are visualized in Sections 11–14. The lingula (B20) is only incompletely seen in Section 11. The LLL bronchus (B23) is seen on the trispiral tomogram of Sections 13–15 and on the specimen radiographs of Sections 14 and 15. The origins of the anteromedial (B25) and lateral basal (B26) segments to the LLL are seen in Section 13. The segmental bronchus to the posterior basal segment (B27) of the LLL is clearly seen in Sections 16 and 17.

Only a few bronchi are not favorably oriented in the coronal plane. The segmental bronchi to the anterior segments of the RUL (B5) and the LUL (B19), as well as the superior segments of the RLL (B11) and the LLL (B24) are all oriented nearly perpendicular to the coronal plane and thus cut in cross-section instead of profile. Because of its anterior, oblique course the lingula (B20) with its segmental divisions is not optimally imaged in the coronal plane. These 5 bronchial segments are best seen in the sagittal and the axial planes. The right middle lobe bronchus (B7) is oblique to the coronal plane and is characteristically seen as a small oval or comma-shaped bronchial segment (Section 9) as it takes-off from the anterior surface of the intermediate bronchus. The RML with its lateral and medial segments is best seen in the right posterior oblique and the sagittal planes.

**Fissures**

The fissures of both lungs are seen throughout the coronal series in all of the gross anatomic specimens and specimen radiographs, but they are not as well imaged in the trispiral tomograms.

**Neural Structures**

The brachial plexus (n1) is well seen on both sides emerging between the anterior scalene (M38) and middle scalene (M39) muscles in the anatomic specimen of Section 10. The course of the plexus and its major divisions can be traced from its origin through the supraclavicular portions in both the gross anatomic specimens and the specimen radiographs from Sections 9–13.

**Vascular Structures**

The cardiac structures and chambers, mediastinal vessels, pulmonary arteries and veins, together with the peripheral arterial and venous structures are well seen throughout the coronal series in familiar presentation. No anomalies were encountered. Sections 15–19 fortuitously show the beginning anterior course

## Table 3.1
### Index to Optimal Bronchial Imaging Coronal Projection
#### Specimen Radiograph and Trispiral Tomogram
Right       Bronchus       Left

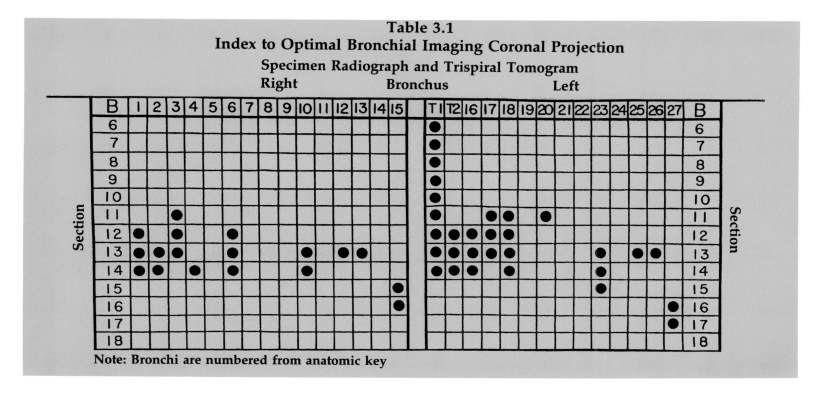

| Section (B) | 1 | 2 | 3 | 4 | 5 | 6 | 7 | 8 | 9 | 10 | 11 | 12 | 13 | 14 | 15 | T1 | T2 | 16 | 17 | 18 | 19 | 20 | 21 | 22 | 23 | 24 | 25 | 26 | 27 | Section (B) |
|---|---|---|---|---|---|---|---|---|---|---|---|---|---|---|---|---|---|---|---|---|---|---|---|---|---|---|---|---|---|---|
| 6 | | | | | | | | | | | | | | | | ● | | | | | | | | | | | | | | 6 |
| 7 | | | | | | | | | | | | | | | | ● | | | | | | | | | | | | | | 7 |
| 8 | | | | | | | | | | | | | | | | ● | | | | | | | | | | | | | | 8 |
| 9 | | | | | | | | | | | | | | | | ● | | | | | | | | | | | | | | 9 |
| 10 | | | | | | | | | | | | | | | | ● | | | | | | | | | | | | | | 10 |
| 11 | | | ● | | | | | | | | | | | | | ● | | | ● | ● | | ● | | | | | | | | 11 |
| 12 | ● | | ● | | | ● | | | | | | | | | | ● | ● | ● | ● | ● | | | | | | | | | | 12 |
| 13 | ● | ● | ● | | | ● | | | ● | | ● | ● | | | | ● | ● | ● | ● | ● | | | | | ● | | ● | ● | | 13 |
| 14 | ● | ● | | ● | | ● | | | ● | | | | | | | ● | ● | ● | | ● | | | | | | ● | | | | 14 |
| 15 | | | | | | | | | | | | | | ● | | | | | | | | | | | | ● | | | | 15 |
| 16 | | | | | | | | | | | | | | ● | | | | | | | | | | | | | | | ● | 16 |
| 17 | | | | | | | | | | | | | | | | | | | | | | | | | | | | | ● | 17 |
| 18 | | | | | | | | | | | | | | | | | | | | | | | | | | | | | | 18 |

Note: Bronchi are numbered from anatomic key

of the highest left intercostal vein (V14), which appears on chest radiographs as a "nipple" as it crosses adjacent to the aortic arch.

### Other Structures
The esophagus (E1) and its relationships are nicely seen throughout most of its course through the mediastinum in gross anatomic sections and specimen radiographs (Sections 8–16), and the esophagogastric junction (E2) is seen passing through the crura of the diaphragm (M11 and M12) in Sections 8–12.

The anterior pleural stripe (anterior junction line) is nicely seen in Sections 1–5 in the anatomic specimens and specimen radiographs.

Several fascial structures are identified in the coronal projection: the infradiaphragmatic (transversalis) (F5) and supradiaphragmatic fascia (F10) are seen in coronal anatomic Sections 2 and 3; Sibson's fascia (F8) is seen in anatomic Section 9; the supradiaphragmatic fascia (F10) is also seen in anatomic Section 10; and the prevertebral fascia (F7) is seen in anatomic Section 16.

### Magnetic Resonance Images
The coronal MR images match well to the gross anatomic specimens and specimen radiographs throughout. The correlation to the muscle groups and cardiac and vascular structures is "textbook." Special attention is drawn to the internal thoracic vessels in MR Sections 1–4 (not labeled in this chapter), the venous system draining from the neck and supraclavicular vessels into the brachiocephalic veins in Section 8, the divisions of the aortic arch in Sections 9 and 10, and the pulmonary trunk and proximal pulmonary arteries in Sections 6–14. Fat structures reference well to the MR images.

# Coronal

## Section 1

### Anatomic Key

**Fat**
f   3  extrapleural

**Fissures**
F   1  left major (oblique)
F   2  right major (oblique)
F   3  right minor (horizontal)

**Lobes**
L   4  RUL, anterior segment
L   6  RML, lateral segment
L   7  RML, medial segment
L  11  RLL, anterior basal segment
L  16  LUL, anterior segment
L  18  LUL, superior lingular segment
L  19  LUL, inferior lingular segment
L  22  LLL, anteromedial basal
       segment

**Muscles**
M   1  abdominal, external oblique
M   9  diaphragm
M  10  diaphragm, central tendon
M  16  intercostal
M  31  pectoralis, major
M  32  pectoralis, minor
M  44  serratus, anterior

**Organs**
O   3  liver

**Pleural Structures**
P   1  anterior junction line

**Skeletal Structures**
S   4  rib, body, anterior (7th)
S   5  rib, body, lateral (3rd)
S   7  rib, costal cartilage (2nd)
S  17  sternum, body
S  18  sternum, manubrium

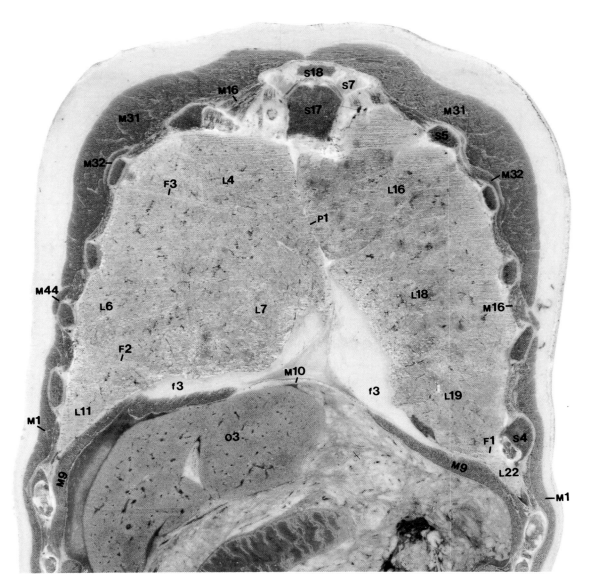

**Anatomic Specimen**

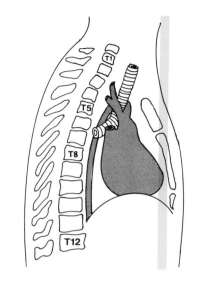

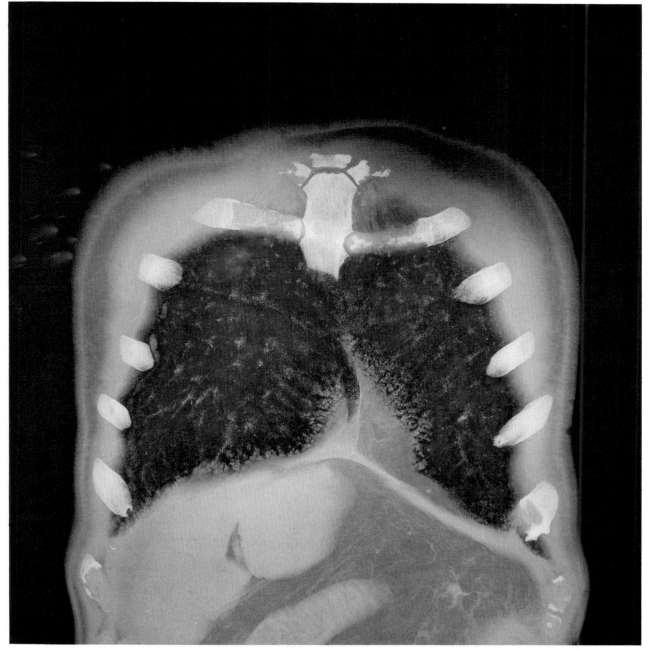

**Specimen Radiograph**

**Magnetic Resonance**

**Trispiral Tomogram**

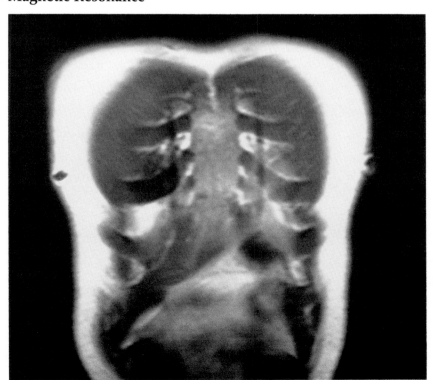

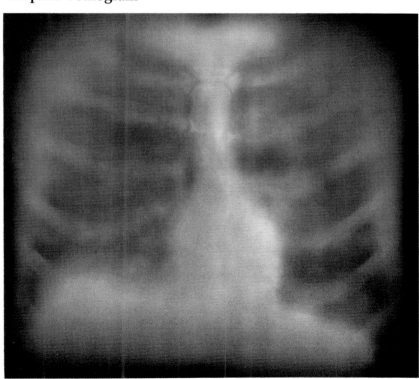

# Coronal

## Section 2

### Anatomic Key

**Fascia**

F  5  infradiaphragmatic (transversalis)
F 10  supradiaphragmatic

**Fat**

f  2  epicardial
f  3  extrapleural

**Fissures**

F  1  left major (oblique)
F  2  right major (oblique)
F  3  right minor (horizontal)

**Heart**

H 21  ventricle, left

**Lobes**

L  4  RUL, anterior segment
L  6  RML, lateral segment
L  7  RML, medial segment
L 11  RLL, anterior basal segment
L 16  LUL, anterior segment
L 18  LUL, superior lingular segment
L 19  LUL, inferior lingular segment
L 22  LLL, anteromedial basal segment

**Muscles**

M  1  abdominal, external oblique
M  9  diaphragm
M 10  diaphragm, central tendon
M 16  intercostal
M 31  pectoralis, major
M 32  pectoralis, minor
M 44  serratus, anterior

**Organs**

O  3  liver

**Skeletal Structures**

S  4  rib, body, anterior (7th)
S  7  rib, costal cartilage (2nd)
S 17  sternum, body
S 18  sternum, manubrium

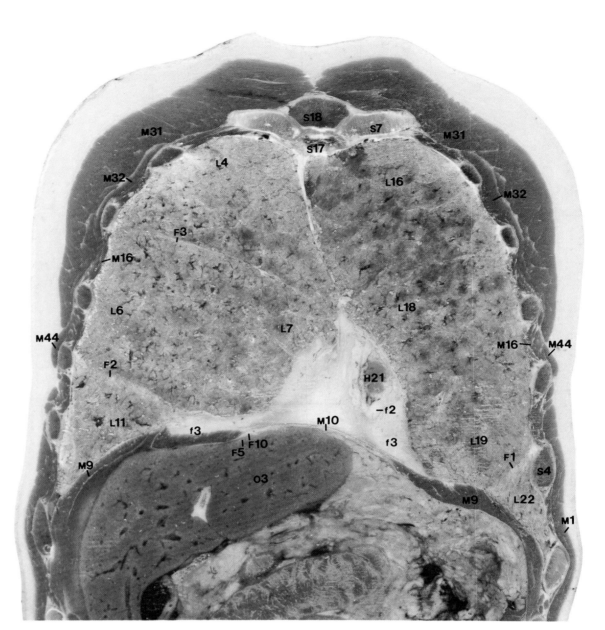

**Anatomic Specimen**

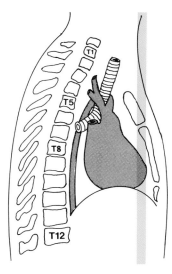

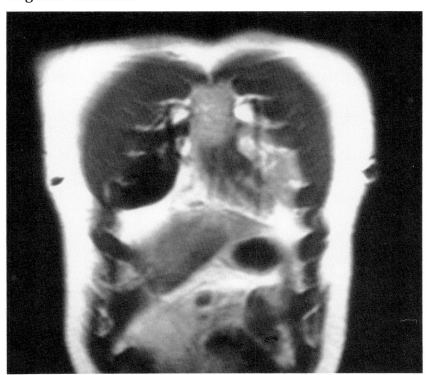

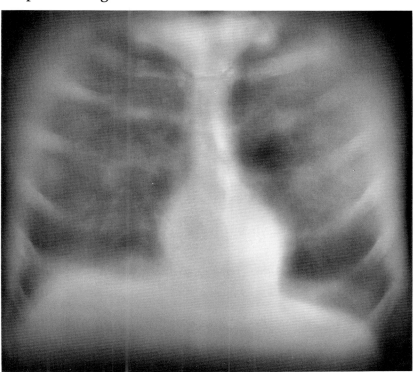

**Specimen Radiograph**

**Magnetic Resonance**

**Trispiral Tomogram**

# Coronal

## Section 3

## Anatomic Key

**Fascia**
F  5  infradiaphragmatic (transversalis)
F 10  supradiaphragmatic

**Fat**
f  2  epicardial
f  3  extrapleural
f  7  subcutaneous

**Fissures**
F  1  left major (oblique)
F  2  right major (oblique)
F  3  right minor (horizontal)

**Heart**
H 13  pericardium, parietal serous
H 14  pericardium, visceral serous
H 21  ventricle, left
H 22  ventricle, right

**Lobes**
L  4  RUL, anterior segment
L  6  RML, lateral segment
L  7  RML, medial segment
L 11  RLL, anterior basal segment
L 16  LUL, anterior segment
L 18  LUL, superior lingular segment
L 19  LUL, inferior lingular segment
L 22  LLL, anteromedial basal segment

**Muscles**
M  1  abdominal, external oblique
M  9  diaphragm
M 10  diaphragm, central tendon
M 16  intercostal
M 31  pectoralis, major
M 32  pectoralis, minor
M 44  serratus, anterior

**Organs**
O  3  liver

**Pleural Structures**
P  1  anterior junction line

**Skeletal Structures**
S  3  rib (8th)
S  4  rib, body, anterior (5th)
S  7  rib, costal cartilage (2nd)
S 18  sternum, manubrium

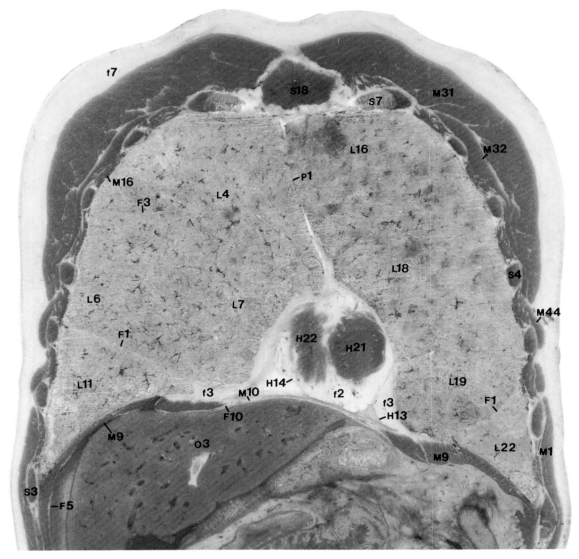

**Anatomic Specimen**

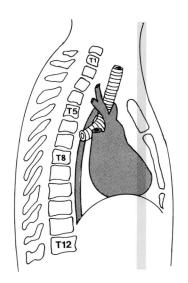

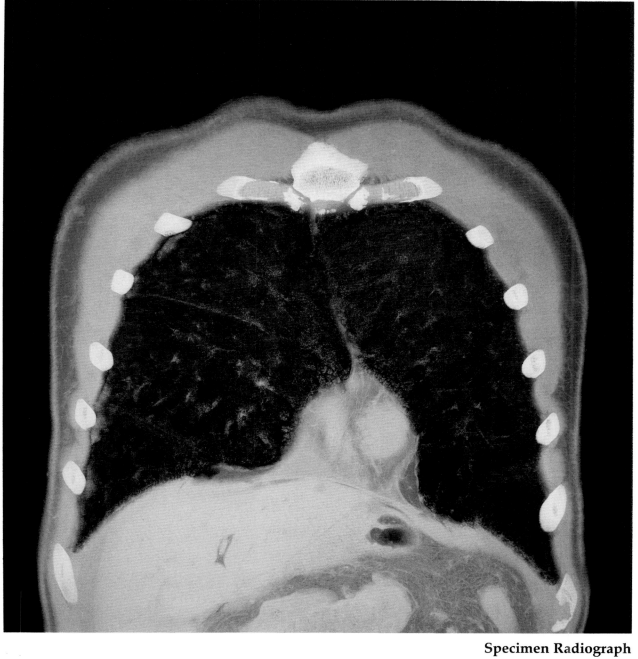

**Specimen Radiograph**

**Magnetic Resonance**

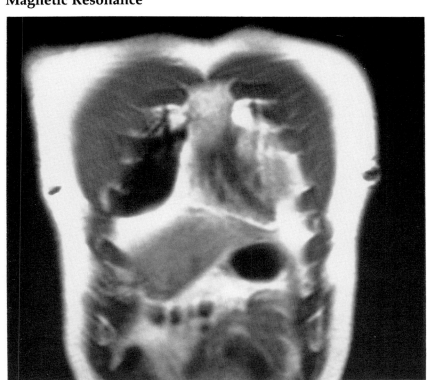

**Trispiral Tomogram**

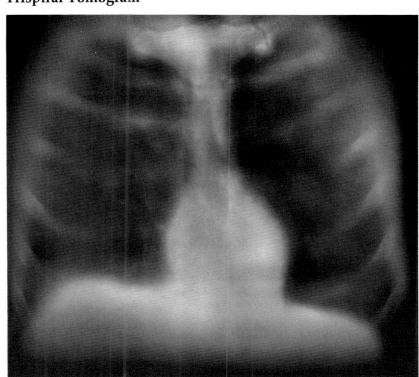

# Coronal

## Section 4

### Anatomic Key

**Fat**
f  2  epicardial
f  3  extrapleural

**Fissures**
F  1  left major (oblique)
F  2  right major (oblique)
F  3  right minor (horizontal)

**Heart**
H 13  pericardium, parietal serous
H 14  pericardium, visceral serous
H 16  septum, interventricular
H 21  ventricle, left
H 22  ventricle, right

**Lobes**
L  4  RUL, anterior segment
L  6  RML, lateral segment
L  7  RML, medial segment
L 11  RLL, anterior basal segment
L 16  LUL, anterior segment
L 18  LUL, superior lingular segment
L 19  LUL, inferior lingular segment
L 22  LLL, anteromedial basal
     segment

**Muscles**
M  1  abdominal, external oblique
M  9  diaphragm
M 10  diaphragm, central tendon
M 16  intercostal
M 31  pectoralis, major
M 32  pectoralis, minor
M 44  serratus, anterior
M 52  sternocleidomastoid

**Organs**
O  3  liver

**Pleural Structures**
P  1  anterior junction line

**Skeletal Structures**
S  1  clavicle
S  3  rib (4th)
S  4  rib, body, anterior (8th)
S  7  rib, costal cartilage (1st)
S 18  sternum, manubrium

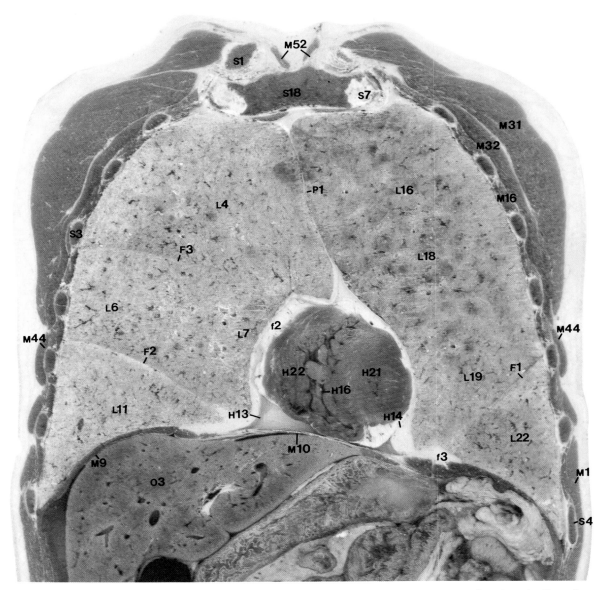

**Anatomic Specimen**

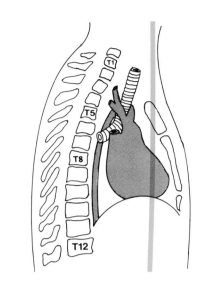

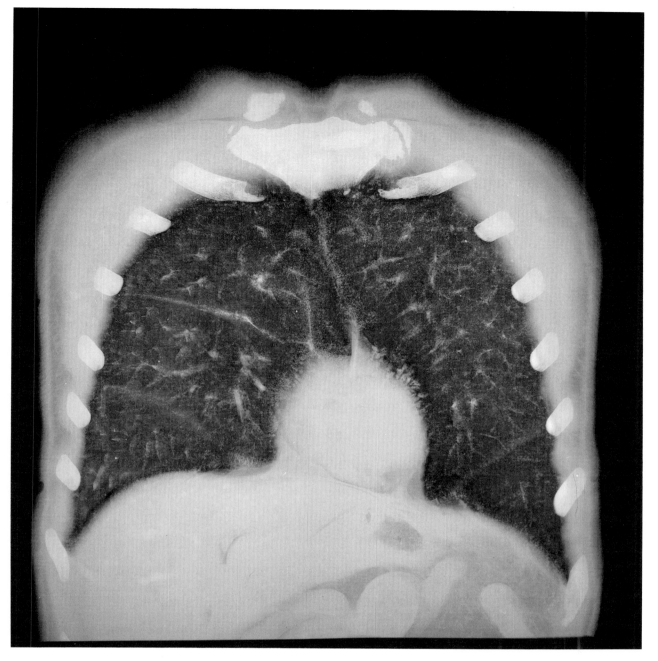

**Specimen Radiograph**

**Magnetic Resonance**

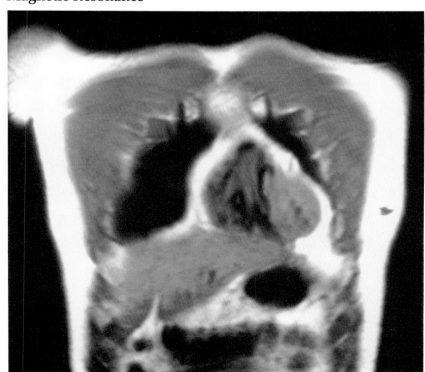

**Trispiral Tomogram**

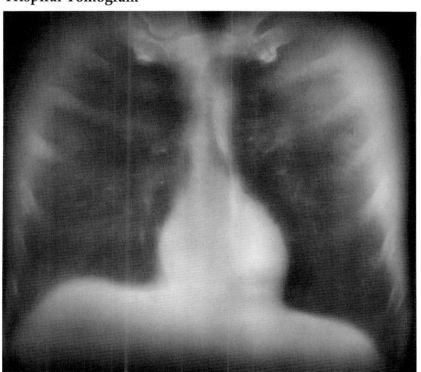

# Coronal

## Section 5

### Anatomic Key

**Bronchi**
B 5 RUL, anterior segment
B 8 RML, lateral segment
B 9 RML, medial segment
B 19 LUL, anterior segment
B 21 LUL, superior lingular segment
B 22 LUL, inferior lingular segment

**Fat**
f 2 epicardial
f 4 mediastinal

**Fissures**
F 1 left major (oblique)
F 2 right major (oblique)
F 3 right minor (horizontal)

**Glands**
G 4 thyroid

**Heart**
H 11 coronary artery, right
H 13 pericardium, parietal serous
H 14 pericardium, visceral serous
H 16 septum, interventricular
H 21 ventricle, left
H 22 ventricle, right

**Lobes**
L 2 RUL, apical segment
L 4 RUL, anterior segment
L 6 RML, lateral segment
L 7 RML, medial segment
L 10 RLL, medial basal segment
L 11 RLL, anterior basal segment
L 16 LUL, anterior segment
L 18 LUL, superior lingular segment
L 19 LUL, inferior lingular segment
L 22 LLL, anteromedial basal
    segment

**Muscles**
M 1 abdominal, external oblique
M 9 diaphragm
M 10 diaphragm, central tendon
M 16 intercostal
M 31 pectoralis, major
M 32 pectoralis, minor
M 33 platysma
M 44 serratus, anterior
M 52 sternocleidomastoid
M 53 sternohyoid
M 54 sternothyroid

**Organs**
O 3 liver
O 6 stomach

**Skeletal Structures**
S 1 clavicle
S 3 rib (8th)
S 7 rib, costal cartilage (1st)

**Trachea**
T 1 trachea

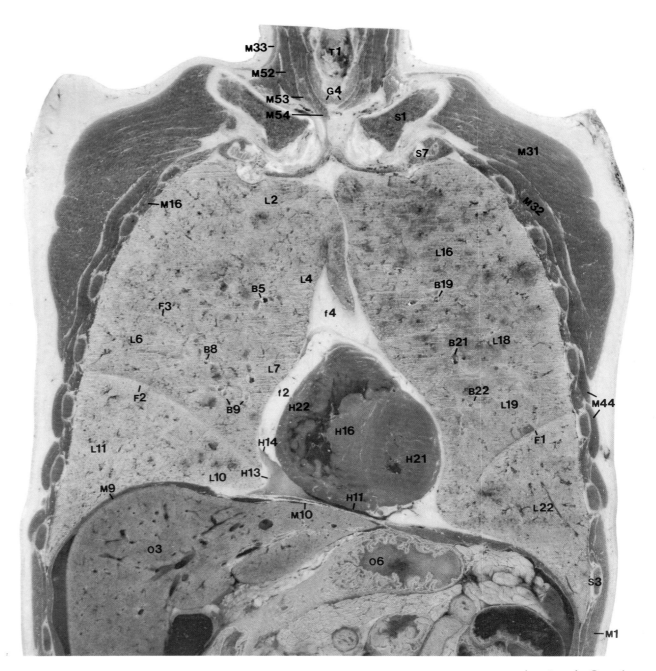

**Anatomic Specimen**

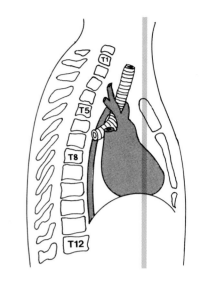

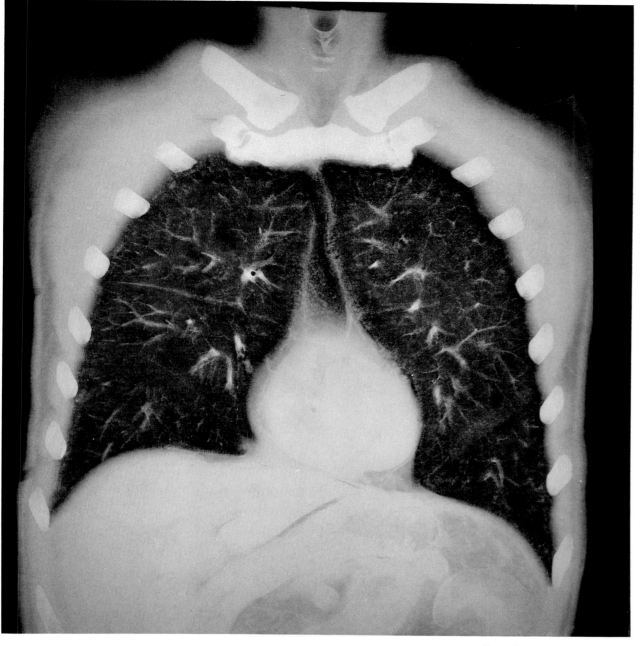

**Specimen Radiograph**

**Magnetic Resonance**

**Trispiral Tomogram**

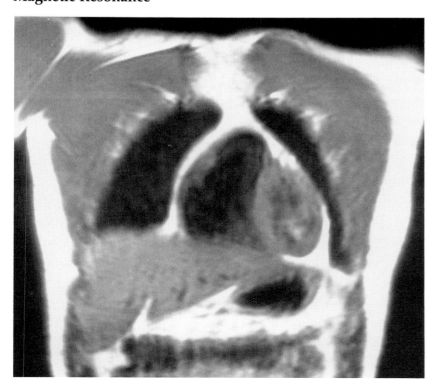

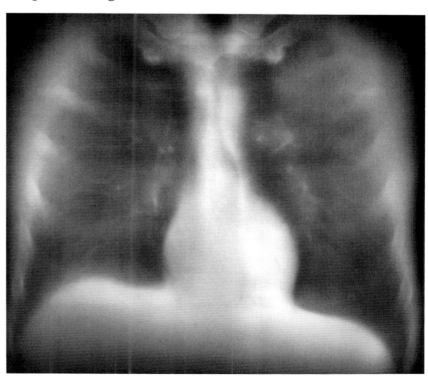

# Coronal

## Section 6

### Anatomic Key

**Arteries**
A 1 pulmonary trunk
A 30 aorta, ascending

**Bronchi**
B 5 RUL, anterior segment
B 8 RML, lateral segment
B 9 RML, medial segment
B 18 LUL, apicoposterior segment
B 19 LUL, anterior segment
B 21 LUL, superior lingular segment
B 22 LUL, inferior lingular segment

**Fat**
f 2 epicardial
f 3 extrapleural
f 4 mediastinal

**Fissures**
F 1 left major (oblique)
F 2 right major (oblique)
F 3 right minor (horizontal)

**Glands**
G 4 thyroid

**Heart**
H 2 atrium, right
H 3 auricle, left
H 4 auricle, right
H 16 septum, interventricular
H 21 ventricle, left
H 22 ventricle, right

**Lobes**
L 2 RUL, apical segment
L 4 RUL, anterior segment
L 6 RML, lateral segment
L 7 RML, medial segment
L 10 RLL, medial basal segment
L 11 RLL, anterior basal segment
L 15 LUL, apicoposterior segment
L 16 LUL, anterior segment
L 18 LUL, superior lingular segment
L 19 LUL, inferior lingular segment
L 22 LLL, anteromedial basal segment

**Muscles**
M 1 abdominal, external oblique
M 8 deltoid
M 9 diaphragm
M 10 diaphragm, central tendon
M 16 intercostal
M 21 latissimus dorsi
M 31 pectoralis, major
M 32 pectoralis, minor
M 33 platysma
M 44 serratus, anterior
M 52 sternocleidomastoid
M 54 sternothyroid
M 55 subclavius

**Organs**
O 3 liver
O 5 spleen
O 6 stomach

**Skeletal Structures**
S 1 clavicle
S 3 rib (8th)
S 4 rib, body, anterior (1st)
S 5 rib, body, lateral (5th)
S 12 scapula, coracoid process

**Trachea and Larynx**
T 1 trachea
T 3 larynx

**Veins**
V 13 subclavian, right
V 39 vena cava, superior

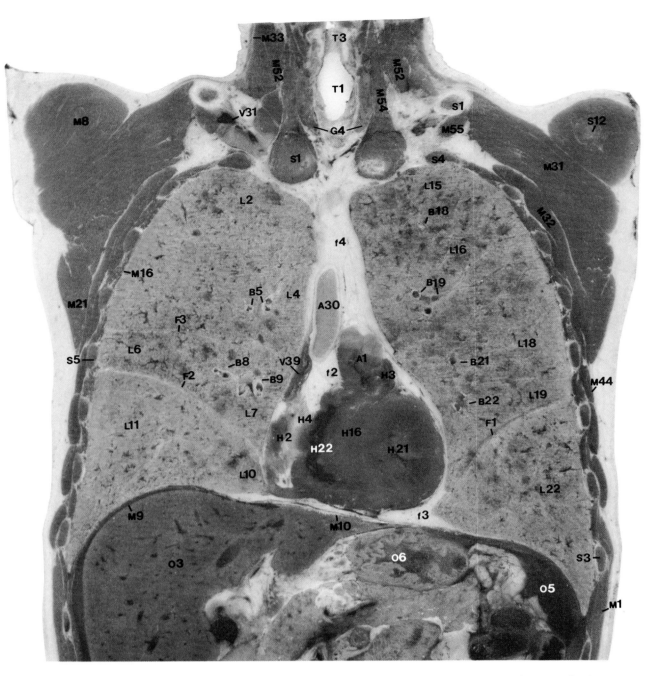

**Anatomic Specimen**

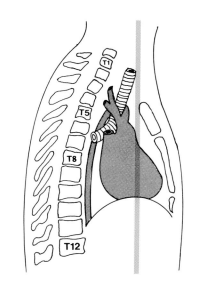

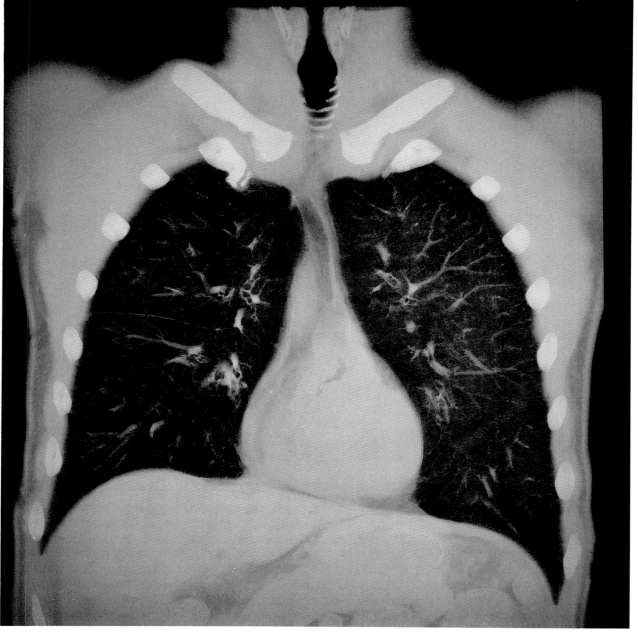

**Specimen Radiograph**

**Magnetic Resonance**

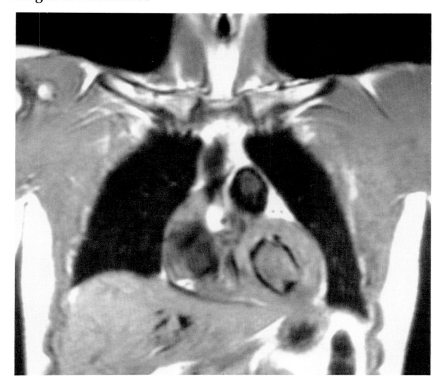

**Trispiral Tomogram**

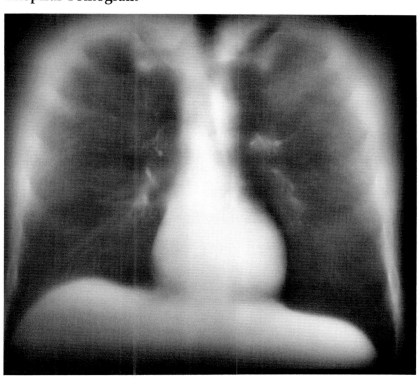

# Coronal

## Section 7

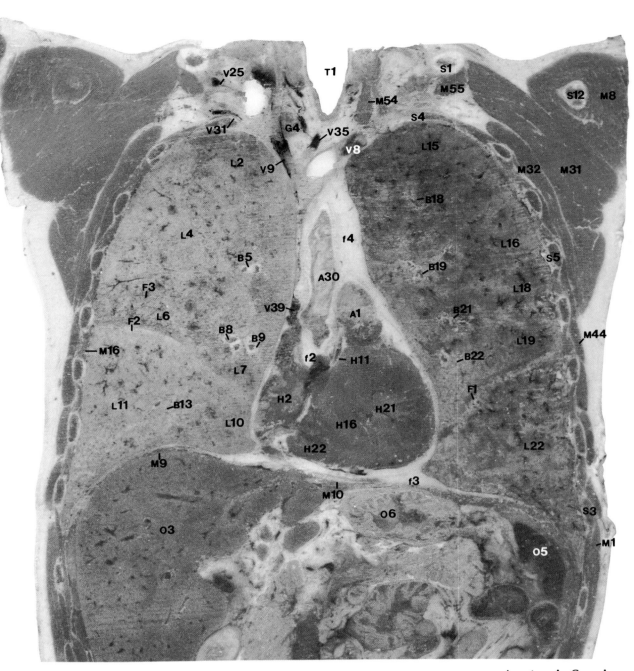

**Anatomic Specimen**

## Anatomic Key

### Arteries
A  1  pulmonary trunk
A 30  aorta, ascending

### Bronchi
B  5  RUL, anterior segment
B  8  RML, lateral segment
B  9  RML, medial segment
B 13  RLL, anterior basal segment
B 18  LUL, apicoposterior segment
B 19  LUL, anterior segment
B 21  LUL, superior lingular segment
B 22  LUL, inferior lingular segment

### Fat
f  2  epicardial
f  3  extrapleural
f  4  mediastinal

### Fissures
F  1  left major (oblique)
F  2  right major (oblique)
F  3  right minor (horizontal)

### Glands
G  4  thyroid

### Heart
H  2  atrium, right
H 11  coronary artery, right
H 16  septum, interventricular
H 21  ventricle, left
H 22  ventricle, right

### Lobes
L  2  RUL, apical segment
L  4  RUL, anterior segment
L  6  RML, lateral segment
L  7  RML, medial segment
L 10  RLL, medial basal segment
L 11  RLL, anterior basal segment
L 15  LUL, apicoposterior segment
L 16  LUL, anterior segment
L 18  LUL, superior lingular segment
L 19  LUL, inferior lingular segment
L 22  LLL, anteromedial basal
       segment

### Muscles
M  1  abdominal, external oblique
M  8  deltoid
M  9  diaphragm
M 10  diaphragm, central tendon
M 16  intercostal
M 31  pectoralis, major
M 32  pectoralis, minor
M 44  serratus, anterior
M 54  sternothyroid
M 55  subclavius

### Organs
O  3  liver
O  5  spleen
O  6  stomach

### Skeletal Structures
S  1  clavicle
S  3  rib (8th)
S  4  rib, body, anterior (1st)
S  5  rib, body, lateral (4th)
S 12  scapula, coracoid process

### Trachea
T  1  trachea

### Veins
V  8  brachiocephalic (innominate),
       left
V  9  brachiocephalic (innominate),
       right
V 25  jugular, external, right
V 31  subclavian, right
V 35  thyroid, inferior, right
V 39  vena cava, superior

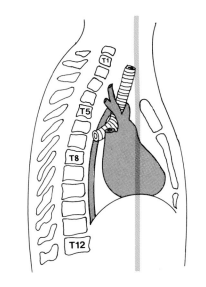

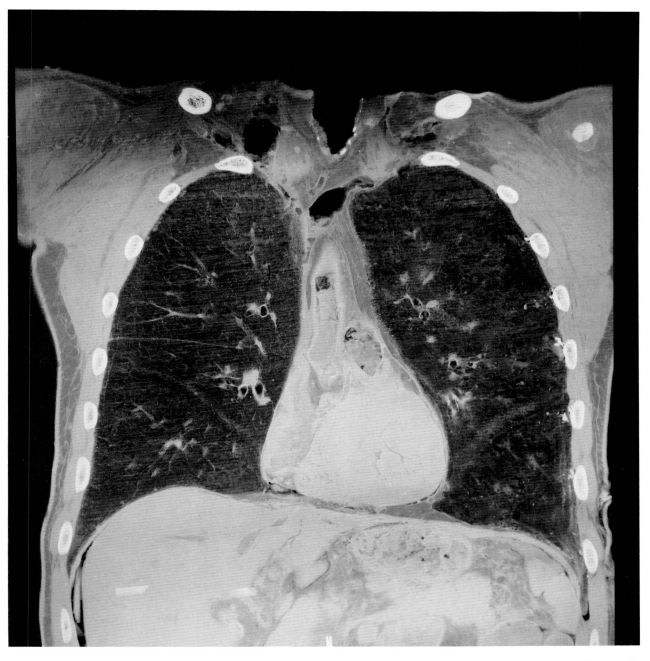

Specimen Radiograph

**Magnetic Resonance**

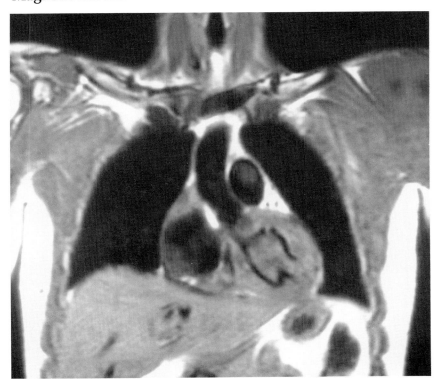

**Trispiral Tomogram**

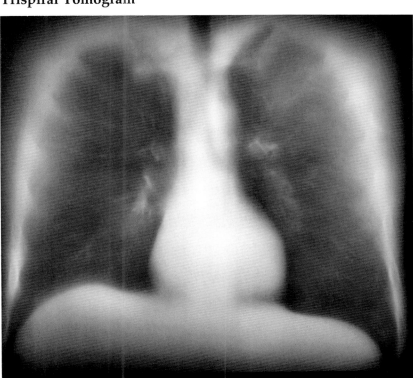

# Coronal

## Section 8

## Anatomic Key

### Arteries
A   1   pulmonary trunk
A  29   aortic arch
A  30   aorta, ascending
A  34   brachiocephalic (innominate) trunk
A  37   carotid, common, left
A  38   carotid, common, right
A  50   subclavian, right

### Bronchi
B   5   RUL, anterior segment
B   8   RML, lateral segment
B   9   RML, medial segment
B  13   RLL, anterior basal segment
B  18   LUL, apicoposterior segment
B  19   LUL, anterior segment
B  21   LUL, superior lingular segment
B  22   LUL, inferior lingular segment
B  25   LLL, anteromedial basal segment

### Esophagus
E   2   esophagogastric junction

### Fat
f   4   mediastinal

### Fissures
F   1   left major (oblique)
F   2   right major (oblique)
F   3   right minor (horizontal)

### Glands
G   4   thyroid

### Heart
H   2   atrium, right
H  10   coronary artery, left
H  21   ventricle, left

### Lobes
L   2   RUL, apical segment
L   4   RUL, anterior segment
L   6   RML, lateral segment
L   7   RML, medial segment
L  10   RLL, medial basal segment
L  11   RLL, anterior basal segment
L  15   LUL, apicoposterior segment
L  16   LUL, anterior segment
L  18   LUL, superior lingular segment
L  19   LUL, inferior lingular segment
L  22   LLL, anteromedial basal segment
L  23   LLL, lateral basal segment

### Muscles
M   1   abdominal, external oblique
M   6   biceps brachii, short head
M   7   coracobrachialis
M   8   deltoid
M   9   diaphragm
M  16   intercostal
M  21   latissimus dorsi
M  28   longus colli
M  31   pectoralis, major
M  32   pectoralis, minor
M  38   scalene, anterior
M  44   serratus, anterior
M  52   sternocleidomastoid
M  55   subclavius

### Organs
O   3   liver
O   5   spleen
O   6   stomach

### Skeletal Structures
S   1   clavicle
S   5   rib, body, lateral (8th)
S   6   rib, body, posterior (1st)
S  12   scapula, coracoid process

### Trachea and Larynx
T   1   trachea
T   3   larynx

### Veins
V   0   pulmonary, superior, RUL
V   1   pulmonary, superior, RML
V   2   pulmonary, superior, LUL
V   3   pulmonary, superior, LUL, lingula
V   8   brachiocephalic (innominate), left
V   9   brachiocephalic (innominate), right
V  24   jugular, external, left
V  25   jugular, external, right
V  26   jugular, internal, left
V  27   jugular, internal, right
V  30   subclavian, left
V  31   subclavian, right
V  38   vena cava, inferior
V  39   vena cava, superior

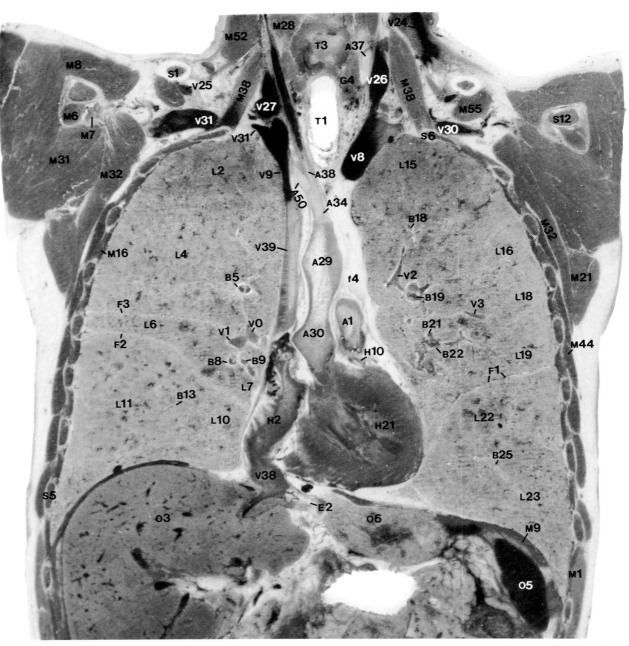

**Anatomic Specimen**

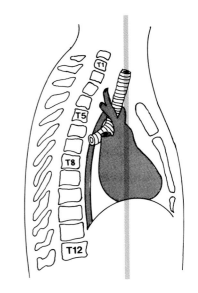

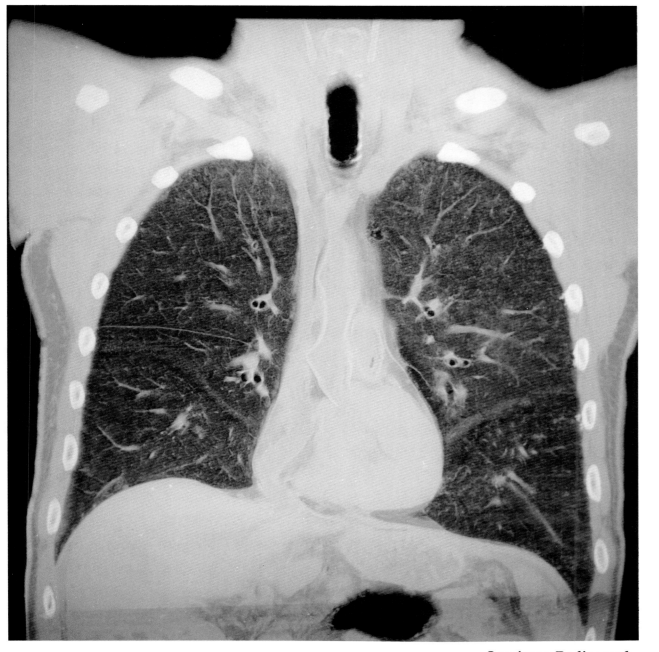

**Specimen Radiograph**

**Magnetic Resonance**

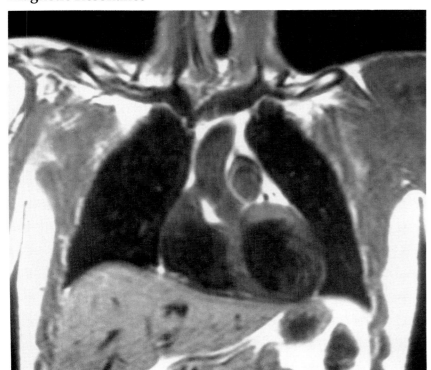

**Trispiral Tomogram**

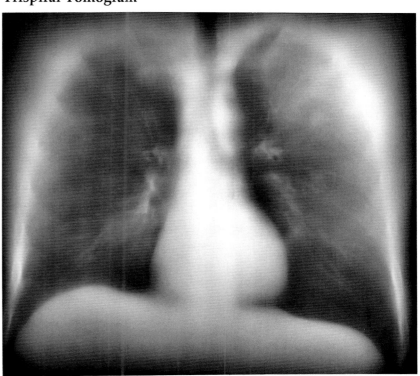

# Coronal
## Section 9

### Anatomic Key

**Arteries**
A 1 pulmonary trunk
A 29 aortic arch
A 30 aorta, ascending
A 34 brachiocephalic (innominate) trunk
A 37 carotid, common, left
A 38 carotid, common, right
A 50 subclavian, right

**Bronchi**
B 5 RUL, anterior segment
B 6 intermediate, right
B 7 right middle lobe (RML)
B 12 RLL, medial basal segment
B 13 RLL, anterior basal segment
B 18 LUL, apicoposterior segment
B 19 LUL, anterior segment
B 21 LUL, superior lingular segment
B 22 LUL, inferior lingular segment
B 25 LLL, anteromedial basal segment
B 26 LLL, lateral basal segment

**Esophagus**
E 1 esophagus
E 2 esophagogastric junction

**Fascia**
F 8 Sibson's

**Fat**
f 2 epicardial
f 3 extrapleural
f 4 mediastinal
f 7 subcutaneous

**Fissures**
F 1 left major (oblique)
F 2 right major (oblique)
F 3 right minor (horizontal)

**Heart**
H 1 atrium, left
H 2 atrium, right
H 12 coronary sinus
H 19 valve, mitral
H 20 valve, aortic
H 21 ventricle, left

**Lobes**
L 2 RUL, apical segment
L 4 RUL, anterior segment
L 6 RML, lateral segment
L 7 RML, medial segment
L 10 RLL, medial basal segment
L 11 RLL, anterior basal segment
L 15 LUL, apicoposterior segment
L 16 LUL, anterior segment
L 18 LUL, superior lingular segment
L 19 LUL, inferior lingular segment
L 22 LLL, anteromedial basal segment
L 23 LLL, lateral basal segment

**Muscles**
M 1 abdominal, external oblique
M 6 biceps brachii, short head
M 7 coracobrachialis
M 8 deltoid
M 9 diaphragm
M 12 diaphragm, crus, right
M 16 intercostal
M 21 latissimus dorsi
M 28 longus colli
M 30 omohyoid
M 31 pectoralis, major
M 32 pectoralis, minor
M 33 platysma
M 38 scalene, anterior
M 39 scalene, middle
M 44 serratus, anterior
M 52 sternocleidomastoid
M 55 subclavius

**Neural Structures**
n 1 brachial plexus

**Organs**
o 3 liver
o 5 spleen
o 6 stomach

**Skeletal Structures**
s 1 clavicle
s 2 humerus
s 3 rib (9th)
s 4 rib, body, anterior (1st)
s 5 rib, body, lateral (5th)
s 12 scapula, coracoid process
s 20 intervertebral disc (C5–C6)
s 21 vertebra, body (C6)
s 29 vertebra, transverse process (C6)

**Trachea**
T 1 trachea

**Veins**
v 0 pulmonary, superior, RUL
v 1 pulmonary, superior, RML
v 3 pulmonary, superior, LUL, lingula
v 4 pulmonary, inferior, RLL
v 8 brachiocephalic (innominate), left
v 31 subclavian, right
v 38 vena cava, inferior
v 39 vena cava, superior

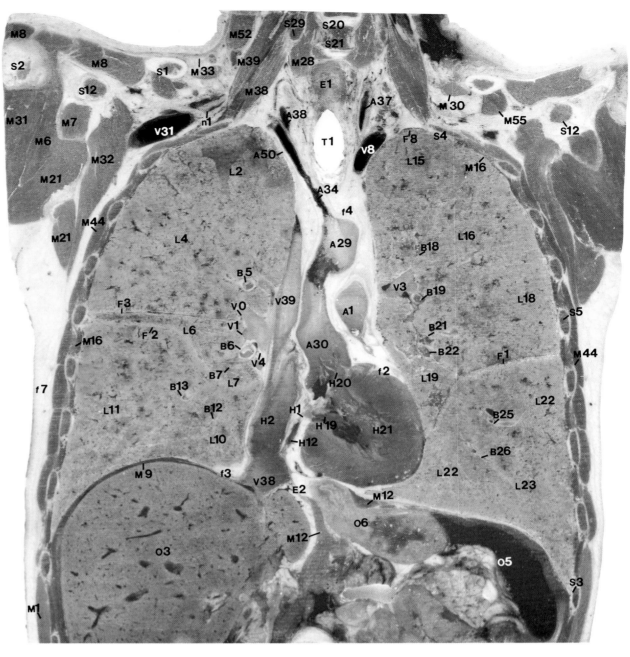

**Anatomic Specimen**

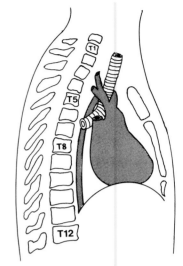

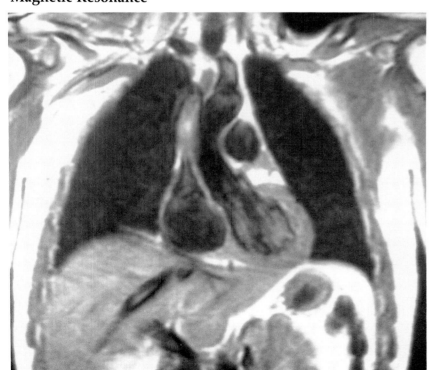

**Specimen Radiograph**

**Magnetic Resonance**

**Trispiral Tomogram**

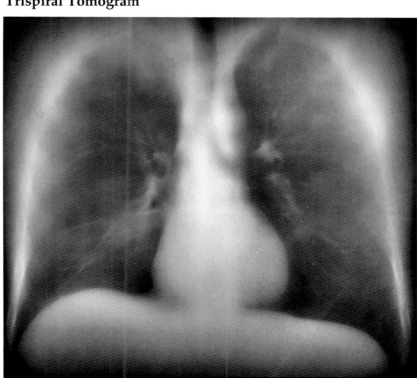

# Coronal

## Section 10

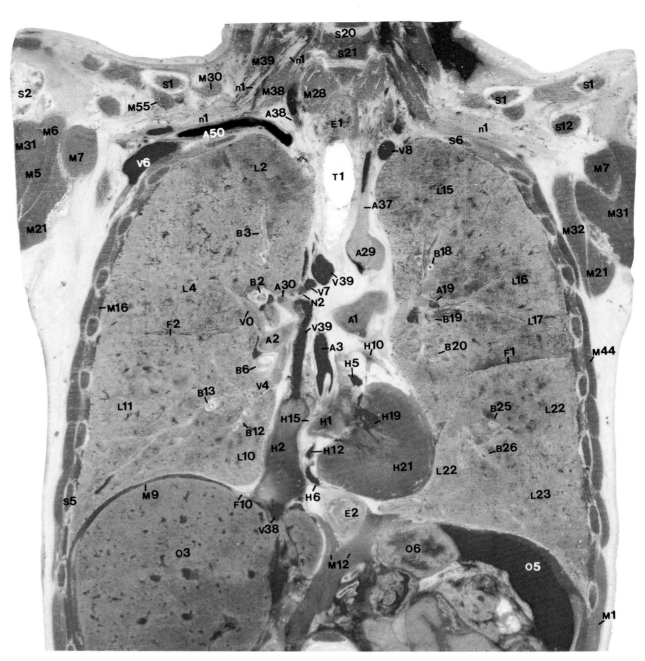

**Anatomic Specimen**

## Anatomic Key

### Arteries
A  1  pulmonary trunk
A  2  main pulmonary, right
A  3  right upper lobe (RUL)
A 19  LUL, anterior branch
A 29  aortic arch
A 30  aorta, ascending
A 37  carotid, common, left
A 38  carotid, common, right
A 50  subclavian, right

### Bronchi
B  2  right upper lobe (RUL)
B  3  RUL, apical segment
B  6  intermediate, right
B 12  RLL, medial basal segment
B 13  RLL, anterior basal segment
B 18  LUL, apicoposterior segment
B 19  LUL, anterior segment
B 20  LUL, lingula
B 25  LLL, anteromedial basal
      segment
B 26  LLL, lateral basal segment

### Esophagus
E  1  esophagus
E  2  esophagogastric junction

### Fascia
F 10  supradiaphragmatic

### Fissures
F  1  left major (oblique)
F  2  right major (oblique)

### Heart
H  1  atrium, left
H  2  atrium, right
H  5  cardiac vein, great
H  6  cardiac vein, middle
H 10  coronary artery, left
H 12  coronary sinus
H 15  septum, interatrial
H 19  valve, mitral
H 21  ventricle, left

### Lobes
L  2  RUL, apical segment
L  4  RUL, anterior segment
L 10  RLL, medial basal segment
L 11  RLL, anterior basal segment
L 15  LUL, apicoposterior segment
L 16  LUL, anterior segment
L 17  LUL, lingula
L 22  LLL, anteromedial basal
      segment
L 23  LLL, lateral basal segment

### Muscles
M  1  abdominal, external oblique
M  5  biceps brachii, long head
M  6  biceps brachii, short head
M  7  coracobrachialis
M  9  diaphragm
M 12  diaphragm, crus, right
M 16  intercostal
M 21  latissimus dorsi
M 28  longus colli
M 30  omohyoid
M 31  pectoralis, major
M 32  pectoralis, minor
M 38  scalene, anterior
M 39  scalene, middle
M 44  serratus, anterior
M 55  subclavius

### Neural Structures
n  1  brachial plexus

### Nodes
N  2  bronchopulmonary (hilar)

### Organs
O  3  liver
O  5  spleen
O  6  stomach

### Skeletal Structures
S  1  clavicle
S  2  humerus
S  5  rib, body, lateral (8th)
S  6  rib, body, posterior (1st)
S 12  scapula, coracoid process
S 20  intervertebral disc (C5–C6)
S 21  vertebra, body (C6)

### Trachea
T  1  trachea

### Veins
V  0  pulmonary, superior, RUL
V  4  pulmonary, inferior, RLL
V  6  axillary, right
V  7  azygos
V  8  brachiocephalic (innominate),
      left
V 38  vena cava, inferior
V 39  vena cava, superior

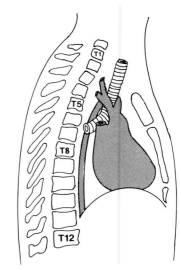

**Specimen Radiograph**

**Magnetic Resonance**

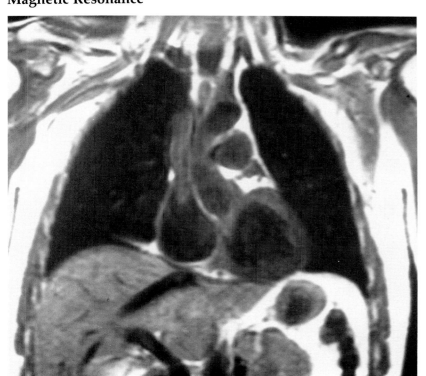

**Trispiral Tomogram**

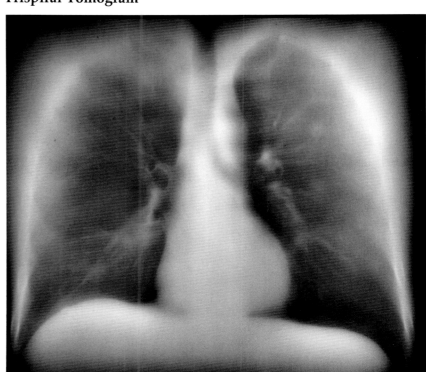

# Coronal

## Section 11

### Anatomic Key

**Arteries**

A  2  main pulmonary, right
A  2* main pulmonary, right (interlobar part)
A  3  right upper lobe (RUL)
A 16  main pulmonary, left
A 29  aortic arch
A 32  axillary, left
A 33  axillary, right
A 37  carotid, common, left
A 49  subclavian, left
A 65  vertebral, left

**Bronchi**

B  2  right upper lobe (RUL)
B  3  RUL, apical segment
B  6  intermediate, right
B 12  RLL, medial basal segment
B 13  RLL, anterior basal segment
B 17  left upper lobe (LUL)
B 18  LUL, apicoposterior segment
B 19  LUL, anterior segment
B 20  LUL, lingula
B 25  LLL, anteromedial basal segment
B 26  LLL, lateral basal segment

**Esophagus**

E  1  esophagus
E  2  esophagogastric junction

**Fat**

f  1  axillary
f  2  epicardial
f  3  extrapleural
f  4  mediastinal

**Fissures**

F  1  left major (oblique)
F  2  right major (oblique)

**Heart**

H  1  atrium, left
H  2  atrium, right
H  8  coronary artery, circumflex
H  9  coronary artery, anterior descending
H 21  ventricle, left

**Lobes**

L  2  RUL, apical segment
L  4  RUL, anterior segment
L  9  RLL, superior segment
L 10  RLL, medial basal segment
L 11  RLL, anterior basal segment
L 15  LUL, apicoposterior segment
L 16  LUL, anterior segment
L 17  LUL, lingula
L 22  LLL, anteromedial basal segment
L 23  LLL, lateral basal segment

**Muscles**

M  1  abdominal, external oblique
M  7  coracobrachialis
M  9  diaphragm
M 11  diaphragm, crus, left
M 12  diaphragm, crus, right
M 16  intercostal
M 28  longus colli
M 30  omohyoid
M 32  pectoralis, minor
M 33  platysma
M 38  scalene, anterior
M 39  scalene, middle
M 44  serratus, anterior
M 55  subclavius
M 63  trapezius

**Neural Structures**

n  1  brachial plexus

**Nodes**

N  2  bronchopulmonary (hilar)

**Organs**

O  3  liver
O  5  spleen
O  6  stomach

**Skeletal Structures**

s  1  clavicle
s  2  humerus
s  3  rib (9th)
s  5  rib, body, lateral (6th)
s  6  rib, body, posterior (1st)
s 11  scapula, acromion
s 12  scapula, coracoid process
s 20  intervertebral disc (C7–T1)
s 21  vertebra, body (C6)

**Trachea**

T  1  trachea

**Veins**

v  0  pulmonary, superior, RUL
v  4  pulmonary, inferior, RLL
v  6  axillary, right
v  7  azygos
v 26  jugular, internal, left
v 38  vena cava, inferior

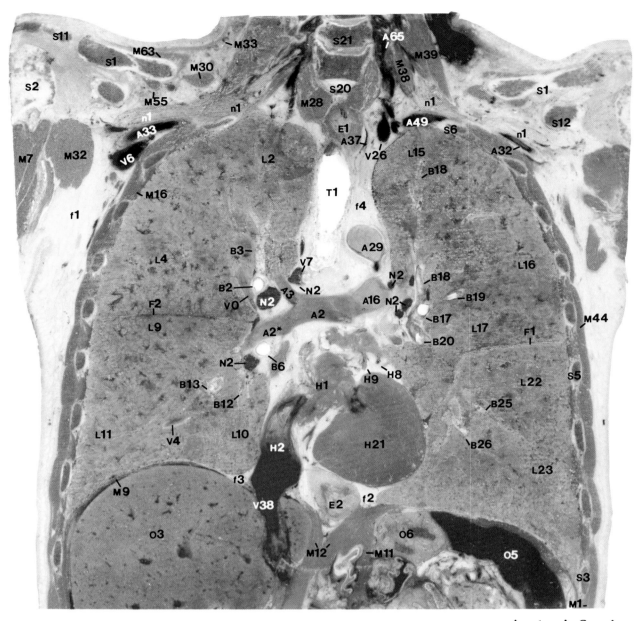

**Anatomic Specimen**

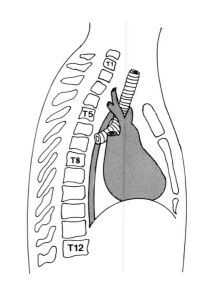

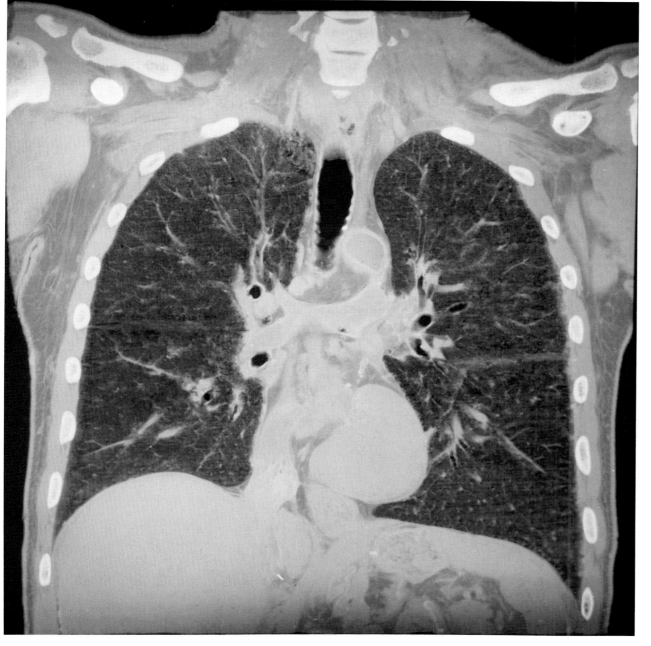

**Specimen Radiograph**

**Magnetic Resonance**

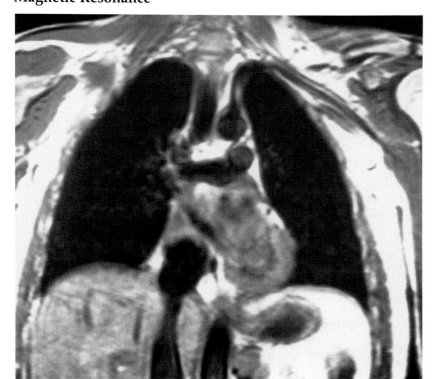

**Trispiral Tomogram**

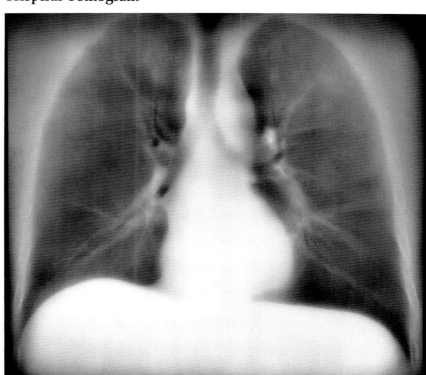

# Coronal
## Section 12

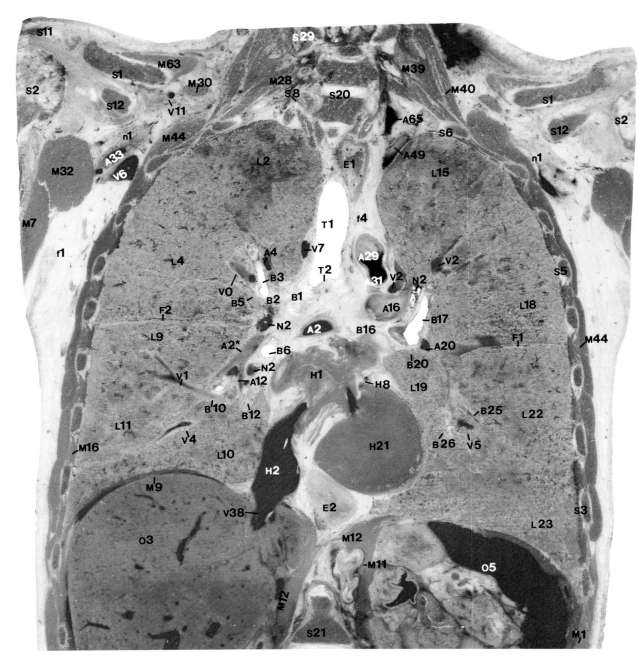

**Anatomic Specimen**

## Anatomic Key

**Arteries**
A  2  main pulmonary, right
A  2* main pulmonary, right (interlobar part)
A  4  RUL, apical branch
A 12 RLL, medial basal branch
A 16 main pulmonary, left
A 20 LUL, lingula
A 29 aortic arch
A 31 aorta, descending
A 33 axillary, right
A 49 subclavian, left
A 65 vertebral, left

**Bronchi**
B  1  main, right
B  2  right upper lobe (RUL)
B  3  RUL, apical segment
B  5  RUL, anterior segment
B  6  intermediate, right
B 10 right lower lobe (RLL)
B 12 RLL, medial basal segment
B 16 main, left
B 17 left upper lobe (LUL)
B 20 LUL, lingula
B 25 LLL, anteromedial basal segment
B 26 LLL, lateral basal segment

**Esophagus**
E  1  esophagus
E  2  esophagogastric junction

**Fat**
f  1  axillary
f  4  mediastinal

**Fissures**
F  1  left major (oblique)
F  2  right major (oblique)

**Heart**
H  1  atrium, left
H  2  atrium, right
H  8  coronary artery, circumflex
H 21 ventricle, left

**Lobes**
L  2  RUL, apical segment
L  4  RUL, anterior segment
L  9  RLL, superior segment
L 10 RLL, medial basal segment
L 11 RLL, anterior basal segment
L 15 LUL, apicoposterior segment
L 18 LUL, superior lingular segment
L 19 LUL, inferior lingular segment
L 22 LLL, anteromedial basal segment
L 23 LLL, lateral basal segment

**Muscles**
M  1  abdominal, external oblique
M  7  coracobrachialis
M  9  diaphragm
M 11 diaphragm, crus, left
M 12 diaphragm, crus, right
M 16 intercostal
M 28 longus colli
M 30 omohyoid
M 32 pectoralis, minor
M 39 scalene, middle
M 40 scalene, posterior
M 44 serratus, anterior
M 63 trapezius

**Neural Structures**
n  1  brachial plexus

**Nodes**
N  2  bronchopulmonary (hilar) (calcified)

**Organs**
O  3  liver
O  5  spleen

**Skeletal Structures**
S  1  clavicle
S  2  humerus
S  3  rib (8th)
S  5  rib, body, lateral (4th)
S  6  rib, body, posterior (1st)
S  8  rib, head and neck (1st)
S 11 scapula, acromion
S 12 scapula, coracoid process
S 20 intervertebral disc (C7–T1)
S 21 vertebra, body (T12)
S 29 vertebra, transverse process (C6)

**Trachea**
T  1  trachea
T  2  carina (bifurcation of trachea)

**Veins**
V  0  pulmonary, superior, RUL
V  1  pulmonary, superior, RML
V  2  pulmonary, superior, LUL
V  4  pulmonary, inferior, RLL
V  5  pulmonary, inferior, LLL
V  6  axillary, right
V  7  azygos
V 11 cephalic, right
V 38 vena cava, inferior

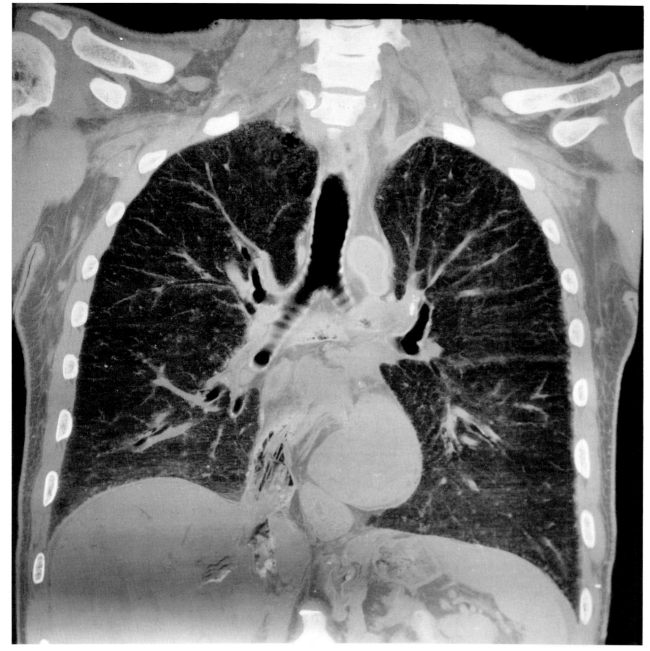

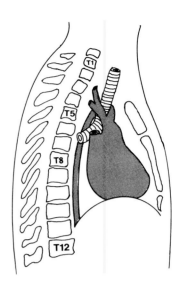

**Specimen Radiograph**

**Magnetic Resonance**

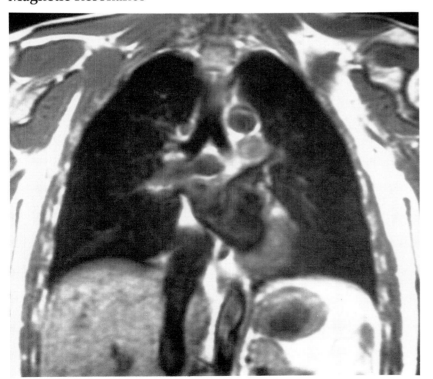

**Trispiral Tomogram**

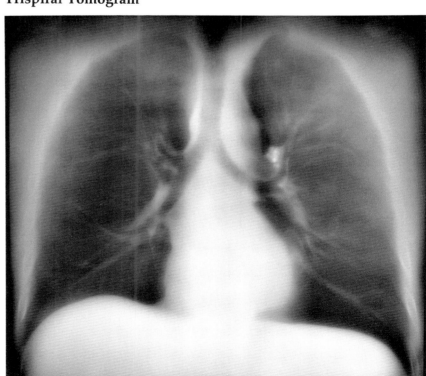

## Anatomic Key

### Arteries
A  2* main pulmonary, right (interlobar part)
A  5 RUL, posterior branch
A 16 main pulmonary, left
A 23 left lower lobe (LLL)
A 31 aorta, descending
A 32 axillary, left
A 33 axillary, right
A 49 subclavian, left
A 65 vertebral, left

### Bronchi
B  1 main, right
B  2 right upper lobe (RUL)
B  4 RUL, posterior segment
B  6 intermediate, right
B 10 right lower lobe (RLL)
B 12 RLL, medial basal segment
B 13 RLL, anterior basal segment
B 16 main, left
B 17 left upper lobe (LUL)
B 18 LUL, apicoposterior segment
B 19 LUL, anterior segment
B 25 LLL, anteromedial basal segment
B 26 LLL, lateral basal segment

### Esophagus
E  1 esophagus

### Fat
f  1 axillary
f  3 extrapleural
f  4 mediastinal

### Fissures
F  1 left major (oblique)
F  2 right major (oblique)

### Heart
H  1 atrium, left
H  2 atrium, right
H  5 cardiac vein, great
H  8 coronary artery, circumflex
H 21 ventricle, left

### Lobes
L  2 RUL, apical segment
L  4 RUL, anterior segment
L  9 RLL, superior segment
L 10 RLL, medial basal segment
L 11 RLL, anterior basal segment
L 15 LUL, apicoposterior segment
L 19 LUL, inferior lingular segment
L 22 LLL, anteromedial basal segment
L 23 LLL, lateral basal segment

### Muscles
M  1 abdominal, external oblique
M  5 biceps brachii, long head
M  6 biceps brachii, short head
M  7 coracobrachialis
M  9 diaphragm
M 11 diaphragm, crus, left
M 12 diaphragm, crus, right
M 16 intercostal
M 28 longus colli
M 30 omohyoid
M 33 platysma
M 39 scalene, middle
M 40 scalene, posterior
M 44 serratus, anterior
M 63 trapezius

### Neural Structures
n  1 brachial plexus
n  5 spinal cord

### Nodes
N  2 bronchopulmonary (hilar) (calcified)
N 11 tracheobronchial, inferior (subcarinal)

### Organs
O  3 liver
O  5 spleen
O  6 stomach

### Skeletal Structures
S  1 clavicle
S  2 humerus
S  6 rib, body, posterior (1st)
S  8 rib, head and neck (1st)
S 11 scapula, acromion
S 12 scapula, coracoid process
S 20 intervertebral disc (T11–T12)
S 21 vertebra, body (T1)
S 26 vertebra, pedicle (C7)

### Trachea
T  1 trachea
T  2 carina (bifurcation of trachea)

### Veins
V  2 pulmonary, superior, LUL
V  4 pulmonary, inferior, RLL
V  5 pulmonary, inferior, LLL
V  6 axillary, right
V  7 azygos
V 11 cephalic, right
V 40 venous plexus, mediastinal

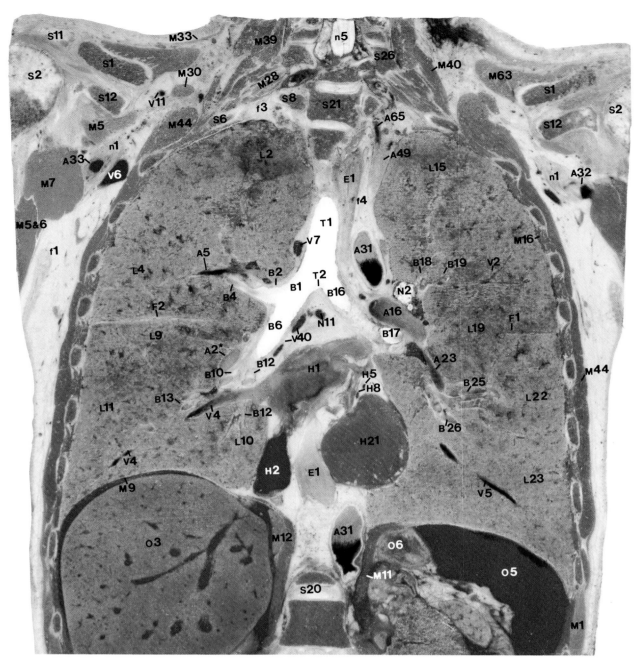

**Anatomic Specimen**

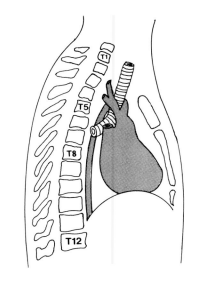

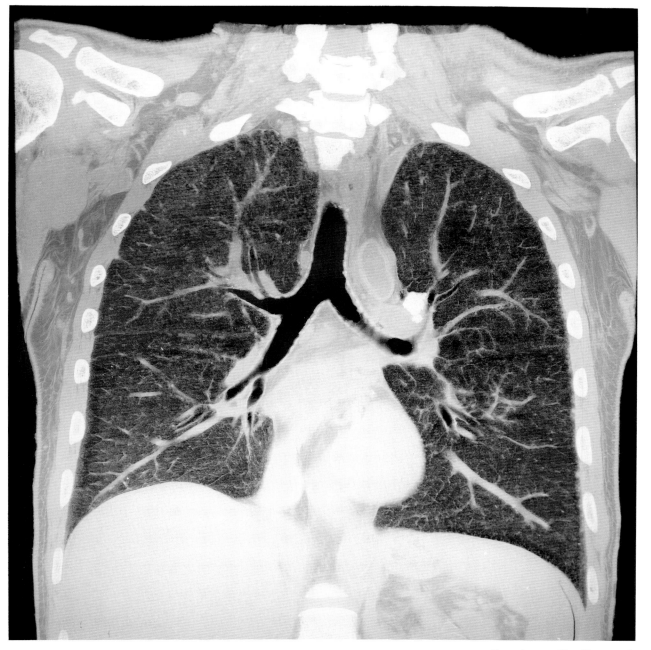

**Specimen Radiograph**

**Magnetic Resonance**

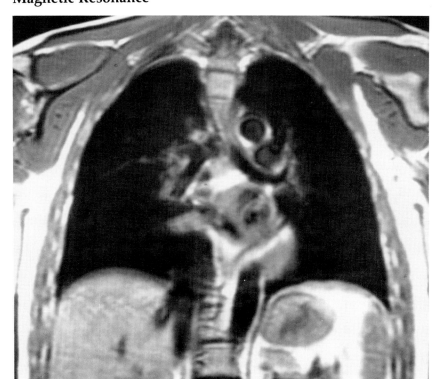

**Trispiral Tomogram**

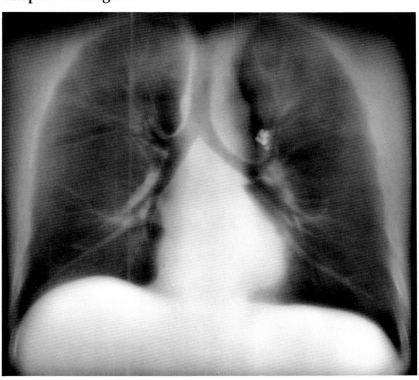

# Coronal

## Section 14

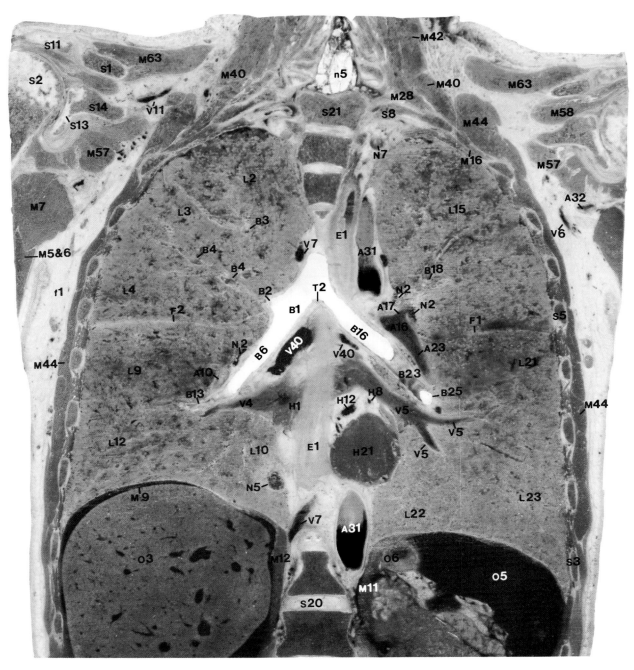

**Anatomic Specimen**

## Anatomic Key

### Arteries
A 10 right lower lobe (RLL)
A 16 main pulmonary, left
A 17 left upper lobe (LUL)
A 23 left lower lobe (LLL)
A 31 aorta, descending
A 32 axillary, left

### Bronchi
B 1 main, right
B 2 right upper lobe (RUL)
B 3 RUL, apical segment
B 4 RUL, posterior segment
B 6 intermediate, right
B 13 RLL, anterior basal segment
B 16 main, left
B 18 LUL, apicoposterior segment
B 23 left lower lobe (LLL)
B 25 LLL, anteromedial basal
    segment

### Esophagus
E 1 esophagus

### Fat
f 1 axillary

### Fissures
F 1 left major (oblique)
F 2 right major (oblique)

### Heart
H 1 atrium, left
H 8 coronary artery, circumflex
H 12 coronary sinus
H 21 ventricle, left

### Lobes
L 2 RUL, apical segment
L 3 RUL, posterior segment
L 4 RUL, anterior segment
L 9 RLL, superior segment
L 10 RLL, medial basal segment
L 12 RLL, lateral basal segment
L 15 LUL, apicoposterior segment
L 21 LLL, superior segment
L 22 LLL, anteromedial basal
    segment
L 23 LLL, lateral basal segment

### Muscles
M 5 biceps brachii, long head
M 6 biceps brachii, short head
M 7 coracobrachialis
M 9 diaphragm
M 11 diaphragm, crus, left
M 12 diaphragm, crus, right
M 16 intercostal
M 28 longus colli
M 40 scalene, posterior
M 42 semispinalis, cervicis
M 44 serratus, anterior
M 57 subscapularis
M 58 supraspinatus
M 63 trapezius

### Neural Structures
n 5 spinal cord

### Nodes
N 2 bronchopulmonary (hilar)
N 5 intrapulmonary
N 7 mediastinal, posterior

### Organs
O 3 liver
O 5 spleen
O 6 stomach

### Skeletal Structures
S 1 clavicle
S 2 humerus
S 3 rib (9th)
S 5 rib, body, lateral (5th)
S 8 rib, head and neck (1st)
S 11 scapula, acromion
S 13 scapula, glenoid cavity
S 14 scapula, neck
S 20 intervertebral disc (T11–T12)
S 21 vertebra, body (T1)

### Trachea
T 2 carina (bifurcation of trachea)

### Veins
V 4 pulmonary, inferior, RLL
V 5 pulmonary, inferior, LLL
V 6 axillary, left
V 7 azygos
V 11 cephalic, right
V 40 venous plexus, mediastinal

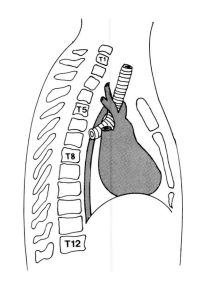

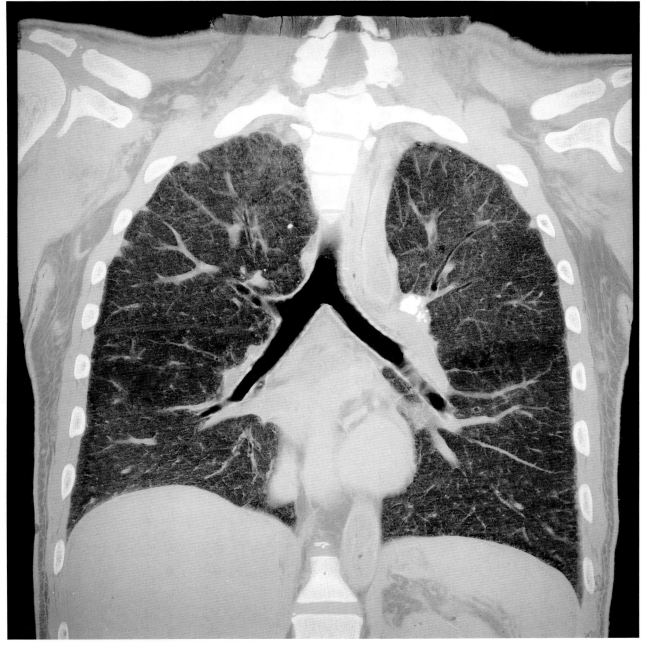

**Specimen Radiograph**

**Magnetic Resonance**

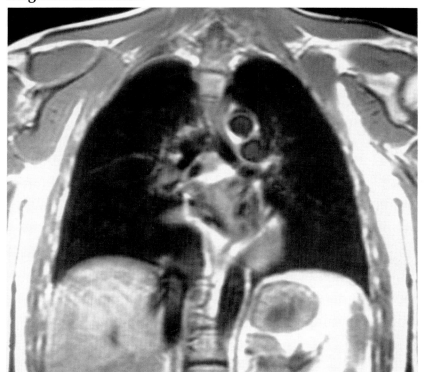

**Trispiral Tomogram**

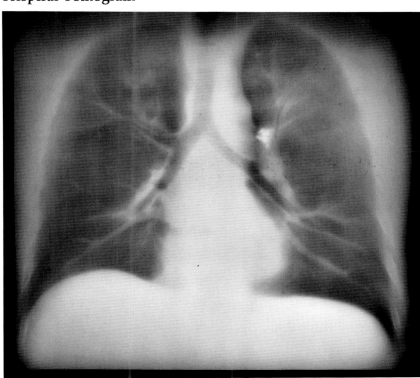

# Coronal

## Section 15

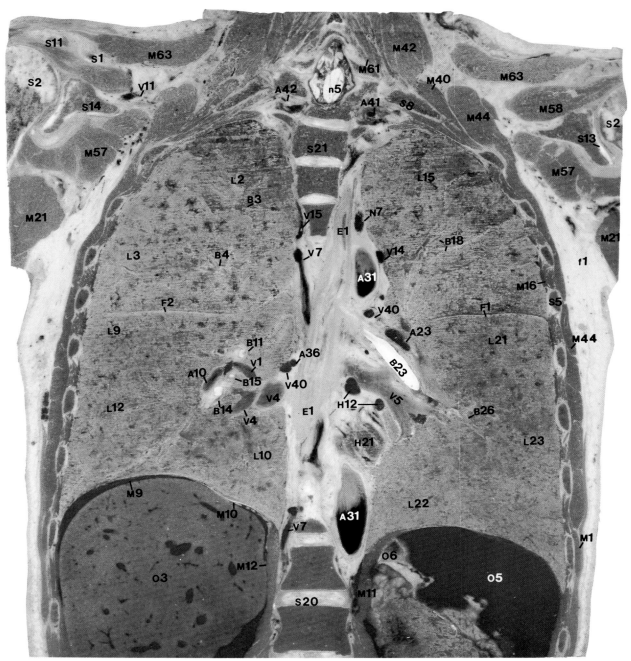

**Anatomic Specimen**

## Anatomic Key

### Arteries
A **10** right lower lobe (RLL)
A **23** left lower lobe (LLL)
A **31** aorta, descending
A **36** bronchial, right
A **41** intercostal, highest (superior), left
A **42** intercostal, highest (superior), right

### Bronchi
B **3** RUL, apical segment
B **4** RUL, posterior segment
B **11** RLL, superior segment
B **14** RLL, lateral basal segment
B **15** RLL, posterior basal segment
B **18** LUL, apicoposterior segment
B **23** left lower lobe (LLL)
B **26** LLL, lateral basal segment

### Esophagus
E **1** esophagus

### Fat
f **1** axillary

### Fissures
F **1** left major (oblique)
F **2** right major (oblique)

### Heart
H **12** coronary sinus
H **21** ventricle, left

### Lobes
L **2** RUL, apical segment
L **3** RUL, posterior segment
L **9** RLL, superior segment
L **10** RLL, medial basal segment
L **12** RLL, lateral basal segment
L **15** LUL, apicoposterior segment
L **21** LLL, superior segment
L **22** LLL, anteromedial basal segment
L **23** LLL, lateral basal segment

### Muscles
M **1** abdominal, external oblique
M **9** diaphragm
M **10** diaphragm, central tendon
M **11** diaphragm, crus, left
M **12** diaphragm, crus, right
M **16** intercostal
M **21** latissimus dorsi
M **40** scalene, posterior
M **42** semispinalis, cervicis
M **44** serratus, anterior
M **57** subscapularis
M **58** supraspinatus
M **61** transversospinalis
M **63** trapezius

### Neural Structures
n **5** spinal cord

### Nodes
N **7** mediastinal, posterior

### Organs
o **3** liver
o **5** spleen
o **6** stomach

### Skeletal Structures
s **1** clavicle
s **2** humerus
s **5** rib, body, lateral (5th)
s **8** rib, head and neck (1st)
s **11** scapula, acromion
s **13** scapula, glenoid cavity
s **14** scapula, neck
s **20** intervertebral disc (T11–T12)
s **21** vertebra, body (T2)

### Veins
v **1** pulmonary, superior, RML
v **4** pulmonary, inferior, RLL
v **5** pulmonary, inferior, LLL
v **7** azygos
v **11** cephalic, right
v **14** intercostal, highest (superior), left
v **15** intercostal, highest (superior), right
v **40** venous plexus, mediastinal

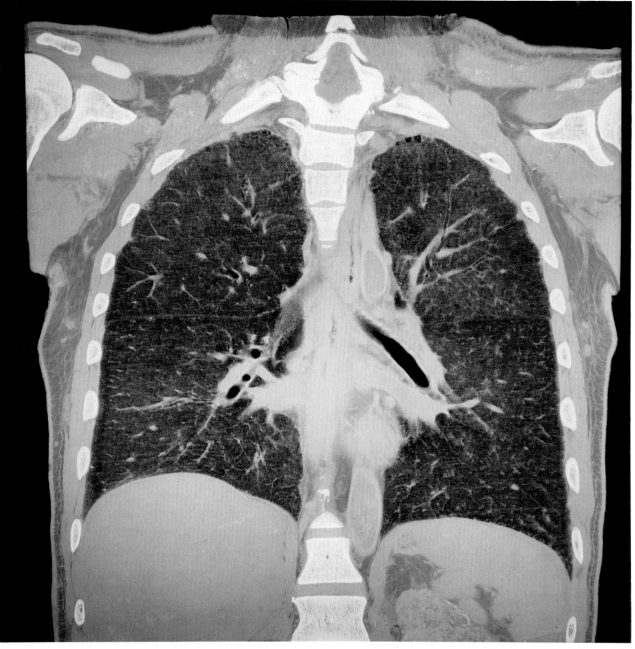

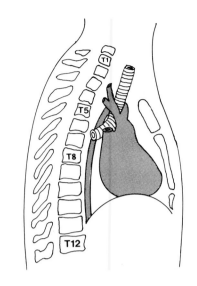

**Specimen Radiograph**

**Magnetic Resonance**

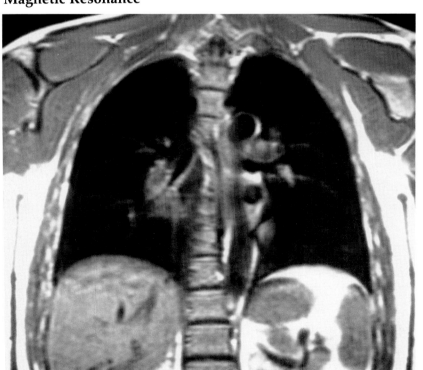

**Trispiral Tomogram**

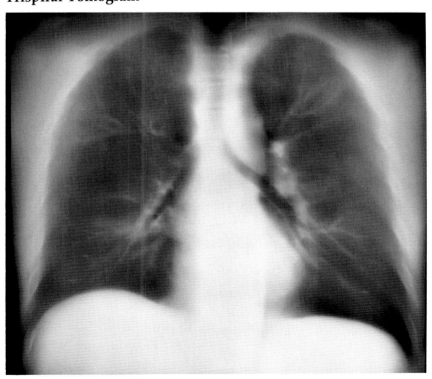

# Coronal

## Section 16

### Anatomic Key

**Arteries**
A 24 LLL, superior branch
A 31 aorta, descending

**Bronchi**
B 3 RUL, apical segment
B 4 RUL, posterior segment
B 11 RLL, superior segment
B 14 RLL, lateral basal segment
B 15 RLL, posterior basal segment
B 18 LUL, apicoposterior segment
B 24 LLL, superior segment
B 26 LLL, lateral basal segment
B 27 LLL, posterior basal segment

**Esophagus**
E 1 esophagus

**Fascia**
F 7 prevertebral

**Fat**
f 1 axillary
f 3 extrapleural
f 7 subcutaneous

**Fissures**
F 1 left major (oblique)
F 2 right major (oblique)

**Lobes**
L 2 RUL, apical segment
L 3 RUL, posterior segment
L 9 RLL, superior segment
L 10 RLL, medial basal segment
L 12 RLL, lateral basal segment
L 15 LUL, apicoposterior segment
L 21 LLL, superior segment
L 22 LLL, anteromedial basal segment
L 23 LLL, lateral basal segment

**Muscles**
M 1 abdominal, external oblique
M 9 diaphragm
M 11 diaphragm, crus, left
M 12 diaphragm, crus, right
M 16 intercostal
M 21 latissimus dorsi
M 40 scalene, posterior
M 42 semispinalis, cervicis
M 44 serratus, anterior
M 57 subscapularis
M 58 supraspinatus
M 61 transversospinalis
M 63 trapezius

**Neural Structures**
n 5 spinal cord

**Nodes**
N 2 bronchopulmonary (hilar)

**Organs**
o 3 liver
o 5 spleen

**Skeletal Structures**
S 2 humerus
S 5 rib, body, lateral (5th)
S 8 rib, head and neck (1st)
S 11 scapula, acromion
S 13 scapula, glenoid cavity
S 14 scapula, neck
S 20 intervertebral disc
S 21 vertebra, body (T3)
S 29 vertebra, transverse process (T1)

**Veins**
V 4 pulmonary, inferior, RLL
V 5 pulmonary, inferior, LLL
V 7 azygos
V 12 hemiazygos
V 14 intercostal, highest (superior), left
V 15 intercostal, highest (superior), right
V 19 intercostal, posterior, right

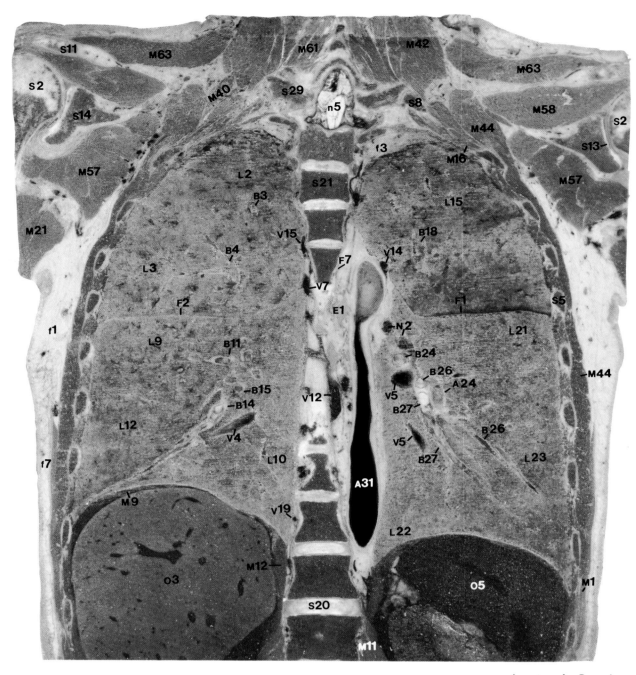

**Anatomic Specimen**

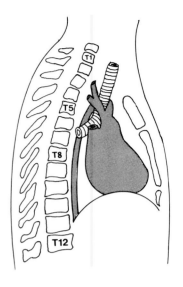

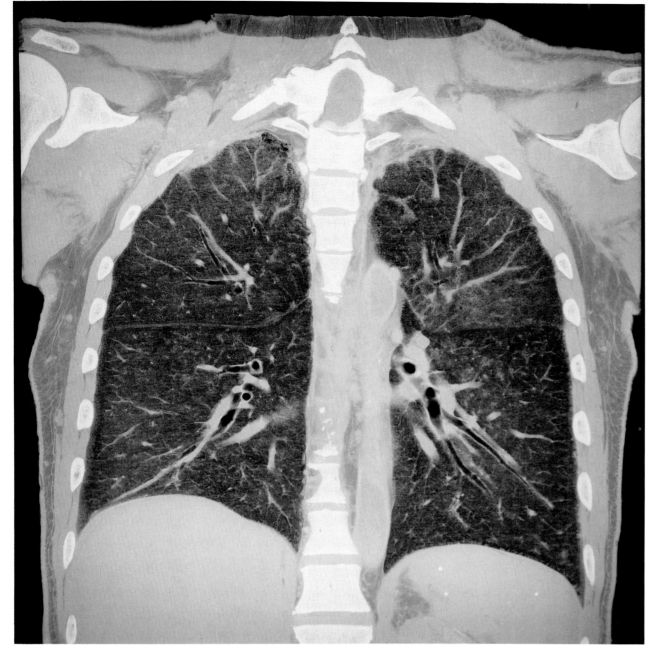

**Specimen Radiograph**

**Magnetic Resonance**

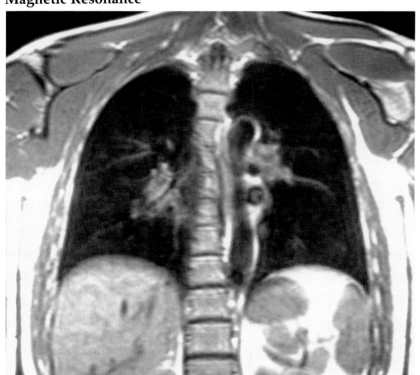

**Trispiral Tomogram**

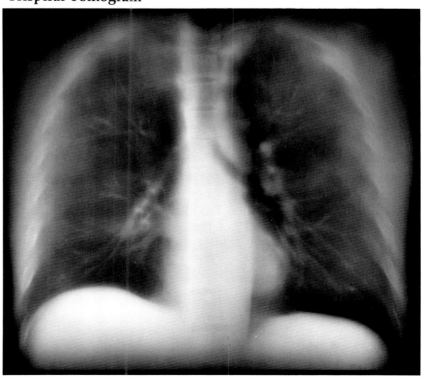

# Coronal
## Section 17

### Anatomic Key

**Arteries**

A 23 left lower lobe (LLL)
A 26 LLL, lateral basal branch
A 27 LLL, posterior basal branch
A 31 aorta, descending
A 46 intercostal, posterior, right

**Bronchi**

B 4 RUL, posterior segment
B 11 RLL, superior segment
B 14 RLL, lateral basal segment
B 15 RLL, posterior basal segment
B 24 LLL, superior segment
B 26 LLL, lateral basal segment
B 27 LLL, posterior basal segment

**Fat**

f 1 axillary

**Fissures**

F 1 left major (oblique)
F 2 right major (oblique)

**Lobes**

L 2 RUL, apical segment
L 3 RUL, posterior segment
L 9 RLL, superior segment
L 10 RLL, medial basal segment
L 12 RLL, lateral basal segment
L 15 LUL, apicoposterior segment
L 21 LLL, superior segment
L 23 LLL, lateral basal segment
L 24 LLL, posterior basal segment

**Muscles**

M 9 diaphragm
M 16 intercostal
M 21 latissimus dorsi
M 40 scalene, posterior
M 42 semispinalis, cervicis
M 44 serratus, anterior
M 50 splenius, capitis
M 57 subscapularis
M 58 supraspinatus
M 61 transversospinalis
M 63 trapezius

**Neural Structures**

n 5 spinal cord

**Organs**

O 3 liver
O 5 spleen

**Skeletal Structures**

s 2 humerus
s 3 rib (9th)
s 5 rib, body, lateral (2nd)
s 8 rib, head and neck (2nd)
s 10 scapula
s 11 scapula, acromion
s 13 scapula, glenoid cavity
s 14 scapula, neck
s 20 intervertebral disc (T12–L1)
s 21 vertebra, body (T2)
s 25 vertebra, lamina (T1)
s 29 vertebra, transverse process (T1)

**Veins**

v 4 pulmonary, inferior, RLL
v 5 pulmonary, inferior, LLL
v 12 hemiazygos
v 14 intercostal, highest (superior), left
v 19 intercostal, posterior, right

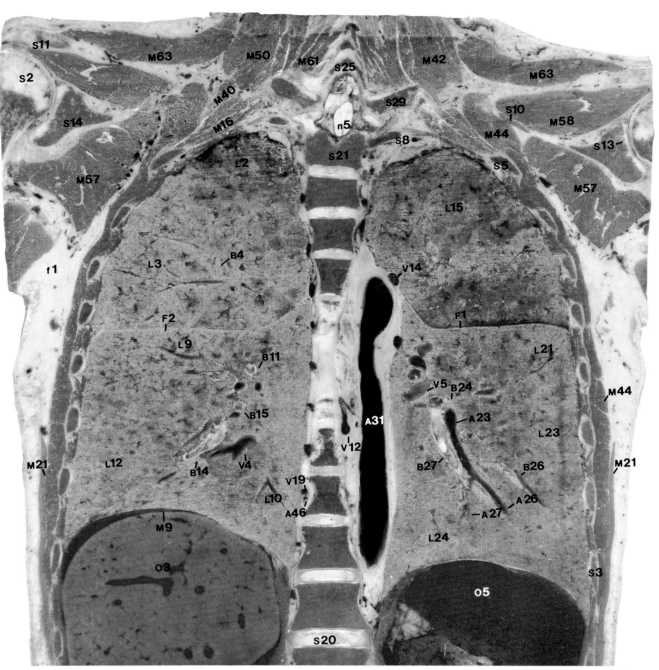

**Anatomic Specimen**

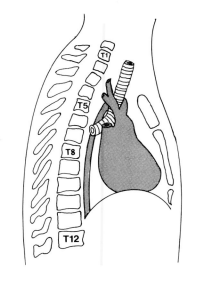

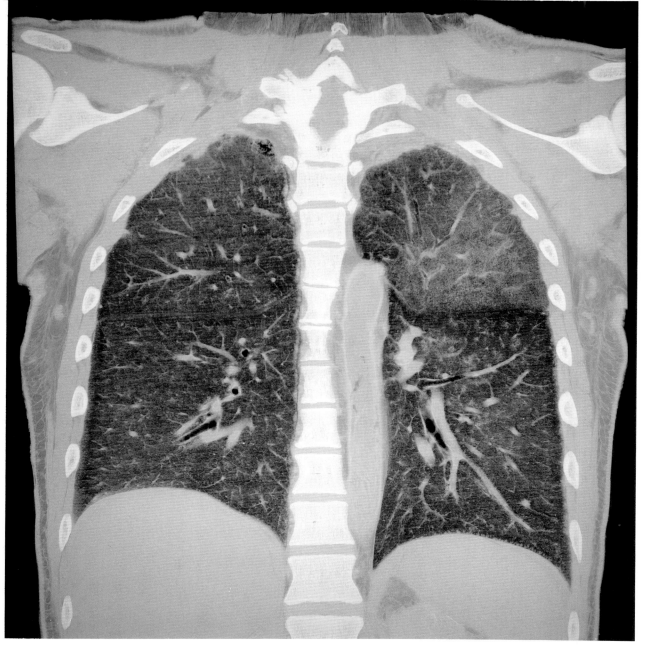

Specimen Radiograph

**Magnetic Resonance**

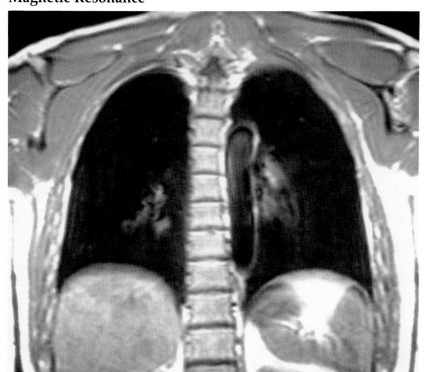

**Trispiral Tomogram**

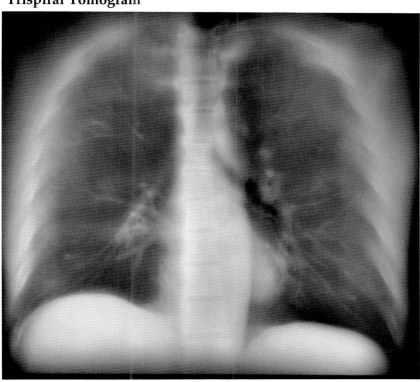

# Coronal

## Section 18

## Anatomic Key

### Arteries
A **24** LLL, superior branch
A **31** aorta, descending

### Bronchi
B **4** RUL, posterior segment
B **11** RLL, superior segment
B **14** RLL, lateral basal segment
B **15** RLL, posterior basal segment
B **26** LLL, lateral basal segment
B **27** LLL, posterior basal segment

### Fissures
F **1** left major (oblique)
F **2** right major (oblique)

### Lobes
L **2** RUL, apical segment
L **3** RUL, posterior segment
L **9** RLL, superior segment
L **10** RLL, medial basal segment
L **12** RLL, lateral basal segment
L **15** LUL, apicoposterior segment
L **21** LLL, superior segment
L **23** LLL, lateral basal segment
L **24** LLL, posterior basal segment

### Muscles
M **1** abdominal, external oblique
M **9** diaphragm
M **16** intercostal
M **42** semispinalis, cervicis
M **44** serratus, anterior
M **50** splenius, capitis
M **57** subscapularis
M **58** supraspinatus
M **61** transversospinalis
M **63** trapezius

### Neural Structures
n **5** spinal cord

### Organs
o **3** liver
o **5** spleen

### Skeletal Structures
s **2** humerus
s **3** rib (9th)
s **5** rib, body, lateral (3rd)
s **6** rib, body, posterior (1st)
s **8** rib, head and neck (2nd)
s **10** scapula
s **11** scapula, acromion
s **13** scapula, glenoid cavity
s **20** intervertebral disc (T11–T12)
s **21** vertebra, body (T3)

### Veins
v **4** pulmonary, inferior, RLL
v **5** pulmonary, inferior, LLL
v **14** intercostal, highest (superior), left
v **15** intercostal, highest (superior), right

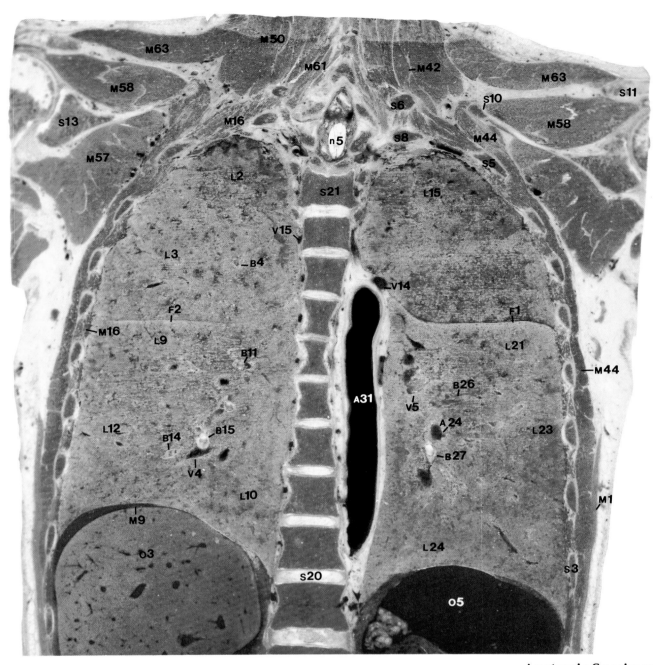

**Anatomic Specimen**

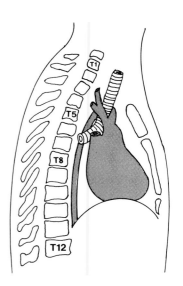

**Specimen Radiograph**

**Magnetic Resonance**

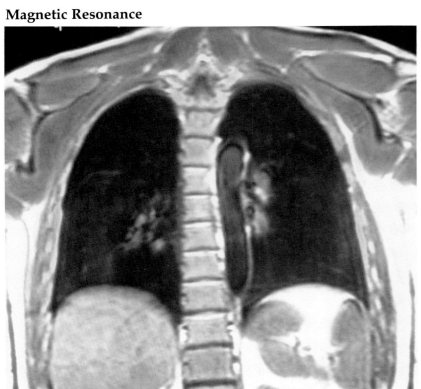

**Trispiral Tomogram**

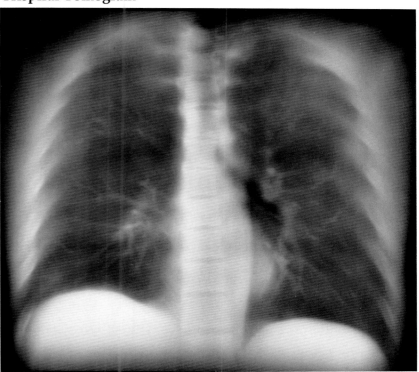

# Coronal

## Section 19

## Anatomic Key

### Arteries
A 24 LLL, superior branch
A 31 aorta, descending

### Bronchi
B 11 RLL, superior segment
B 12 RLL, medial basal segment
B 13 RLL, anterior basal segment
B 14 RLL, lateral basal segment
B 15 RLL, posterior basal segment
B 24 LLL, superior segment
B 26 LLL, lateral basal segment
B 27 LLL, posterior basal segment

### Fissures
F 1 left major (oblique)
F 2 right major (oblique)

### Lobes
L 2 RUL, apical segment
L 3 RUL, posterior segment
L 9 RLL, superior segment
L 10 RLL, medial basal segment
L 12 RLL, lateral basal segment
L 13 RLL, posterior basal segment
L 15 LUL, apicoposterior segment
L 21 LLL, superior segment
L 23 LLL, lateral basal segment
L 24 LLL, posterior basal segment

### Muscles
M 9 diaphragm
M 11 diaphragm, crus, left
M 12 diaphragm, crus, right
M 16 intercostal
M 21 latissimus dorsi
M 40 scalene, posterior
M 42 semispinalis, cervicis
M 44 serratus, anterior
M 51 splenius, cervicis
M 57 subscapularis
M 58 supraspinatus
M 61 transversospinalis
M 63 trapezius

### Neural Structures
n 5 spinal cord

### Organs
O 3 liver
O 5 spleen

### Skeletal Structures
s 2 humerus
s 3 rib (9th)
s 5 rib, body, lateral (3rd)
s 8 rib, head and neck (2nd)
s 10 scapula
s 11 scapula, acromion
s 13 scapula, glenoid cavity
s 14 scapula, neck
s 20 intervertebral disc (T12–L1)
s 21 vertebra, body (T3)
s 25 vertebra, lamina (T1)
s 27 vertebra, spinous process (C7)
s 29 vertebra, transverse process
   (T2)

### Veins
v 4 pulmonary, inferior, RLL
v 14 intercostal, highest (superior),
   left
v 18 intercostal, posterior, left
v 19 intercostal, posterior, right
v 40 venous plexus, mediastinal

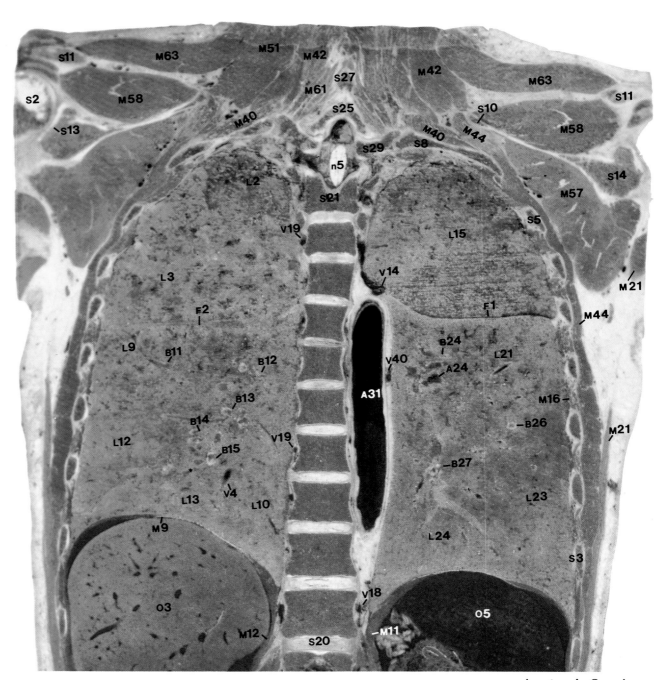

**Anatomic Specimen**

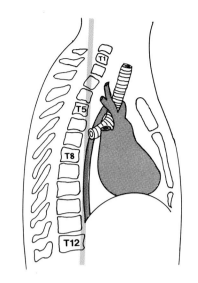

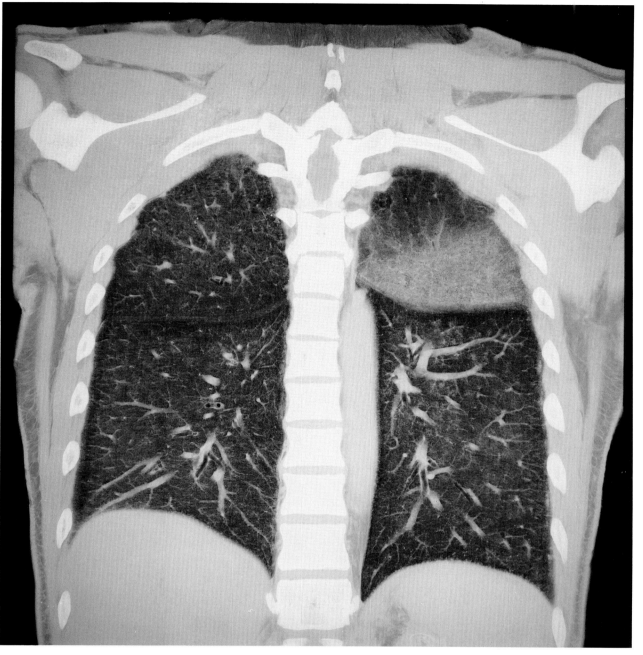

**Specimen Radiograph**

**Magnetic Resonance**

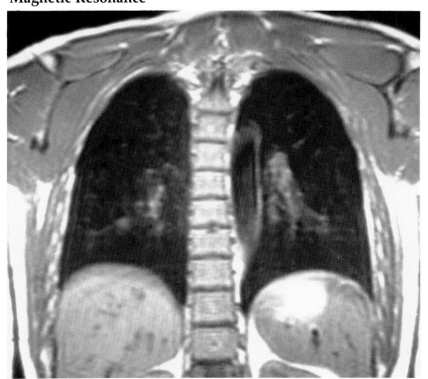

**Trispiral Tomogram**

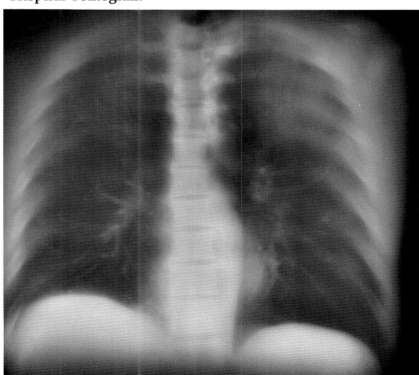

# Coronal

## Section 20

### Anatomic Key

**Arteries**
A 26 LLL, lateral basal branch
A 27 LLL, posterior basal branch
A 31 aorta, descending
A 46 intercostal, posterior, right

**Bronchi**
B 11 RLL, superior segment
B 12 RLL, medial basal segment
B 14 RLL, lateral basal segment
B 15 RLL, posterior basal segment
B 24 LLL, superior segment
B 26 LLL, lateral basal segment
B 27 LLL, posterior basal segment

**Fissures**
F 1 left major (oblique)
F 2 right major (oblique)

**Lobes**
L 3 RUL, posterior segment
L 9 RLL, superior segment
L 10 RLL, medial basal segment
L 13 RLL, posterior basal segment
L 15 LUL, apicoposterior segment
L 21 LLL, superior segment
L 24 LLL, posterior basal segment

**Muscles**
M 8 deltoid
M 9 diaphragm
M 15 infraspinatus
M 16 intercostal
M 21 latissimus dorsi
M 23 levator scapulae
M 42 semispinalis, cervicis
M 44 serratus, anterior
M 51 splenius, cervicis
M 57 subscapularis
M 58 supraspinatus
M 59 teres major
M 63 trapezius

**Neural Structures**
n 5 spinal cord

**Organs**
O 3 liver
O 5 spleen

**Skeletal Structures**
S 2 humerus
S 3 rib (9th)
S 6 rib, body, posterior (2nd)
S 10 scapula
S 11 scapula, acromion
S 15 scapula, spine
S 20 intervertebral disc (T11–T12)
S 21 vertebra, body (T4)
S 25 vertebra, lamina (T1)
S 27 vertebra, spinous process (C7)
S 29 vertebra, transverse process (T2)

**Veins**
v 14 intercostal, highest (superior), left
v 19 intercostal, posterior, right

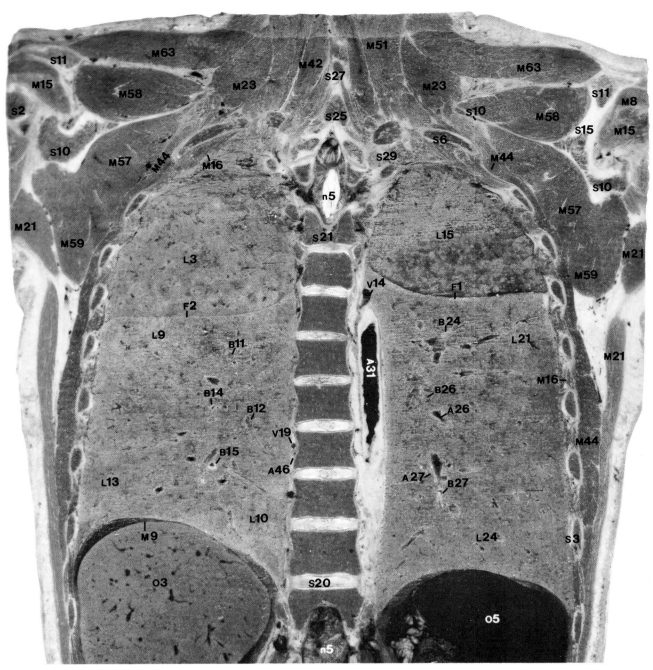

**Anatomic Specimen**

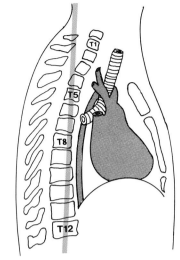

**Specimen Radiograph**

**Magnetic Resonance**

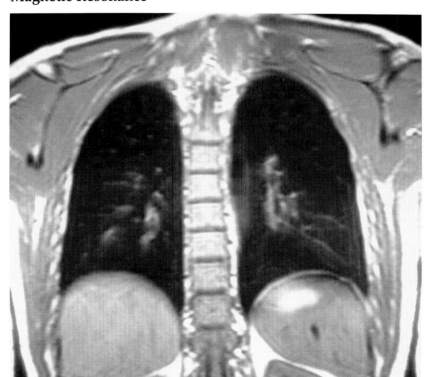

**Trispiral Tomogram**

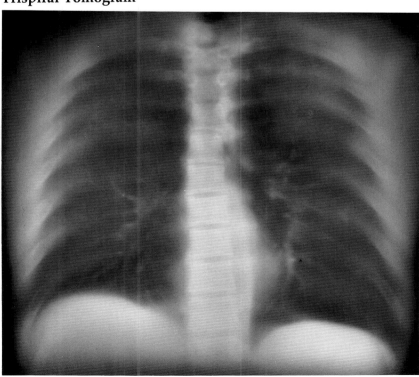

# Coronal
## Section 21

### Anatomic Key

**Fissures**
F 1 left major (oblique)
F 2 right major (oblique)

**Lobes**
L 3 RUL, posterior segment
L 9 RLL, superior segment
L 10 RLL, medial basal segment
L 13 RLL, posterior basal segment
L 15 LUL, apicoposterior segment
L 21 LLL, superior segment
L 24 LLL, posterior basal segment

**Muscles**
M 8 deltoid
M 9 diaphragm
M 15 infraspinatus
M 16 intercostal
M 21 latissimus dorsi
M 23 levator scapulae
M 34 psoas
M 42 semispinalis, cervicis
M 44 serratus, anterior
M 57 subscapularis
M 58 supraspinatus
M 59 teres major
M 61 transversospinalis
M 63 trapezius

**Neural Structures**
n 5 spinal cord

**Organs**
o 3 liver
o 5 spleen

**Skeletal Structures**
s 3 rib (10th)
s 5 rib, body, lateral (7th)
s 6 rib, body, posterior (3rd)
s 8 rib, head and neck (12th)
s 9 rib, tubercle (4th)
s 10 scapula
s 15 scapula, spine
s 27 vertebra, spinous process (T3)
s 29 vertebra, transverse process (T4)

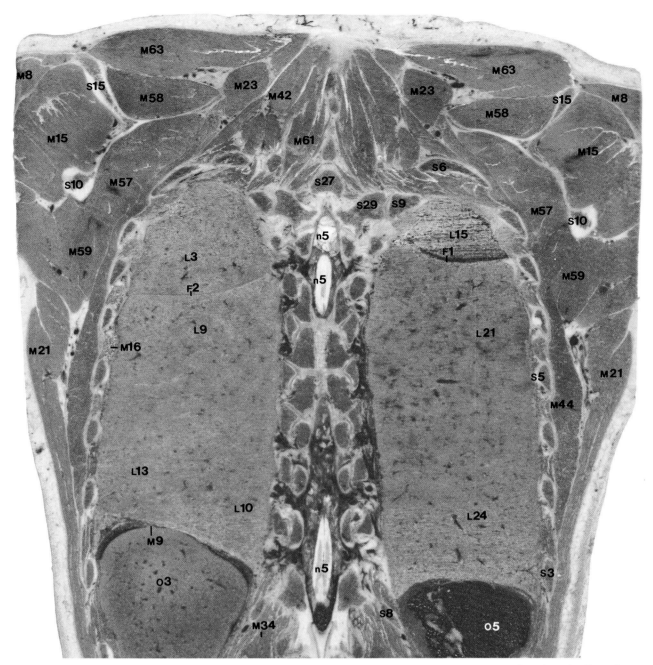

**Anatomic Specimen**

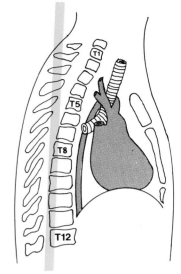

**Specimen Radiograph**

**Magnetic Resonance**

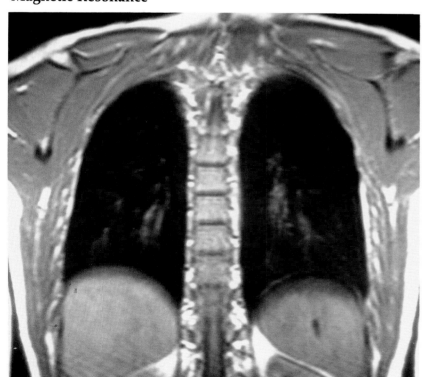

**Trispiral Tomogram**

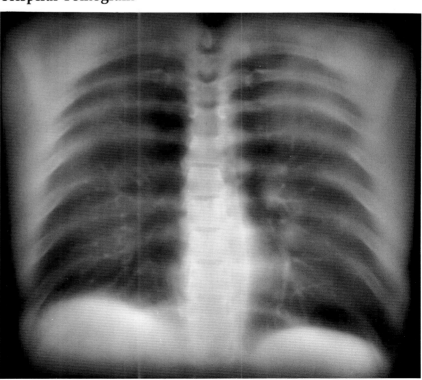

# Coronal

## Section 22

### Anatomic Key

**Fissures**

F  2  right major (oblique)

**Lobes**

L  3  RUL, posterior segment
L  9  RLL, superior segment
L 13  RLL, posterior basal segment
L 21  LLL, superior segment
L 24  LLL, posterior basal segment

**Muscles**

M  8  deltoid
M  9  diaphragm
M 13  erector spinae
M 15  infraspinatus
M 16  intercostal
M 21  latissimus dorsi
M 23  levator scapulae
M 27  longissimus, thoracis
M 34  psoas
M 42  semispinalis, cervicis
M 44  serratus, anterior
M 57  subscapularis
M 58  supraspinatus
M 59  teres major
M 61  transversospinalis
M 63  trapezius

**Neural Structures**

n  5  spinal cord

**Organs**

O  3  liver
O  5  spleen

**Skeletal Structures**

S  3  rib (9th)
S  5  rib, body, lateral (12th)
S  6  rib, body, posterior (4th)
S  8  rib, head and neck (8th)
S 10  scapula
S 15  scapula, spine
S 26  vertebra, pedicle (T8)
S 27  vertebra, spinous process (T4)
S 29  vertebra, transverse process (5th)

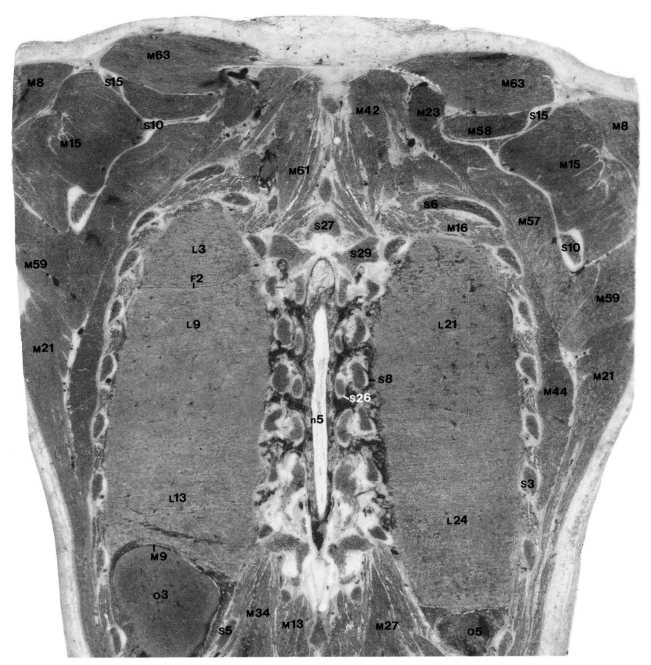

**Anatomic Specimen**

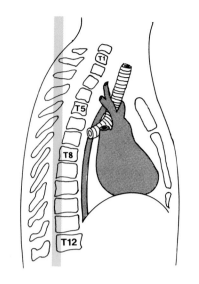

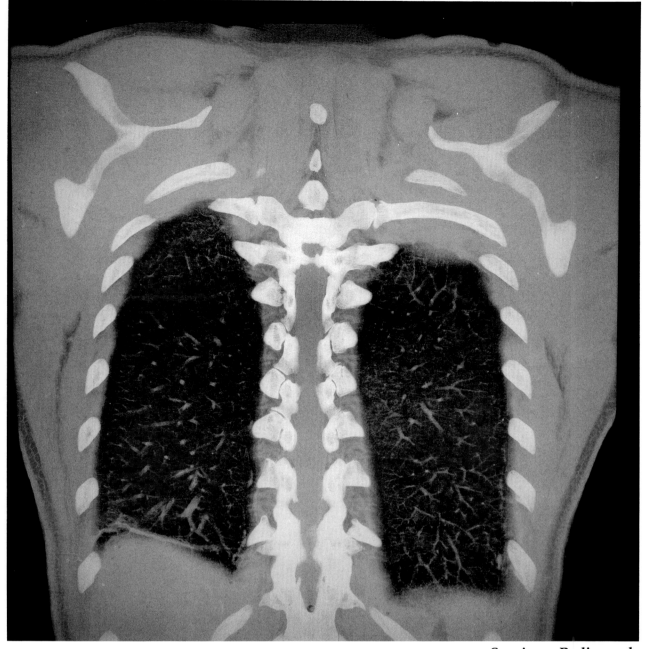

**Specimen Radiograph**

**Magnetic Resonance**

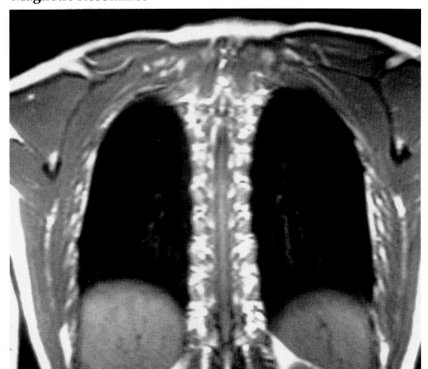

**Trispiral Tomogram**

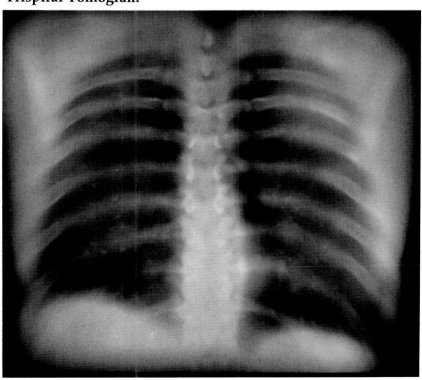

# Coronal

## Section 23

### Anatomic Key

**Fissures**
F  2  right major (oblique)

**Lobes**
L  3  RUL, posterior segment
L  9  RLL, superior segment
L  13  RLL, posterior basal segment
L  21  LLL, superior segment
L  24  LLL, posterior basal segment

**Muscles**
M  8  deltoid
M  13  erector spinae
M  15  infraspinatus
M  21  latissimus dorsi
M  23  levator scapulae
M  42  semispinalis, cervicis
M  44  serratus, anterior
M  57  subscapularis
M  58  supraspinatus
M  59  teres major
M  60  teres minor
M  61  transversospinalis
M  63  trapezius

**Neural Structures**
n  5  spinal cord

**Skeletal Structures**
S  3  rib (9th)
S  6  rib, body, posterior (4th)
S  10  scapula
S  15  scapula, spine
S  27  vertebra, spinous process (T6)

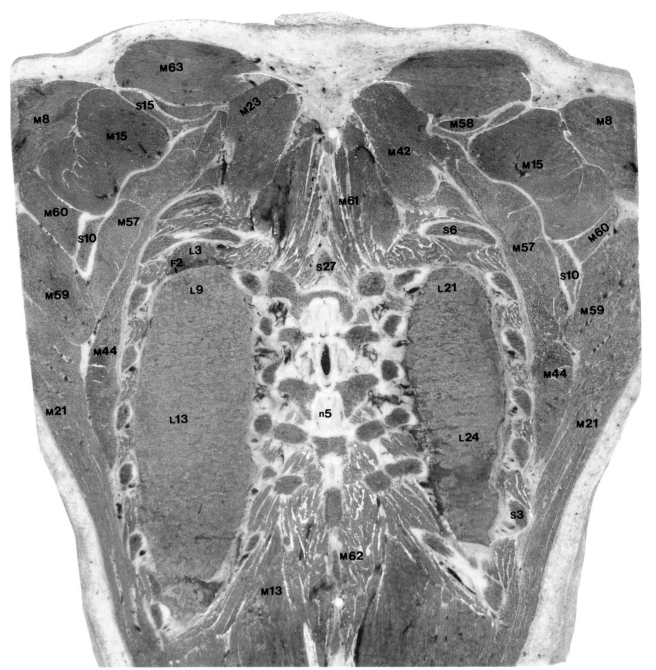

**Anatomic Specimen**

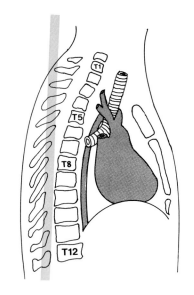

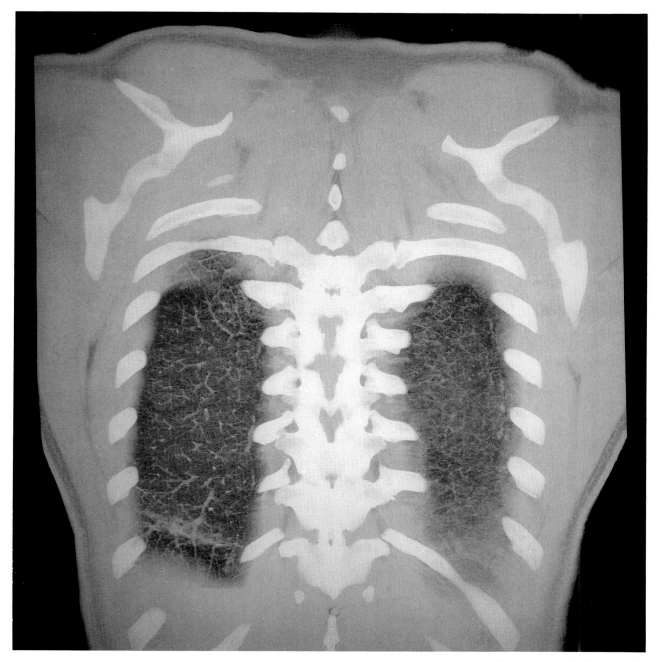

**Specimen Radiograph**

**Magnetic Resonance**

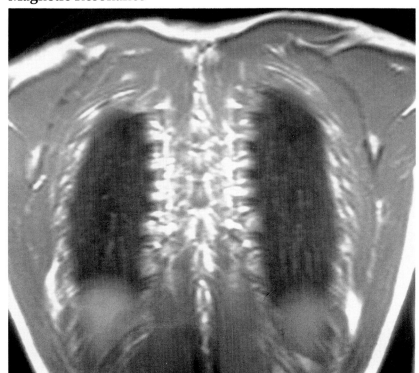

**Trispiral Tomogram**

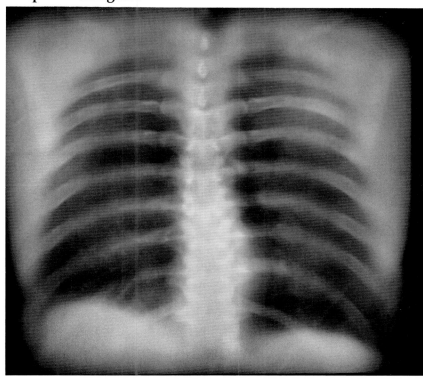

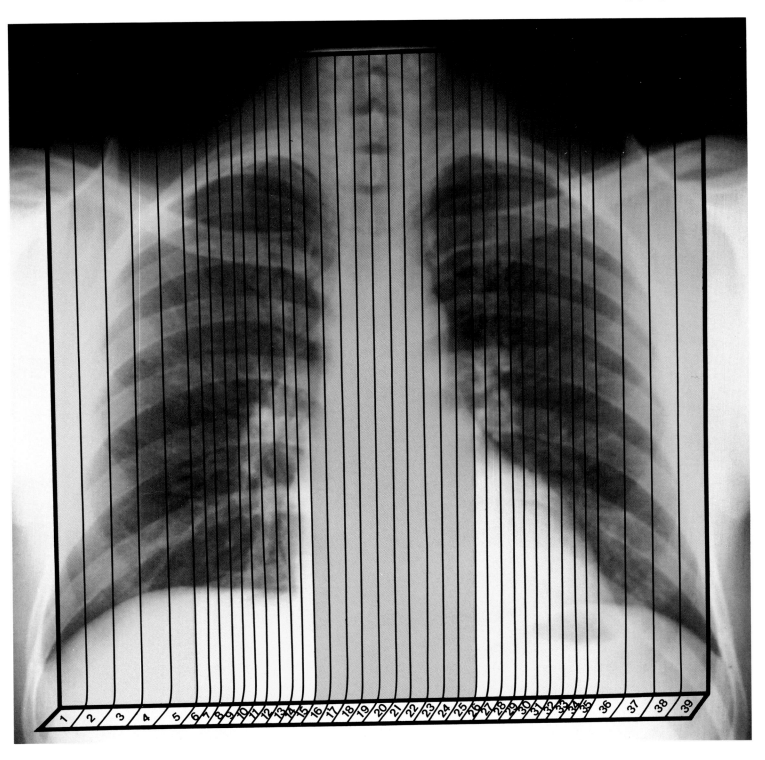

# Chapter 4

# Sagittal Plane

## The Anatomic Specimen

The cadaver for the sagittal sections was that of a 65-year-old white male who died suddenly during physical exertion. Cardiac arrest secondary to acute myocardial ischemia was assumed. The lungs were reexpanded and the chest radiographed within 2 hours following death. The thoracic structures were deemed normal for a subject of the stated age. The cadaver was placed in the freezer in a neutral supine position. The body remained in the freezer for 49 months prior to isolation of the chest and sectioning.

In advance of filming the trispiral tomograms and sawing the sections, the thorax was carefully mounted on a plastic frame and aligned to ensure that a true sagittal plane was sectioned. Satisfactory alignment is confirmed from midline Anatomic Section 19, its matching tomogram, and the specimen radiograph where midplane vertebral sections are present from C7 to T12. Note that the diameter of the spinal cord (n5) within the neural canal is normal. Anteriorly, this section passes through the midplane of the trachea (T1), manubrium (s18), body of the sternum (s17), and xiphoid (s19). Since there was no detectable variation in section thickness of the anatomic slices that were measured, it can be assumed that all sections are representative of an accurate sagittal plane.

Moderate postmortem sludging, as described in Chapter 1, can be noted in the major vessels and intracardiac chambers throughout the sagittal sections. Attempts were made to improve the appearance of those vascular structures where sludging left only clear serum to outline the vessel lumen. These vessels were overlain with "thawed" blood recovered from abdominal vessels. The time consumed and the degree of thawing required to get reasonable filling of these vessels resulted in jeopardizing the appearance of the remaining structures, especially the lung fields. It was therefore decided to present the vessels as they appeared naturally, assuming that the reader will soon become accustomed to the sludging process within the major vascular structures. Opacified blood remained in most of the smaller vessels.

### Anatomic Variant

A single minor anatomic variant is noted in Sections 9 and 31, where the external jugular veins on both sides (v24 and v25) are seen to run posterior to the belly of the omohyoid muscles (M30). No reference to this anatomic variant could be identified. All other thoracic structures appear normal.

## Section Plan

The plan for the sagittal sections (facing page) is shown in colored strips over a routine anteroposterior radiograph of the chest and represents a precise charting of the position of the gross anatomic sections. The sections are color-coded according to section thickness: **blue**, 10 mm; **pink**, 5 mm; **yellow**, 3 mm. The level and thickness of section is indicated throughout the chapter by the color bar overlying the miniature reference key at the top right of each double-page layout.

The sectioning process began on the right side of the cadaver at the level of the lateral portion of the right lung, right axilla, scapula, and lateral portion of the right clavicle. The sections progressed from right to left.

Sections 1–5 are 10 mm thick and extend to the lateral margin of the right hilum. Sections 6–15 are 3 mm thick to include thin-section detail within the right hilum. Sections 16–25 are 5 mm thick to include the mediastinal, major vascular, and cardiac structures. Sections 26–35 are again 3 mm thick to provide thin-section detail through the left hilum. Sections 36–39 are 10 mm thick and conclude in the lateral portion of the left lung and left axilla corresponding to the starting level on the right side.

## Orientation of Illustrations

All of the anatomic and radiographic images are oriented with the anterior structures to the reader's right and the posterior structures to the reader's left, simulating the orientation of a left lateral chest radiograph. The sections are progressive from the lateral portion of the right lung field across to the lateral portion of the left lung field as seen in the section plan.

## Anatomic–Radiographic Correlation

Prior to the sawing process, trispiral tomograms were made in the sagittal projection on the Exatome (General Electric-CGR) at 1-mm increments through the entire diameter of the chest. Trispiral tomograms from this series were subsequently matched to the low-kilovolt specimen radiographs and the gross anatomic specimens. The Exatome is an adjustable-fulcrum device with progressive change in magnification as the level of section is raised above the table top. To keep magnification to a minimum and to obtain essentially equal magnification on both sides, the "down side" of the specimen was reversed at the midsagittal plane. Minimal magnification (1.11×) in the trispiral tomograms was present in the most lateral sections (Sections 1 and 39) and progressed to 1.23× in the midplane of the chest (Section 19).

## Bronchial Structures

The mainstem and segmental bronchi that are optimally visualized in the sagittal projection are identified in Table 4.1.

**Right Lung**: The posterior basal bronchus to the RLL (B15), so difficult to see well in other projections, is clearly seen in specimen radiographs and tomograms in Sections 7–9. The posterior bronchus to the RUL (B4) is clearly seen in the trispiral tomogram in Section 9. In Sections 11 and 12 there is an unusual profile display of the apical (B3) and anterior (B5) segmental bronchi to the RUL and the RML bronchus (B7) with its lateral (B8) and medial (B9) segmental bronchi. The superior segment to the RLL (B11) is normally well seen in lateral projection. In this specimen the superior bronchus is seen at its origin in the anatomic specimen and specimen radiographs (Section 11), but only as cross-section images in more lateral sections. The proximal RML bronchus (B7) is very well seen in Section 13. The intermediate bronchus (B6) is seen in longitudinal profile in Sections 14 and 15, while the right mainstem bronchus (B1) is well seen in Sections 15–17.

The trachea (T1) is clearly outlined in Sections 19–21. The left mainstem bronchus (B16) is seen only in cross section in sagittal projection.

**Left Lung**: The apicoposterior (B18) and the proximal portion of the anterior bronchus to the LUL (B19) are well seen in Sections 30 and 31. The superior (B21) and inferior (B22) lingular bronchi are particularly well seen in the specimen radiographs and trispiral tomograms in Sections 30 and 31. The LLL bronchus (B23) is clearly seen in Sections 29–31, the superior segment of the LLL (B24) in Sections 29 and 30, and the posterior basal segment of the LLL (B27) in Sections 30–32.

Attention is drawn to the detail of the mucosal surfaces of the bronchi in the trispiral tomograms. Very small intrabronchial masses can be identified in the trispiral tomogram with proper orientation of the bronchus.

## Neural Structures

The spinal cord (n5) is seen throughout in Section 19. The brachial plexus (n1) is well visualized in the anatomic specimens and the specimen radiographs in Sections 1–15 on the right side and Sections 25–38 on the left side. The components of the brachial plexus (n1) (lateral, posterior, and medial cords) are clearly seen in Sections 12 (right) and 28 (left) emerging between the anterior scalene (M38) and middle scalene (M39) muscles. The trunks and cords of the brachial plexus extend with the vascular structures into the axilla.

The brachial plexus was not clearly seen in MR images, possibly due to the intense signal arising from adjacent axillary fat. There is insufficient contrast discrimination to see the soft tissues (muscles, vessels, and neural structures) in the trispiral tomograms.

## Skeletal Structures

Skeletal structures are well seen throughout the sagittal series. In Section 19, the thoracic spine in its entirety (T1–T12) and the sternum are clearly seen in the gross anatomic specimen and specimen radiograph, and there is excellent correlation in the trispiral tomogram.

## Vascular Structures

This cadaver was the oldest of the five cadavers. Extensive arteriosclerosis is noted throughout the arterial system, especially the aorta, no other vascular abnormalities were noted.

## Extrapleural Fat

The relationship of a large mass of extrapleural fat (f3) above the anterior portion of the diaphragm is clearly seen in Section 36. Projection of this fat mass over the lower portion of the left lower lung in poorly positioned portable radiographic studies can produce the illusion of consolidation of the left lower lobe, as previously described (49).

## Other Structures

**Fascia**: Sibson's fascia (F8) is seen with assurance in Sections 11 and 12. The perivisceral fascia (F6) and prevertebral fascia (F7) can be identified in Sections 19 and 20 and the superficial (investing) fascia (F9) in Section 20.

## Fissures

The fissures are well seen in both lungs, and especially well defined in the specimen radiographs.

## Table 4.1
### Index to Optimal Bronchial Imaging Sagittal Projection
#### Specimen Radiograph and Trispiral Tomogram
#### Right      Bronchus      Left

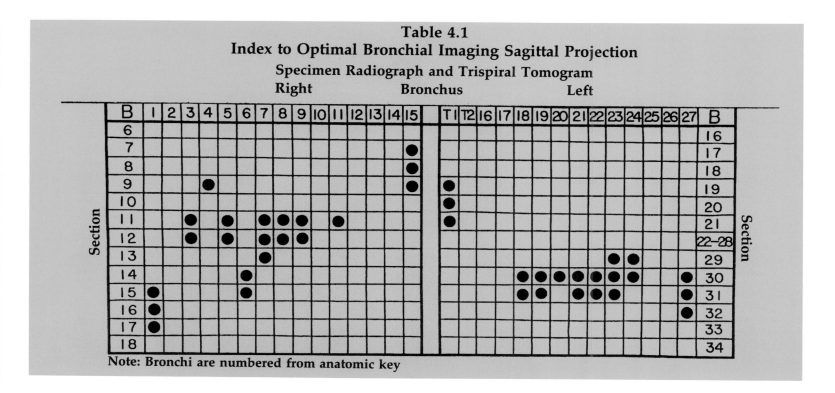

| B | 1 | 2 | 3 | 4 | 5 | 6 | 7 | 8 | 9 | 10 | 11 | 12 | 13 | 14 | 15 | T1 | T2 | 16 | 17 | 18 | 19 | 20 | 21 | 22 | 23 | 24 | 25 | 26 | 27 | B |
|---|---|---|---|---|---|---|---|---|---|----|----|----|----|----|----|----|----|----|----|----|----|----|----|----|----|----|----|----|----|---|
| 6 |  |  |  |  |  |  |  |  |  |  |  |  |  |  |  |  |  |  |  |  |  |  |  |  |  |  |  |  |  | 16 |
| 7 |  |  |  |  |  |  |  |  |  |  |  |  |  |  | ● |  |  |  |  |  |  |  |  |  |  |  |  |  |  | 17 |
| 8 |  |  |  |  |  |  |  |  |  |  |  |  |  |  | ● |  |  |  |  |  |  |  |  |  |  |  |  |  |  | 18 |
| 9 |  |  | ● |  |  |  |  |  |  |  |  |  |  |  | ● | ● |  |  |  |  |  |  |  |  |  |  |  |  |  | 19 |
| 10 |  |  |  |  |  |  |  |  |  |  |  |  |  |  |  | ● |  |  |  |  |  |  |  |  |  |  |  |  |  | 20 |
| 11 |  |  | ● |  | ● |  |  | ● | ● | ● |  | ● |  |  |  | ● |  |  |  |  |  |  |  |  |  |  |  |  |  | 21 |
| 12 |  |  | ● |  | ● |  |  | ● | ● | ● |  |  |  |  |  |  |  |  |  |  |  |  |  |  |  |  |  |  |  | 22–28 |
| 13 |  |  |  |  |  |  |  | ● |  |  |  |  |  |  |  |  |  |  |  |  |  |  |  |  |  | ● | ● |  |  | 29 |
| 14 |  |  |  |  |  | ● |  |  |  |  |  |  |  |  |  |  |  |  |  | ● | ● | ● | ● | ● | ● | ● |  |  | ● | 30 |
| 15 | ● |  |  |  |  | ● |  |  |  |  |  |  |  |  |  |  |  |  |  | ● | ● |  | ● | ● | ● |  |  |  | ● | 31 |
| 16 | ● |  |  |  |  |  |  |  |  |  |  |  |  |  |  |  |  |  |  |  |  |  |  |  |  |  |  |  | ● | 32 |
| 17 | ● |  |  |  |  |  |  |  |  |  |  |  |  |  |  |  |  |  |  |  |  |  |  |  |  |  |  |  |  | 33 |
| 18 |  |  |  |  |  |  |  |  |  |  |  |  |  |  |  |  |  |  |  |  |  |  |  |  |  |  |  |  |  | 34 |

Note: Bronchi are numbered from anatomic key

### Magnetic Resonance Images

Specific correlations are found in a review of the MR images. Although the sagittal MR images do not always precisely correlate to the sagittal anatomic specimen sections, a functional match can be obtained by comparison to adjacent images. The minor mismatch is partly explained by a slight difference in size between the cadaver and the living volunteer.

Several prominent comparisons are noted:

1. Muscles correlate well throughout.
2. The external jugular vein in MR Sections 7 and 8 is seen to descend anterior to the omohyoid muscle in this normal subject, as opposed to the anomalous course seen in Anatomic Sections 9 (V25) and 31 (V24), where the external jugular descends posterior to the omohyoid (M30).
3. A reasonably close match is noted between Anatomic Sections 12–33 and MR images 15–33, where intracardiac and major venous and arterial structures are encountered from right to left.

The superior (V39) and inferior (V38) vena cava, with their major branches, are well seen in Anatomic Sections 15–19. Comparable structures in the MR images are seen in MR Sections 16 and 17. The ascending aorta (A30), aortic arch (A29), and descending aorta (A31) are seen in Anatomic Sections 18–27, while similar segments are seen in MR Sections 20–27. The aortic arch is especially well seen in MR Sections 22–24.

# Sagittal

## Section 1

### Anatomic Key

**Arteries**
A 33 axillary, right
A 46 intercostal, posterior, right
A 58 thoracoacromial, right
A 64 transverse cervical, right

**Fat**
f 1 axillary

**Fissures**
F 2 right major (oblique)

**Lobes**
L 3 RUL, posterior segment
L 4 RUL, anterior segment
L 9 RLL, superior segment
L 11 RLL, anterior basal segment
L 12 RLL, lateral basal segment
L 13 RLL, posterior basal segment

**Muscles**
M 1 abdominal, external oblique
M 7 coracobrachialis
M 8 deltoid
M 9 diaphragm
M 15 infraspinatus
M 16 intercostal
M 21 latissimus dorsi
M 31 pectoralis, major
M 32 pectoralis, minor
M 33 platysma
M 44 serratus, anterior
M 57 subscapularis
M 58 supraspinatus
M 59 teres major
M 63 trapezius

**Neural Structures**
n 1 brachial plexus

**Nodes**
N 1 axillary

**Organs**
O 3 liver

**Pleural Structures**
P 2 costal (parietal)
P 3 costodiaphragmatic recess
P 6 visceral

**Skeletal Structures**
s 1 clavicle
s 3 rib (3rd)
s 4 rib, body, anterior (7th)
s 6 rib, body, posterior (7th)
s 10 scapula
s 12 scapula, coracoid process
s 15 scapula, spine

**Veins**
v 4 pulmonary, inferior, RLL
v 6 axillary, right
v 19 intercostal, posterior, right
v 32 subscapular, right
v 37 transverse cervical, right

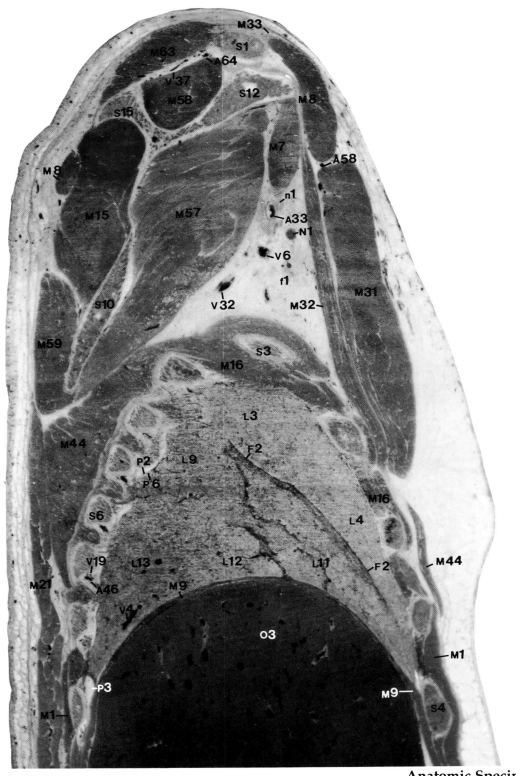

**Anatomic Specimen**

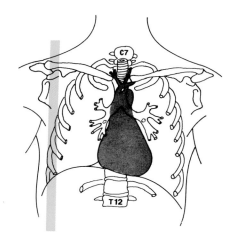

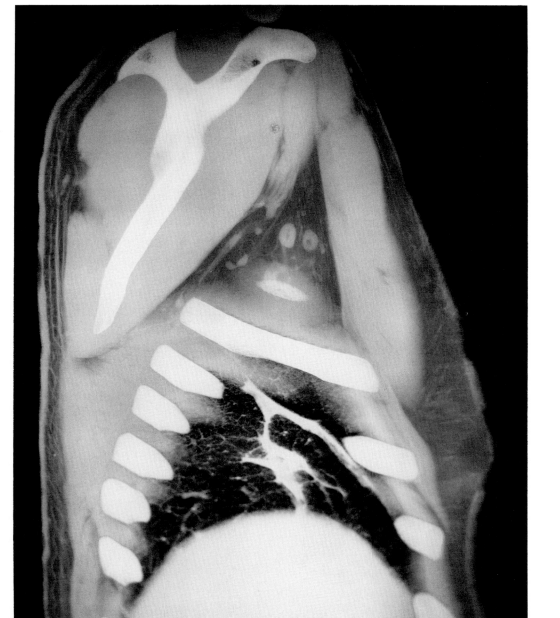

**Specimen Radiograph**

**Magnetic Resonance**

**Trispiral Tomogram**

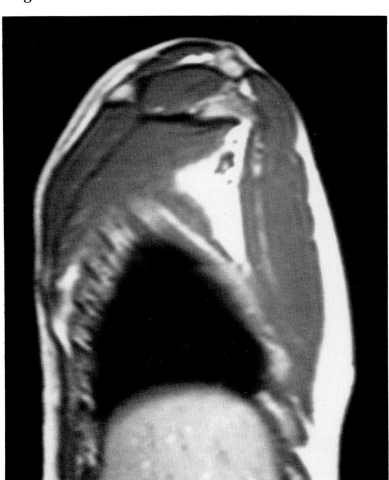

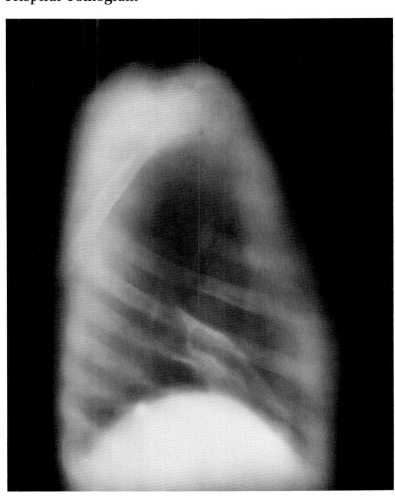

# Sagittal

## Section 2

### Anatomic Key

**Arteries**
A 33 axillary, right
A 46 intercostal, posterior, right
A 52 subscapular, right
A 58 thoracoacromial, right
A 64 transverse cervical, right

**Bronchi**
B 4 RUL, posterior segment
B 5 RUL, anterior segment
B 11 RLL, superior segment
B 13 RLL, anterior basal segment
B 15 RLL, posterior basal segment

**Fat**
f 1 axillary

**Fissures**
F 2 right major (oblique)
F 3 right minor (horizontal)

**Lobes**
L 3 RUL, posterior segment
L 4 RUL, anterior segment
L 6 RML, lateral segment
L 9 RLL, superior segment
L 10 RLL, medial basal segment
L 11 RLL, anterior basal segment
L 13 RLL, posterior basal segment

**Muscles**
M 1 abdominal, external oblique
M 8 deltoid
M 9 diaphragm
M 15 infraspinatus
M 16 intercostal
M 21 latissimus dorsi
M 31 pectoralis, major
M 32 pectoralis, minor
M 33 platysma
M 44 serratus, anterior
M 57 subscapularis
M 58 supraspinatus
M 59 teres major
M 63 trapezius

**Neural Structures**
n 1 brachial plexus
n 2 intercostal nerve

**Organs**
O 3 liver

**Pleural Structures**
P 2 costal (parietal)
P 3 costodiaphragmatic recess
P 6 visceral

**Skeletal Structures**
S 1 clavicle
S 3 rib (3rd)
S 4 rib, body, anterior (7th)
S 6 rib, body, posterior (7th)
S 7 rib, costal cartilage (8th)
S 10 scapula
S 12 scapula, coracoid process
S 15 scapula, spine

**Veins**
V 4 pulmonary, inferior, RLL
V 6 axillary, right
V 19 intercostal, posterior, right
V 29 lateral thoracic, right
V 32 subscapular, right
V 37 transverse cervical, right

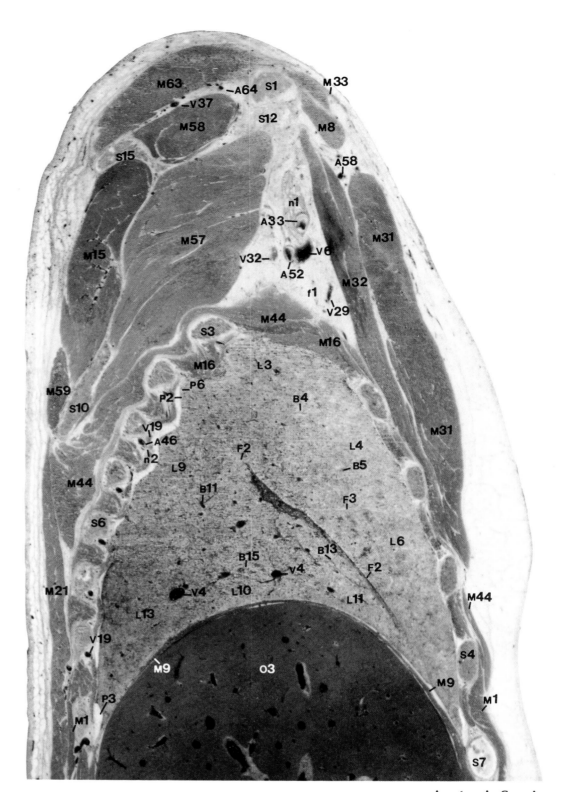

**Anatomic Specimen**

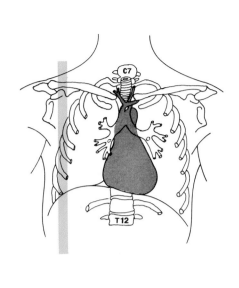

**Specimen Radiograph**

**Magnetic Resonance**

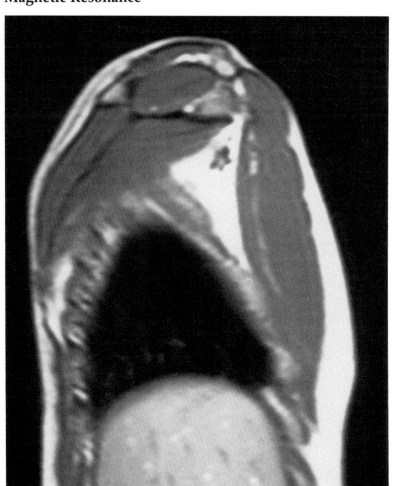

**Trispiral Tomogram**

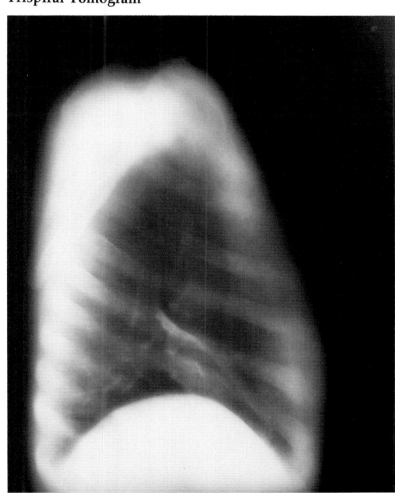

# Sagittal

## Section 3

### Anatomic Key

**Arteries**
A 33 axillary, right
A 46 intercostal, posterior, right
A 52 subscapular, right
A 54 suprascapular, right
A 58 thoracoacromial, right
A 64 transverse cervical, right

**Bronchi**
B 3 RUL, apical segment
B 5 RUL, anterior segment
B 8 RML, lateral segment
B 11 RLL, superior segment
B 13 RLL, anterior basal segment
B 15 RLL, posterior basal segment

**Fat**
f 1 axillary
f 7 subcutaneous

**Fissures**
F 2 right major (oblique)
F 3 right minor (horizontal)

**Lobes**
L 2 RUL, apical segment
L 3 RUL, posterior segment
L 4 RUL, anterior segment
L 6 RML, lateral segment
L 9 RLL, superior segment
L 10 RLL, medial basal segment
L 11 RLL, anterior basal segment
L 13 RLL, posterior basal segment

**Muscles**
M 1 abdominal, external oblique
M 9 diaphragm
M 15 infraspinatus
M 16 intercostal
M 21 latissimus dorsi
M 31 pectoralis, major
M 32 pectoralis, minor
M 33 platysma
M 44 serratus, anterior
M 55 subclavius
M 57 subscapularis
M 58 supraspinatus
M 63 trapezius

**Neural Structures**
n 1 brachial plexus
n 2 intercostal nerve

**Organs**
O 3 liver

**Pleural Structures**
P 2 costal (parietal)
P 3 costodiaphragmatic recess
P 6 visceral

**Skeletal Structures**
S 1 clavicle
S 3 rib (2nd)
S 4 rib, body, anterior (6th)
S 6 rib, body, posterior (7th)
S 7 rib, costal cartilage (7th)
S 10 scapula
S 15 scapula, spine

**Veins**
V 4 pulmonary, inferior, RLL
V 6 axillary, right
V 19 intercostal, posterior, right
V 33 suprascapular, right
V 34 thoracoacromial, right

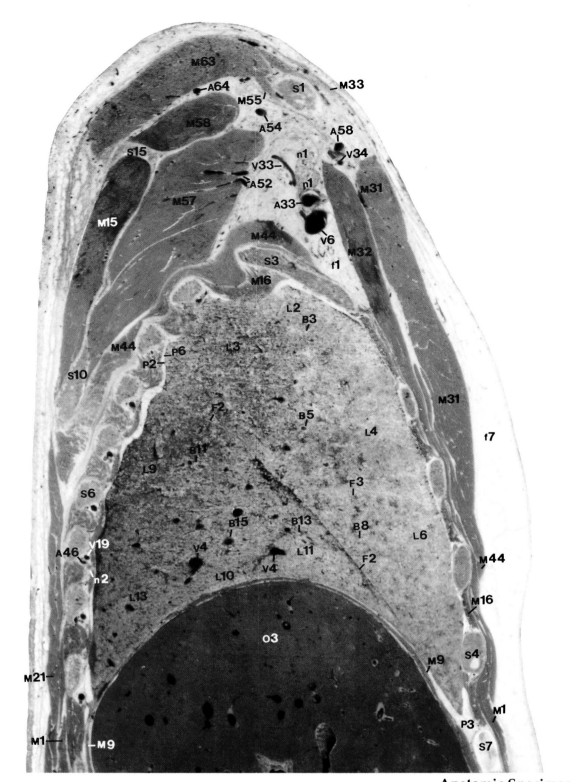

**Anatomic Specimen**

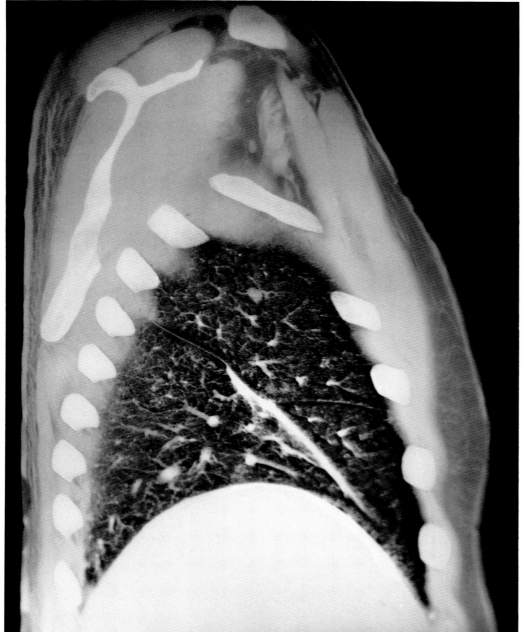

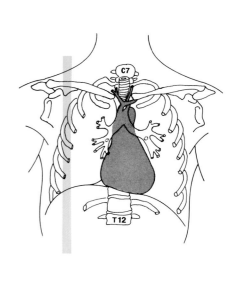

**Specimen Radiograph**

**Magnetic Resonance**

**Trispiral Tomogram**

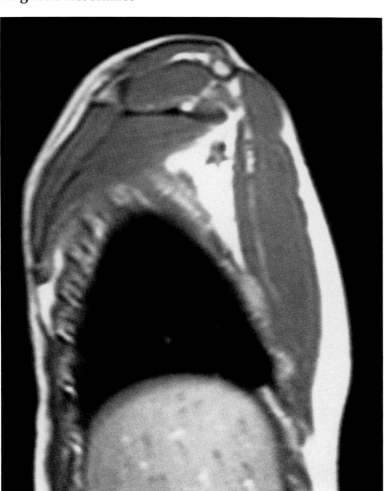

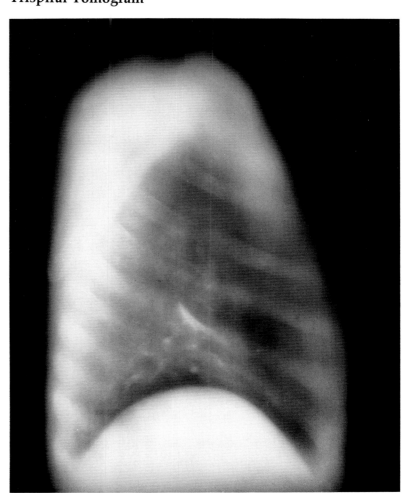

# Sagittal

## Section 4

### Anatomic Key

**Arteries**
A 33 axillary, right
A 46 intercostal, posterior, right
A 52 subscapular, right
A 60 thyrocervical trunk, right

**Bronchi**
B 3 RUL, apical segment
B 4 RUL, posterior segment
B 5 RUL, anterior segment
B 8 RML, lateral segment
B 11 RLL, superior segment
B 13 RLL, anterior basal segment
B 15 RLL, posterior basal segment

**Fat**
f 1 axillary
F 3 extrapleural

**Fissures**
F 2 right major (oblique)
F 3 right minor (horizontal)

**Lobes**
L 2 RUL, apical segment
L 3 RUL, posterior segment
L 4 RUL, anterior segment
L 6 RML, lateral segment
L 9 RLL, superior segment
L 10 RLL, medial basal segment
L 11 RLL, anterior basal segment
L 13 RLL, posterior basal segment

**Muscles**
M 1 abdominal, external oblique
M 9 diaphragm
M 14 iliocostalis, thoracis
M 15 infraspinatus
M 16 intercostal
M 21 latissimus dorsi
M 31 pectoralis, major
M 32 pectoralis, minor
M 33 platysma
M 44 serratus, anterior
M 57 subscapularis
M 58 supraspinatus
M 63 trapezius

**Neural Structures**
n 1 brachial plexus
n 2 intercostal nerve

**Organs**
O 3 liver

**Pleural Structures**
P 2 costal (parietal)
P 3 costodiaphragmatic recess
P 6 visceral

**Skeletal Structures**
S 1 clavicle
S 3 rib (2nd)
S 4 rib, body, anterior (5th)
S 6 rib, body, posterior (9th)
S 7 rib, costal cartilage (6th)
S 10 scapula
S 15 scapula, spine

**Veins**
V 4 pulmonary, inferior, RLL
V 6 axillary, right
V 19 intercostal, posterior, right
V 25 jugular, external, right
V 33 suprascapular, right
V 37 transverse cervical, right

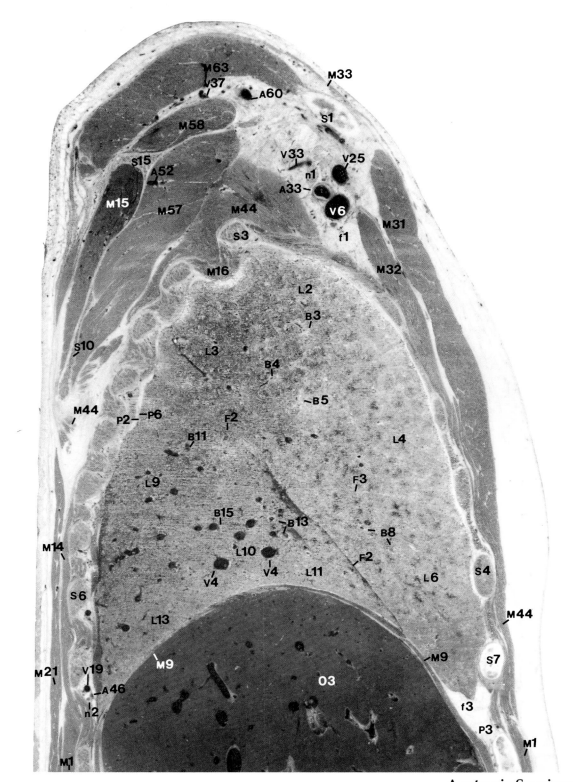

**Anatomic Specimen**

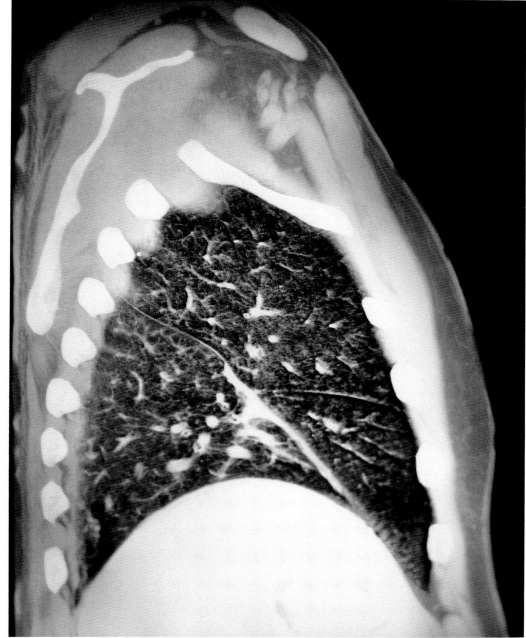

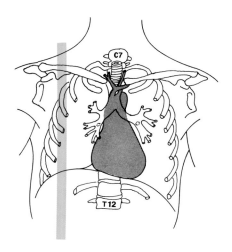

**Specimen Radiograph**

**Magnetic Resonance**

**Trispiral Tomogram**

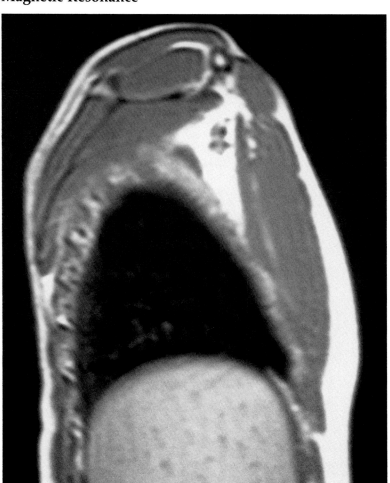

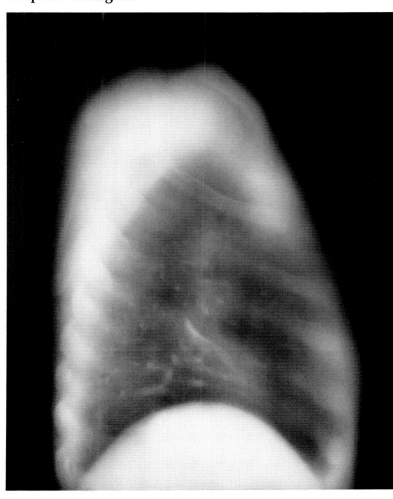

# Sagittal

## Section 5

### Anatomic Key

**Arteries**

A 46 intercostal, posterior, right
A 50 subclavian, right
A 60 thyrocervical trunk, right

**Bronchi**

B 3 RUL, apical segment
B 4 RUL, posterior segment
B 5 RUL, anterior segment
B 8 RML, lateral segment
B 11 RLL, superior segment
B 13 RLL, anterior basal segment
B 14 RLL, lateral basal segment

**Fat**

f 3 extrapleural

**Fissures**

F 2 right major (oblique)
B 3 right minor (horizontal)

**Lobes**

L 2 RUL, apical segment
L 3 RUL, posterior segment
L 4 RUL, anterior segment
L 6 RML, lateral segment
L 9 RLL, superior segment
L 10 RLL, medial basal segment
L 11 RLL, anterior basal segment
L 13 RLL, posterior basal segment

**Muscles**

M 1 abdominal, external oblique
M 9 diaphragm
M 13 erector spinae
M 14 iliocostalis, thoracis
M 15 infraspinatus
M 16 intercostal
M 21 latissimus dorsi
M 30 omohyoid
M 31 pectoralis, major
M 32 pectoralis, minor
M 33 platysma
M 35 rhomboid, major
M 44 serratus, anterior
M 55 subclavius
M 57 subscapularis
M 58 supraspinatus
M 63 trapezius

**Neural Structures**

n 1 brachial plexus
n 2 intercostal nerve

**Organs**

O 3 liver

**Pleural Structures**

P 2 costal (parietal)
P 3 costodiaphragmatic recess
P 6 visceral

**Skeletal Structures**

S 1 clavicle
S 3 rib (2nd)
S 4 rib, body, anterior (1st)
S 6 rib, body, posterior (10th)
S 7 rib, costal cartilage (6th)
S 15 scapula, spine

**Veins**

V 4 pulmonary, inferior, RLL
V 19 intercostal, posterior, right
V 31 subclavian, right
V 37 transverse cervical, right

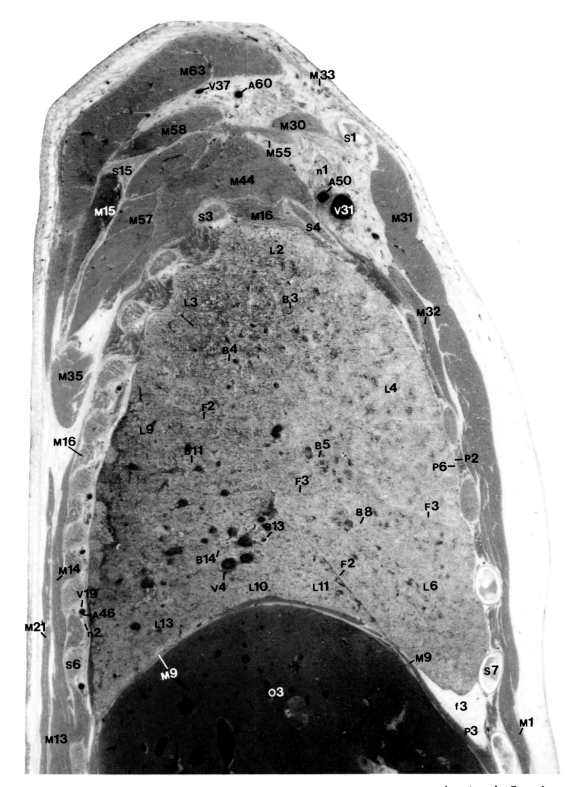

**Anatomic Specimen**

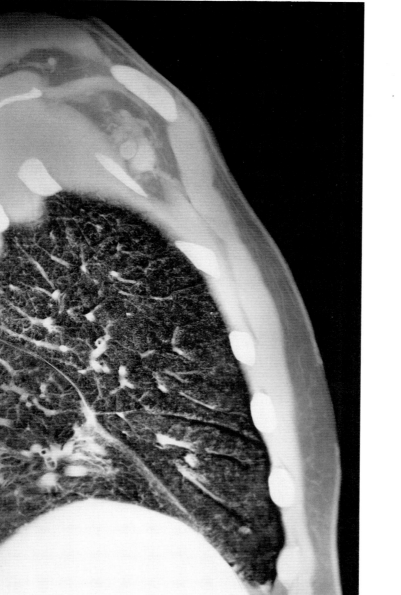

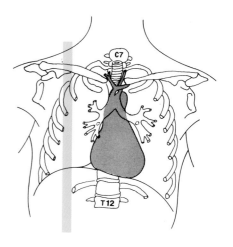

Specimen Radiograph

**Magnetic Resonance**

**Trispiral Tomogram**

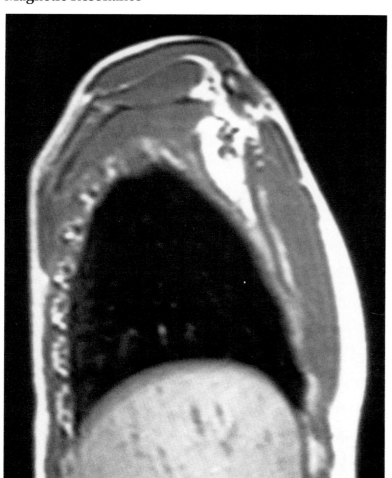

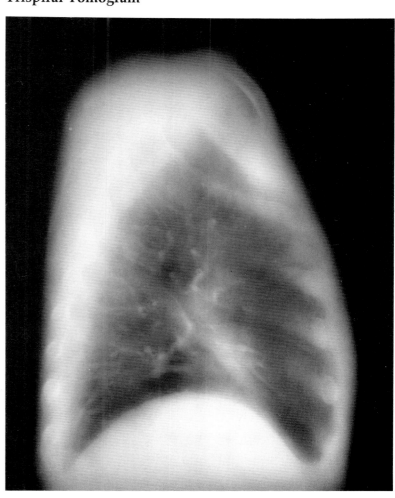

# Sagittal

## Section 6

### Anatomic Key

**Arteries**
A  6  RUL, anterior branch
A 15  RLL, posterior basal branch
A 46  intercostal, posterior, right
A 50  subclavian, right
A 60  thyrocervical trunk, right

**Bronchi**
B  3  RUL, apical segment
B  4  RUL, posterior segment
B  5  RUL, anterior segment
B  8  RML, lateral segment
B 13  RLL, anterior basal segment
B 14  RLL, lateral basal segment
B 15  RLL, posterior basal segment

**Fat**
f  3  extrapleural

**Fissures**
F  2  right major (oblique)
F  3  right minor (horizontal)

**Lobes**
L  2  RUL, apical segment
L  3  RUL, posterior segment
L  4  RUL, anterior segment
L  6  RML, lateral segment
L  9  RLL, superior segment
L 10  RLL, medial basal segment
L 11  RLL, anterior basal segment
L 13  RLL, posterior basal segment

**Muscles**
M  1  abdominal, external oblique
M  9  diaphragm
M 13  erector spinae
M 14  iliocostalis, thoracis
M 15  infraspinatus
M 16  intercostal
M 21  latissimus dorsi
M 30  omohyoid
M 31  pectoralis, major
M 32  pectoralis, minor
M 33  platysma
M 35  rhomboid, major
M 44  serratus, anterior
M 55  subclavius
M 57  subscapularis
M 58  supraspinatus
M 63  trapezius

**Neural Structures**
n  1  brachial plexus
n  2  intercostal nerve

**Organs**
O  3  liver

**Pleural Structures**
P  2  costal (parietal)
P  3  costodiaphragmatic recess
P  6  visceral

**Skeletal Structures**
S  1  clavicle
S  3  rib (2nd)
S  4  rib, body, anterior (1st)
S  6  rib, body, posterior (10th)
S  7  rib, costal cartilage (6th)
S 15  scapula, spine

**Veins**
V  4  pulmonary, inferior, RLL
V 19  intercostal, posterior, right
V 25  jugular, external, right
V 31  subclavian, right
V 37  transverse cervical, right

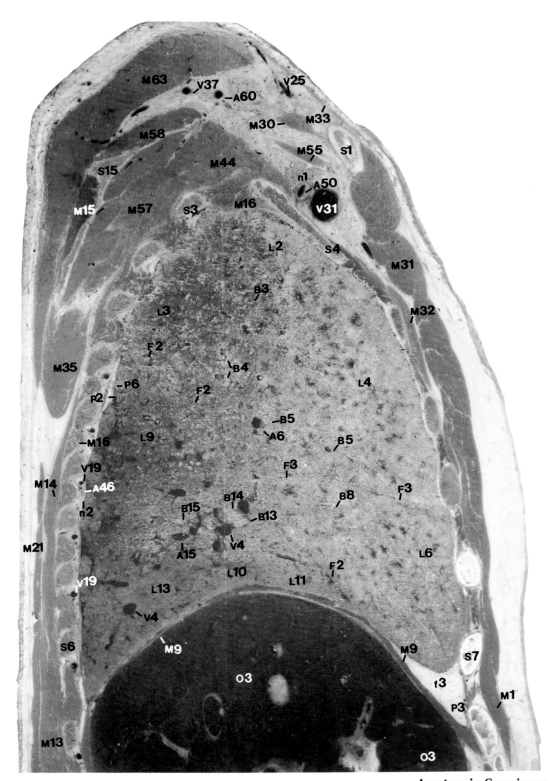

**Anatomic Specimen**

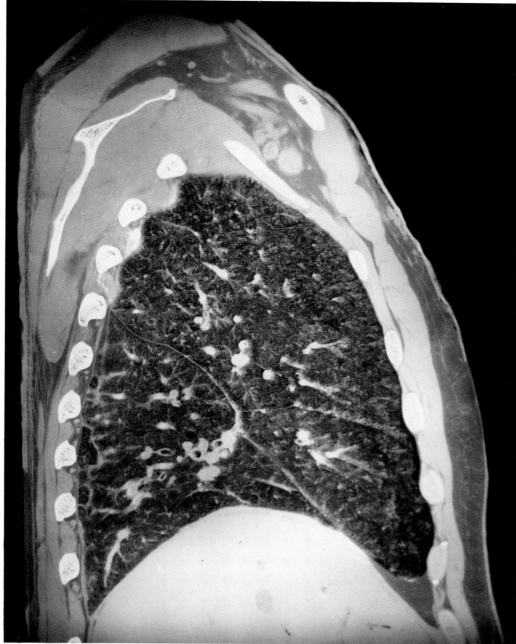

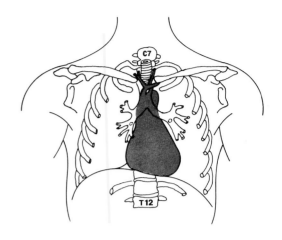

Specimen Radiograph

**Magnetic Resonance**

**Trispiral Tomogram**

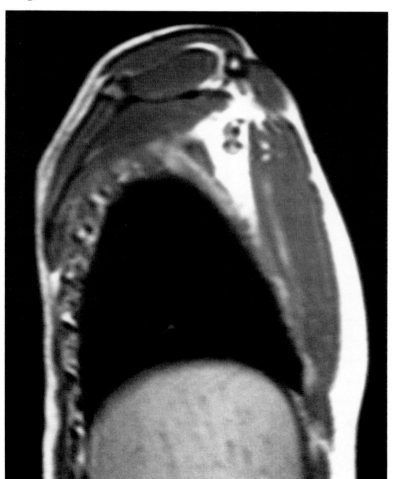

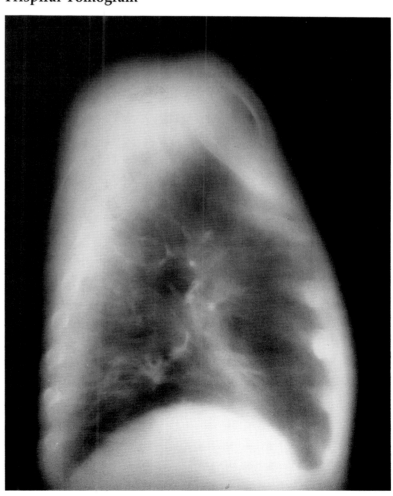

# Sagittal

## Section 7

### Anatomic Key

**Arteries**

A   6  RUL, anterior branch
A  11  RLL, superior branch
A  13  RLL, anterior basal branch
A  14  RLL, lateral basal branch
A  15  RLL, posterior basal branch
A  46  intercostal, posterior, right
A  50  subclavian, right

**Bronchi**

B   3  RUL, apical segment
B   4  RUL, posterior segment
B   5  RUL, anterior segment
B   8  RML, lateral segment
B  11  RLL, superior segment
B  13  RLL, anterior basal segment
B  14  RLL, lateral basal segment
B  15  RLL, posterior basal segment

**Fat**

f   3  extrapleural

**Fissures**

F   2  right major (oblique)
F   3  right minor (horizontal)

**Lobes**

L   2  RUL, apical segment
L   3  RUL, posterior segment
L   4  RUL, anterior segment
L   6  RML, lateral segment
L   9  RLL, superior segment
L  10  RLL, medial basal segment
L  11  RLL, anterior basal segment
L  13  RLL, posterior basal segment

**Muscles**

M   3  abdominis, rectus
M   9  diaphragm
M  13  erector spinae
M  14  iliocostalis, thoracis
M  15  infraspinatus
M  16  intercostal
M  21  latissimus dorsi
M  30  omohyoid
M  31  pectoralis, major
M  32  pectoralis, minor
M  33  platysma
M  35  rhomboid, major
M  39  scalene, middle
M  44  serratus, anterior
M  45  serratus, posterior inferior
M  57  subscapularis
M  58  supraspinatus
M  63  trapezius

**Neural Structures**

n   1  brachial plexus
n   2  intercostal nerve

**Organs**

O   3  liver

**Pleural Structures**

P   2  costal (parietal)
P   3  costodiaphragmatic recess
P   6  visceral

**Skeletal Structures**

S   1  clavicle
S   3  rib (2nd)
S   5  rib, body, lateral (1st)
S   6  rib, body, posterior (10th)
S   7  rib, costal cartilage (5th)
S  10  scapula

**Veins**

v   0  pulmonary, superior, RUL
v   4  pulmonary, inferior, RLL
v  19  intercostal, posterior, right
v  25  jugular, external, right
v  31  subclavian, right
v  37  transverse cervical, right

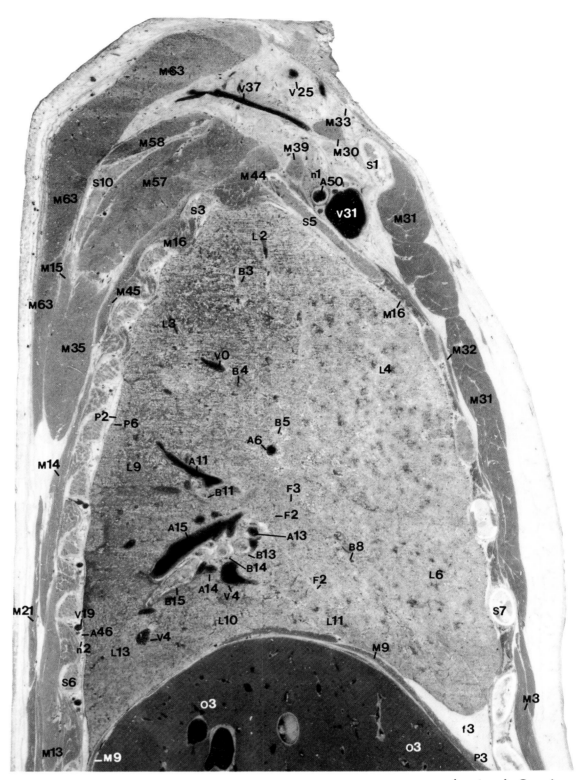

**Anatomic Specimen**

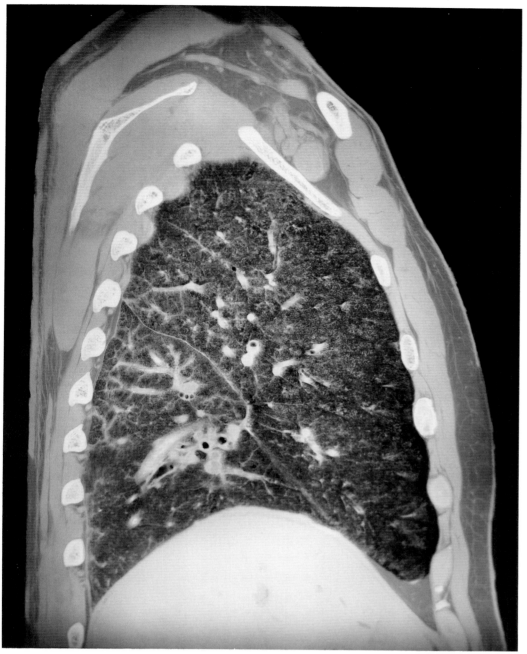

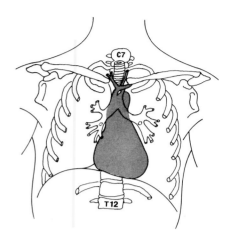

**Specimen Radiograph**

**Magnetic Resonance**

**Trispiral Tomogram**

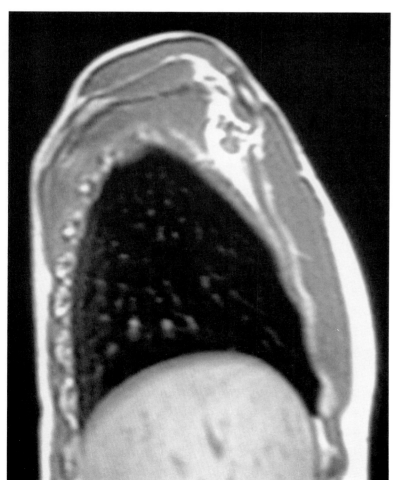

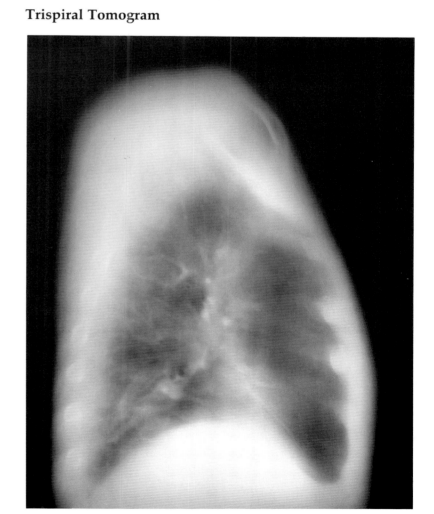

# Sagittal
## Section 8

## Anatomic Key

### Arteries
A **6** RUL, anterior branch
A **10** right lower lobe (RLL)
A **11** RLL, superior branch
A **13** RLL, anterior basal branch
A **15** RLL, posterior basal branch
A **50** subclavian, right

### Bronchi
B **3** RUL, apical segment
B **4** RUL, posterior segment
B **5** RUL, anterior segment
B **8** RML, lateral segment
B **11** RLL, superior segment
B **13** RLL, anterior basal segment
B **14** RLL, lateral basal segment
B **15** RLL, posterior basal segment

### Fat
f **3** extrapleural

### Fissures
F **2** right major (oblique)
F **3** right minor (horizontal)

### Lobes
L **2** RUL, apical segment
L **3** RUL, posterior segment
L **4** RUL, anterior segment
L **6** RML, lateral segment
L **9** RLL, superior segment
L **10** RLL, medial basal segment
L **11** RLL, anterior basal segment
L **13** RLL, posterior basal segment

### Muscles
M **3** abdominis, rectus
M **9** diaphragm
M **13** erector spinae
M **14** iliocostalis, thoracis
M **16** intercostal
M **21** latissimus dorsi
M **30** omohyoid
M **31** pectoralis, major
M **33** platysma
M **35** rhomboid, major
M **39** scalene, middle
M **40** scalene, posterior
M **46** serratus, posterior superior
M **57** subscapularis
M **58** supraspinatus
M **62** transversus thoracis
M **63** trapezius

### Neural Structures
n **1** brachial plexus

### Organs
O **3** liver

### Pleural Structures
P **2** costal (parietal)
P **3** costodiaphragmatic recess
P **6** visceral

### Skeletal Structures
S **1** clavicle
S **3** rib (2nd)
S **4** rib, body, anterior (1st)
S **6** rib, body, posterior (8th)
S **7** rib, costal cartilage (5th)
S **10** scapula

### Veins
V **4** pulmonary, inferior, RLL
V **19** intercostal, posterior, right
V **25** jugular, external, right
V **31** subclavian, right
V **37** transverse cervical, right

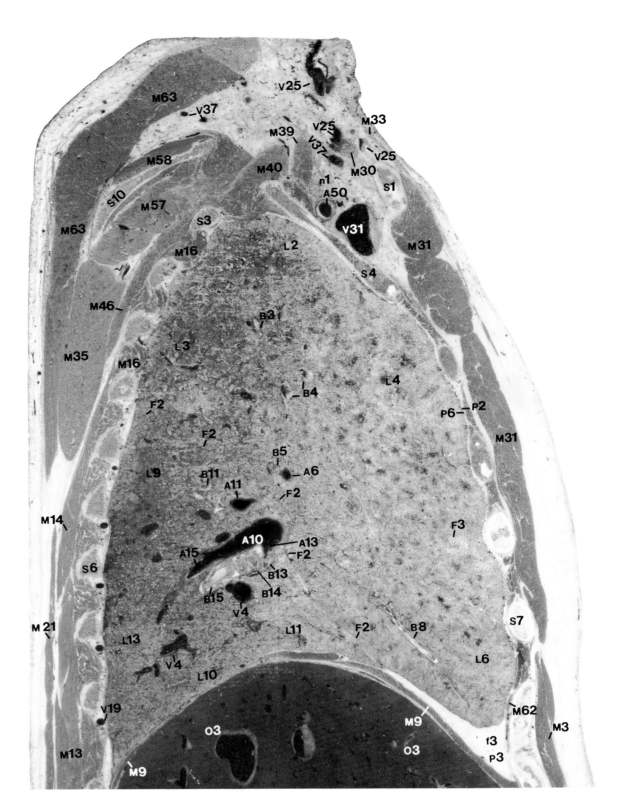

**Anatomic Specimen**

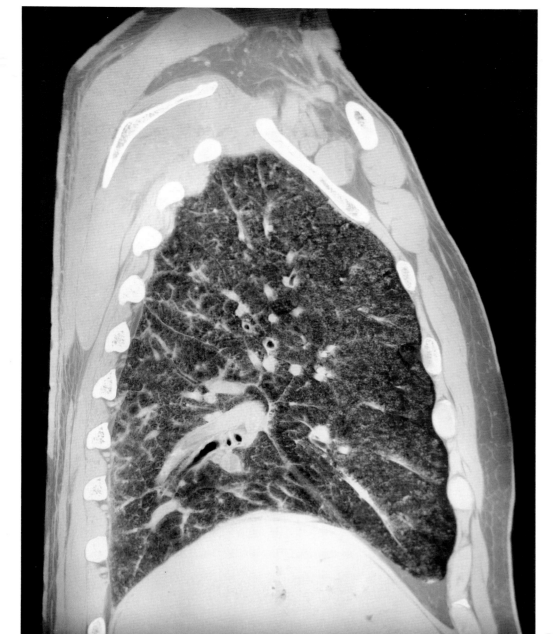

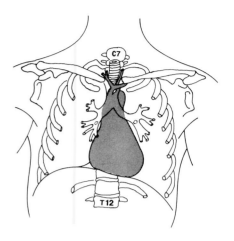

**Specimen Radiograph**

**Magnetic Resonance**

**Trispiral Tomogram**

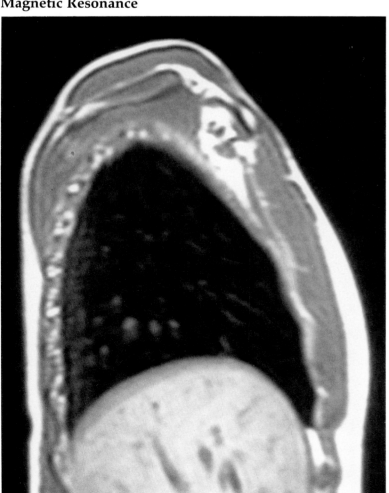

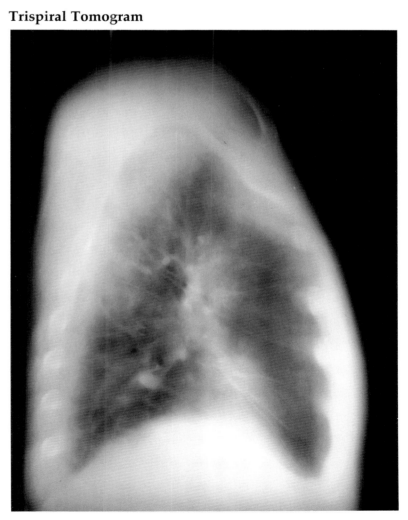

# Sagittal

## Section 9

### Anatomic Key

**Arteries**

A 7 right middle lobe (RML)
A 8 RML, lateral branch
A 10 right lower lobe (RLL)
A 11 RLL, superior branch
A 46 intercostal, posterior, right
A 50 subclavian, right

**Bronchi**

B 3 RUL, apical segment
B 4 RUL, posterior segment
B 5 RUL, anterior segment
B 8 RML, lateral segment
B 10 right lower lobe (RLL)
B 11 RLL, superior segment

**Fat**

f 3 extrapleural

**Fissures**

F 2 right major (oblique)
F 3 right minor (horizontal)

**Lobes**

L 2 RUL, apical segment
L 3 RUL, posterior segment
L 4 RUL, anterior segment
L 6 RML, lateral segment
L 9 RLL, superior segment
L 10 RLL, medial basal segment
L 11 RLL, anterior basal segment
L 13 RLL, posterior basal segment

**Muscles**

M 3 abdominis, rectus
M 9 diaphragm
M 10 diaphragm, central tendon
M 13 erector spinae
M 14 iliocostalis, thoracis
M 21 latissimus dorsi
M 30 omohyoid
M 31 pectoralis, major
M 33 platysma
M 35 rhomboid, major
M 38 scalene, anterior
M 39 scalene, middle
M 40 scalene, posterior
M 46 serratus, posterior superior
M 50 splenius, capitis
M 52 sternocleidomastoid
M 57 subscapularis
M 58 supraspinatus
M 62 transversus thoracis
M 63 trapezius

**Neural Structures**

n 1 brachial plexus
n 2 intercostal nerve

**Organs**

O 3 liver

**Pleural Structures**

P 2 costal (parietal)
P 3 costodiaphragmatic recess
P 6 visceral

**Skeletal Structures**

s 1 clavicle
s 3 rib (1st)
s 6 rib, body, posterior (5th)
s 7 rib, costal cartilage (1st)
s 10 scapula

**Veins**

v 0 pulmonary, superior, RUL
v 1 pulmonary, superior, RML
v 4 pulmonary, inferior, RLL
v 19 intercostal, posterior, right
v 25 jugular, external, right
v 31 subclavian, right

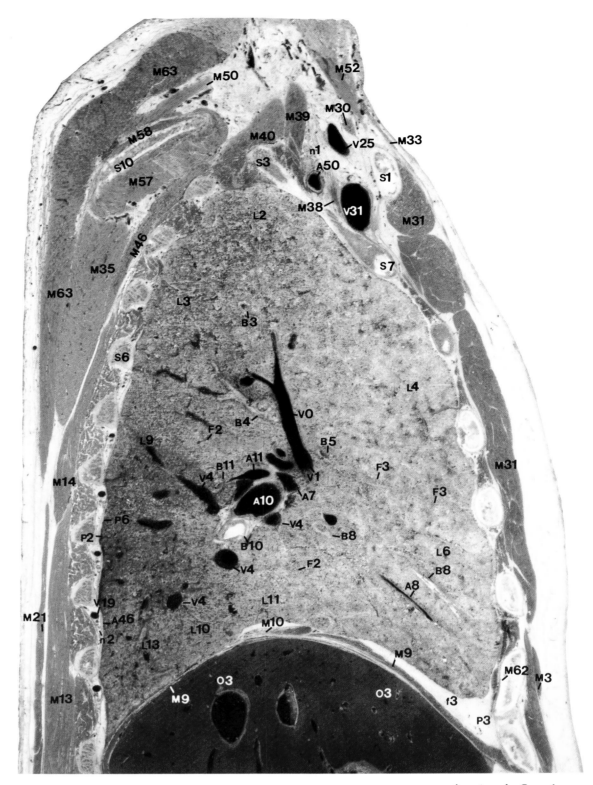

**Anatomic Specimen**

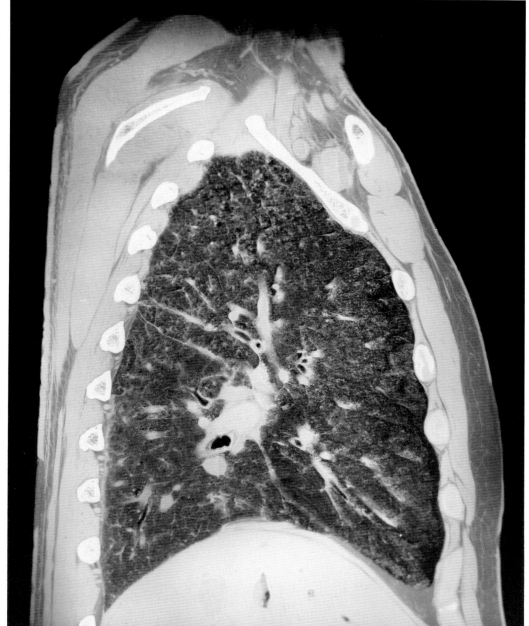

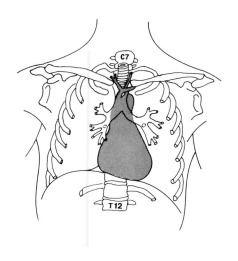

**Specimen Radiograph**

**Magnetic Resonance**

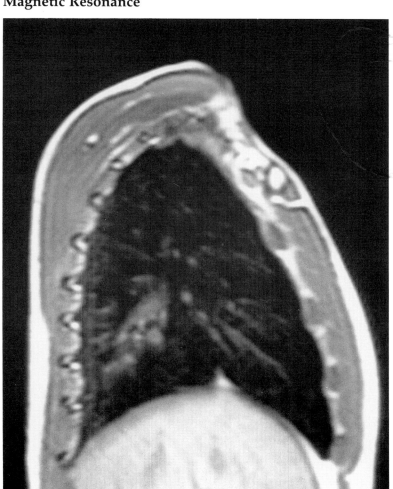

**Trispiral Tomogram**

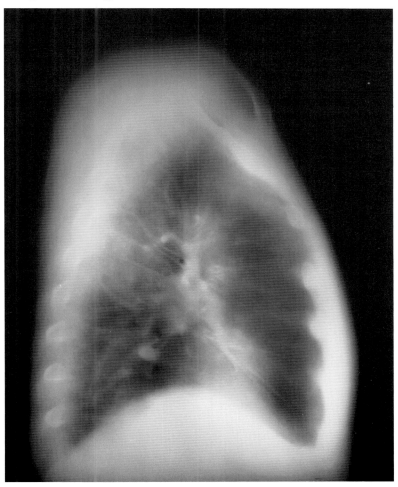

# Sagittal

## Section 10

## Anatomic Key

### Arteries
A 3 right upper lobe (RUL)
A 7 right middle lobe (RML)
A 10 right lower lobe (RLL)
A 11 RLL, superior branch
A 50 subclavian, right

### Bronchi
B 3 RUL, apical segment
B 4 RUL, posterior segment
B 5 RUL, anterior segment
B 8 RML, lateral segment
B 9 RML, medial segment
B 10 right lower lobe (RLL)

### Fat
f 3 extrapleural

### Fissures
F 2 right major (oblique)

### Lobes
L 2 RUL, apical segment
L 3 RUL, posterior segment
L 4 RUL, anterior segment
L 7 RML, medial segment
L 9 RLL, superior segment
L 10 RLL, medial basal segment
L 13 RLL, posterior basal segment

### Muscles
M 3 abdominis, rectus
M 9 diaphragm
M 10 diaphragm, central tendon
M 13 erector spinae
M 14 iliocostalis, thoracis
M 16 intercostal
M 21 latissimus dorsi
M 30 omohyoid
M 31 pectoralis, major
M 35 rhomboid, major
M 38 scalene, anterior
M 39 scalene, middle
M 40 scalene, posterior
M 46 serratus, posterior superior
M 50 splenius, capitis
M 52 sternocleidomastoid
M 57 subscapularis
M 62 transversus thoracis
M 63 trapezius

### Neural Structures
n 1 brachial plexus

### Organs
O 3 liver

### Pleural Structures
P 2 costal (parietal)
P 3 costodiaphragmatic recess
P 6 visceral

### Skeletal Structures
S 1 clavicle
S 3 rib (1st)
S 6 rib, body, posterior (8th)
S 7 rib, costal cartilage (1st)
S 10 scapula

### Veins
V 0 pulmonary, superior, RUL
V 1 pulmonary, superior, RML
V 4 pulmonary, inferior, RLL
V 19 intercostal, posterior, right
V 25 jugular, external, right
V 31 subclavian, right
V 37 transverse cervical, right
V 38 vena cava, inferior

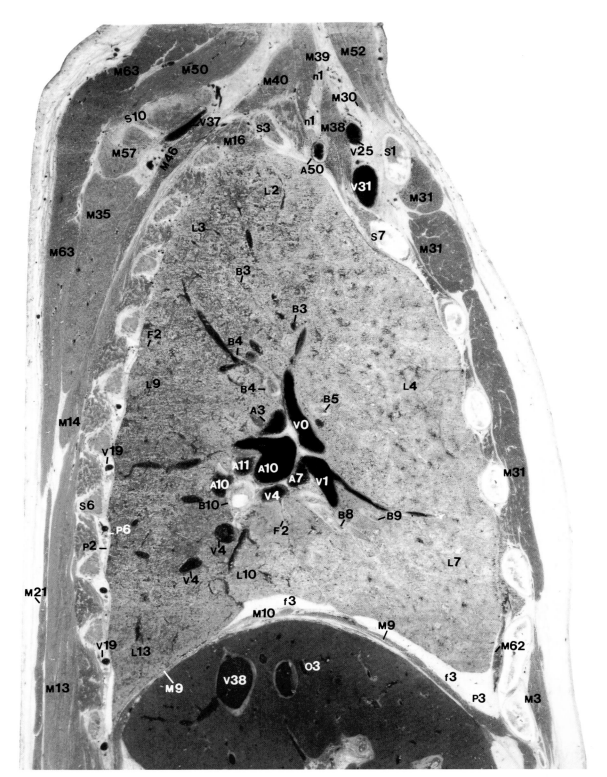

**Anatomic Specimen**

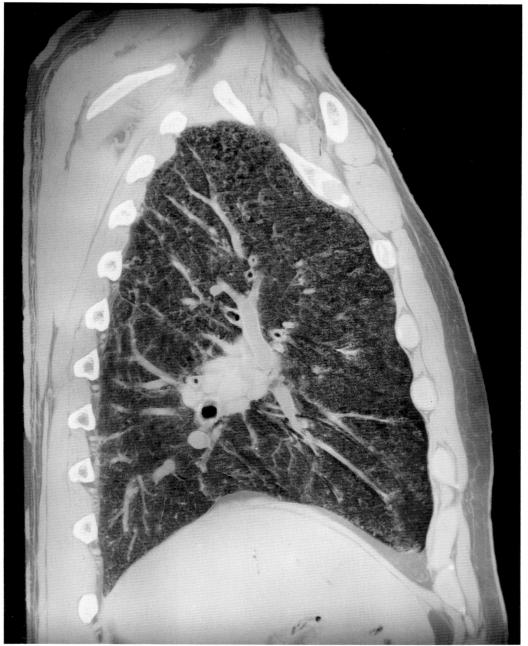

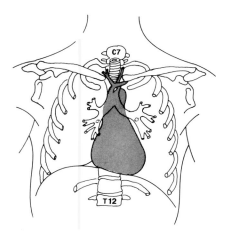

**Specimen Radiograph**

**Magnetic Resonance**

**Trispiral Tomogram**

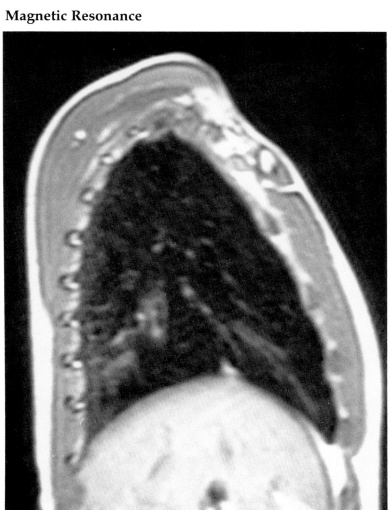

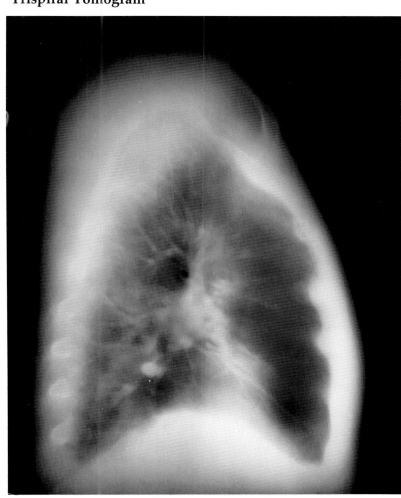

# Sagittal

## Section 11

### Anatomic Key

**Arteries**
A 2* main pulmonary, right (interlobar part)
A 3 right upper lobe (RUL)
A 7 right middle lobe (RML)
A 10 right lower lobe (RLL)
A 11 RLL, superior branch
A 46 intercostal, posterior, right
A 50 subclavian, right

**Bronchi**
B 3 RUL, apical segment
B 4 RUL, posterior segment
B 5 RUL, anterior segment
B 8 RML, lateral segment
B 9 RML, medial segment
B 10 right lower lobe (RLL)
B 11 RLL, superior segment

**Fascia**
F 8 Sibson's

**Fat**
f 2 epicardial
F 3 extrapleural

**Fissures**
F 2 right major (oblique)
F 3 right minor (horizontal)

**Heart**
H 2 atrium, right
H 13 pericardium, parietal serous

**Lobes**
L 2 RUL, apical segment
L 3 RUL, posterior segment
L 4 RUL, anterior segment
L 7 RML, medial segment
L 9 RLL, superior segment
L 10 RLL, medial basal segment

**Muscles**
M 3 abdominis, rectus
M 9 diaphragm
M 10 diaphragm, central tendon
M 13 erector spinae
M 16 intercostal
M 21 latissimus dorsi
M 23 levator scapulae
M 31 pectoralis, major
M 33 platysma
M 35 rhomboid, major
M 36 rhomboid, minor
M 38 scalene, anterior
M 39 scalene, middle
M 40 scalene, posterior
M 46 serratus, posterior superior
M 50 splenius, capitis
M 52 sternocleidomastoid
M 62 transversus thoracis
M 63 trapezius

**Neural Structures**
n 1 brachial plexus
n 2 intercostal nerve

**Nodes**
N 2 bronchopulmonary (hilar)

**Organs**
O 3 liver

**Pleural Structures**
P 3 costodiaphragmatic recess
P 6 visceral

**Skeletal Structures**
S 1 clavicle
S 3 rib (1st)
S 6 rib, body, posterior (5th)
S 7 rib, costal cartilage (1st)

**Veins**
V 0 pulmonary, superior, RUL
V 4 pulmonary, inferior, RLL
V 19 intercostal, posterior, right
V 25 jugular, external, right
V 31 subclavian, right
V 38 vena cava, inferior

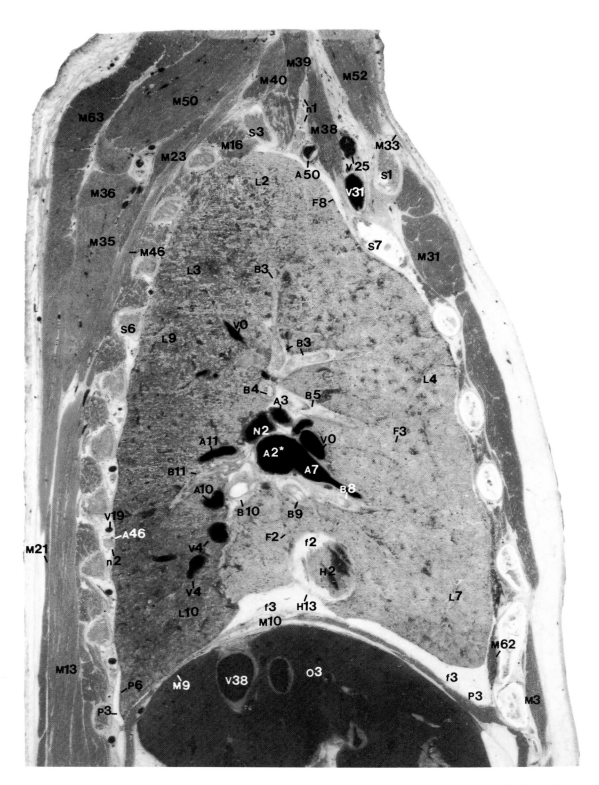

**Anatomic Specimen**

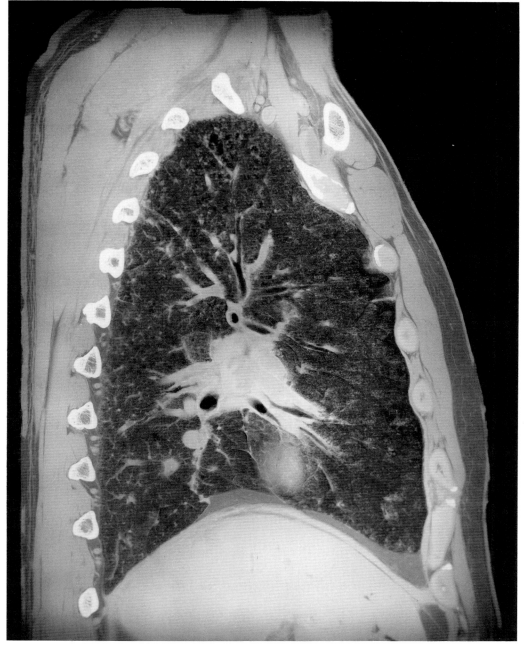

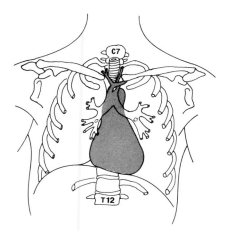

**Specimen Radiograph**

**Magnetic Resonance**

**Trispiral Tomogram**

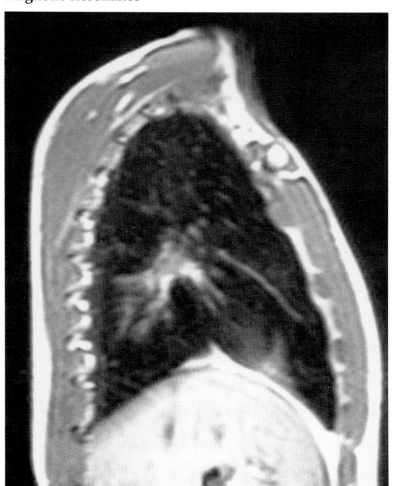

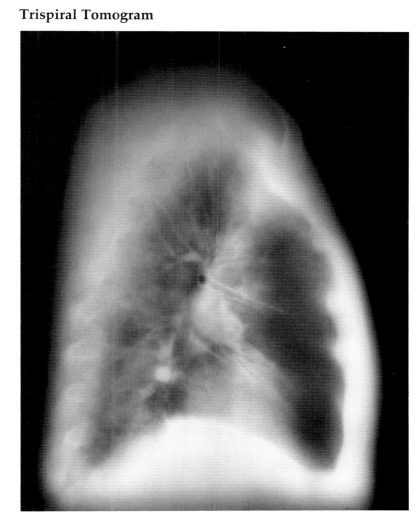

# Sagittal

## Section 12

### Anatomic Key

#### Arteries
A 2* main pulmonary, right (interlobar part)
A 3 right upper lobe (RUL)
A 4 RUL, apical branch
A 5 RUL, posterior branch
A 8 RML, lateral branch
A 12 RLL, medial basal branch
A 46 intercostal, posterior, right
A 50 subclavian, right

#### Bronchi
B 3 RUL, apical segment
B 4 RUL, posterior segment
B 5 RUL, anterior segment
B 7 right middle lobe (RML)
B 8 RML, lateral segment
B 9 RML, medial segment
B 10 right lower lobe (RLL)
B 11 RLL, superior segment

#### Fascia
F 8 Sibson's

#### Fat
f 3 extrapleural
f 7 subcutaneous

#### Fissures
F 3 right minor (horizontal)

#### Heart
H 2 atrium, right
H 13 pericardium, parietal serous

#### Lobes
L 2 RUL, apical segment
L 3 RUL, posterior segment
L 4 RUL, anterior segment
L 7 RML, medial segment
L 9 RLL, superior segment
L 10 RLL, medial basal segment

#### Muscles
M 3 abdominis, rectus
M 9 diaphragm
M 10 diaphragm, central tendon
M 13 erector spinae
M 16 intercostal
M 21 latissimus dorsi
M 23 levator scapulae
M 31 pectoralis, major
M 33 platysma
M 35 rhomboid, major
M 36 rhomboid, minor
M 38 scalene, anterior
M 39 scalene, middle
M 40 scalene, posterior
M 46 serratus, posterior superior
M 50 splenius, capitis
M 52 sternocleidomastoid
M 62 transversus thoracis
M 63 trapezius

#### Neural Structures
n 1 brachial plexus
n 2 intercostal nerve

#### Nodes
N 2 bronchopulmonary (hilar)

#### Organs
O 3 liver

#### Pleural Structures
P 2 costal (parietal)
P 3 costodiaphragmatic recess
P 6 visceral

#### Skeletal Structures
S 1 clavicle
S 3 rib (1st)
S 6 rib, body, posterior (11th)
S 7 rib, costal cartilage (1st)

#### Veins
V 0 pulmonary, superior, RUL
V 4 pulmonary, inferior, RLL
V 19 intercostal, posterior, right
V 25 jugular, external, right
V 31 subclavian, right
V 38 vena cava, inferior

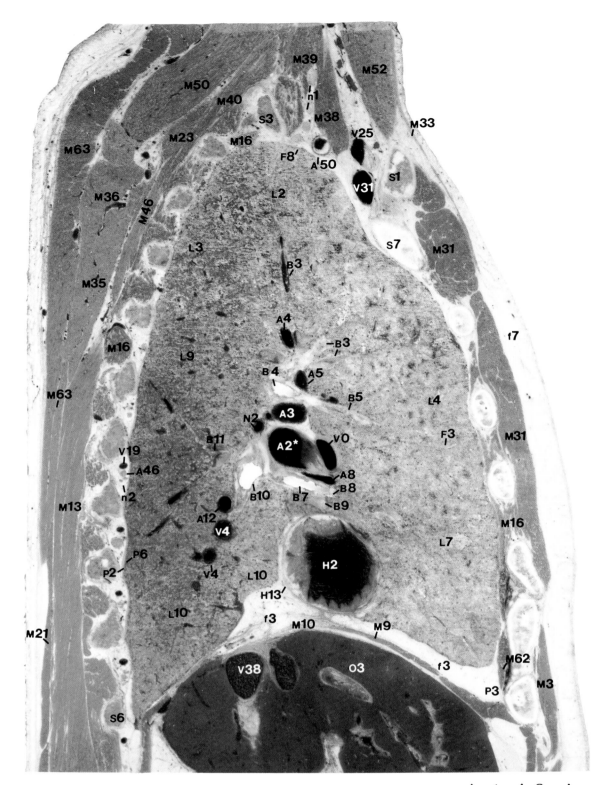

**Anatomic Specimen**

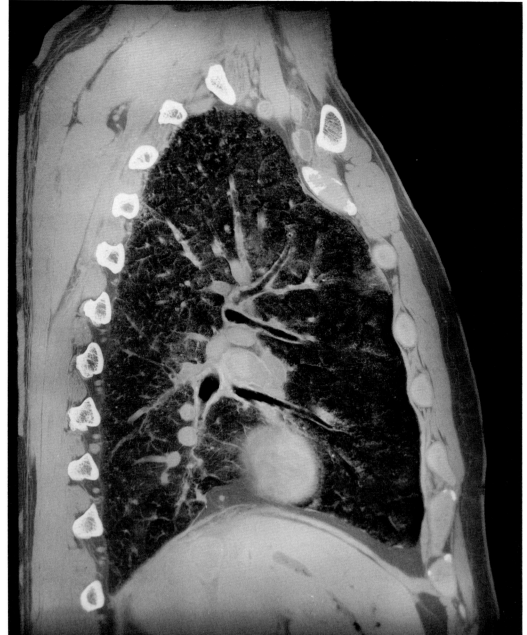

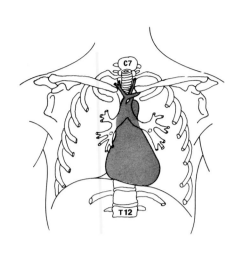

**Specimen Radiograph**

**Magnetic Resonance**

**Trispiral Tomogram**

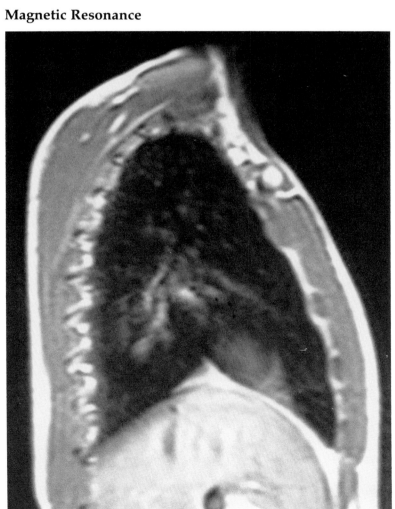

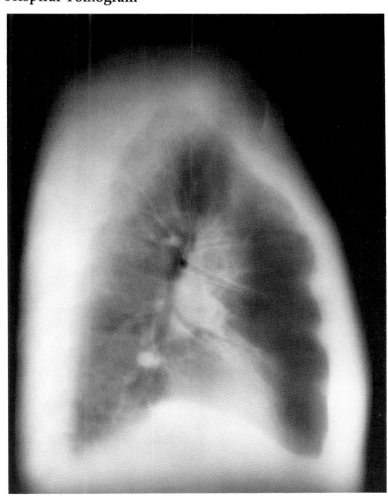

# Sagittal

## Section 13

### Anatomic Key

#### Arteries
A  2* main pulmonary, right
   (interlobar part)
A  3  right upper lobe (RUL)
A  6  RUL, anterior branch
A  9  RML, medial branch
A  12 RLL, medial basal branch
A  44 intercostal, anterior, right
A  46 intercostal, posterior, right
A  50 subclavian, right

#### Bronchi
B  2  right upper lobe (RUL)
B  3  RUL, apical segment
B  4  RUL, posterior segment
B  5  RUL, anterior segment
B  7  right middle lobe (RML)
B  9  RML, medial segment
B  10 right lower lobe (RLL)
B  11 RLL, superior segment
B  12 RLL, medial basal segment
B  15 RLL, posterior basal segment

#### Fat
f  3  extrapleural
f  6  perirenal

#### Fissures
F  3  right minor (horizontal)

#### Heart
H  2  atrium, right
H  4  auricle, right
H  13 pericardium, parietal serous

#### Lobes
L  2  RUL, apical segment
L  3  RUL, posterior segment
L  4  RUL, anterior segment
L  7  RML, medial segment
L  9  RLL, superior segment
L  10 RLL, medial basal segment

#### Muscles
M  3  abdominis, rectus
M  9  diaphragm
M  10 diaphragm, central tendon
M  12 diaphragm, crus, right
M  13 erector spinae
M  16 intercostal
M  21 latissimus dorsi
M  22 levator costarum
M  23 levator scapulae
M  31 pectoralis, major
M  33 platysma
M  35 rhomboid, major
M  36 rhomboid, minor
M  38 scalene, anterior
M  39 scalene, middle
M  40 scalene, posterior
M  46 serratus, posterior superior
M  50 splenius, capitis
M  52 sternocleidomastoid
M  54 sternothyroid
M  62 transversus thoracis
M  63 trapezius

#### Neural Structures
n  1  brachial plexus
n  2  intercostal nerve

#### Nodes
N  2  bronchopulmonary (hilar)

#### Organs
O  3  liver

#### Pleural Structures
P  2  costal (parietal)
P  3  costodiaphragmatic recess
P  6  visceral

#### Skeletal Structures
S  1  clavicle
S  3  rib (1st)
S  6  rib, body, posterior (11th)
S  7  rib, costal cartilage (1st)
S  29 vertebra, transverse process
     (T8)

#### Veins
V  0  pulmonary, superior, RUL
V  4  pulmonary, inferior, RLL
V  19 intercostal, posterior, right
V  27 jugular, internal, right
V  31 subclavian, right
V  38 vena cava, inferior

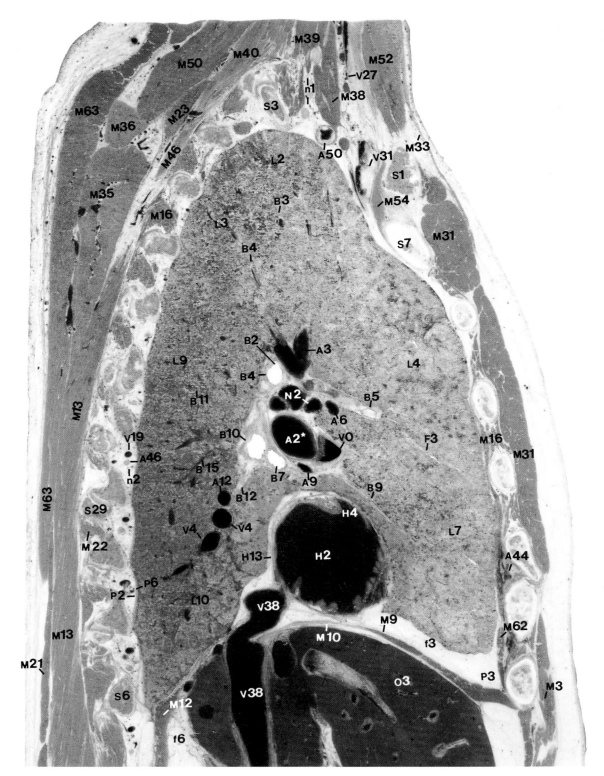

**Anatomic Specimen**

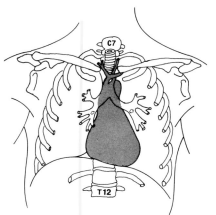

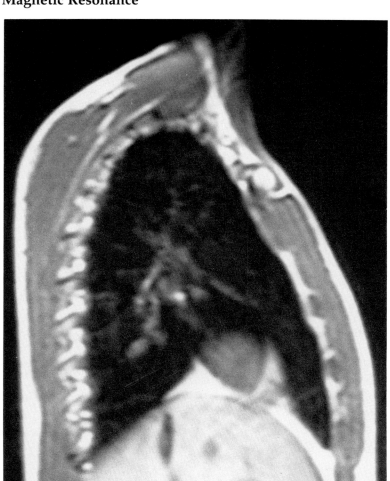

**Magnetic Resonance**

**Specimen Radiograph**

**Trispiral Tomogram**

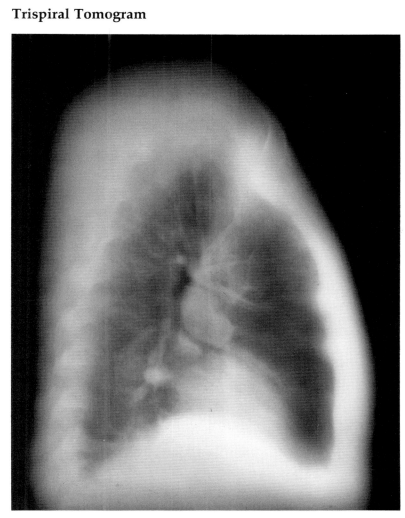

# Sagittal

## Section 14

## Anatomic Key

### Arteries
A  2* main pulmonary, right
   (interlobar part)
A  3  right upper lobe (RUL)
A 12  RLL, medial basal branch
A 44  intercostal, anterior, right
A 46  intercostal, posterior, right
A 48  internal thoracic (mammary),
      right
A 50  subclavian, right
A 64  transverse cervical, right

### Bronchi
B  2  right upper lobe (RUL)
B  3  RUL, apical segment
B  4  RUL, posterior segment
B  5  RUL, anterior segment
B  6  intermediate, right
B  9  RML, medial segment

### Fat
f  3  extrapleural
f  4  mediastinal
f  5  pericardial

### Fissures
F  2  right major (oblique)

### Heart
H  1  atrium, left
H  2  atrium, right
H 13  pericardium, parietal serous
H 14  pericardium, visceral serous

### Lobes
L  2  RUL, apical segment
L  3  RUL, posterior segment
L  4  RUL, anterior segment
L  7  RML, medial segment
L  9  RLL, superior segment
L 10  RLL, medial basal segment

### Muscles
M  3  abdominis, rectus
M  9  diaphragm
M 10  diaphragm, central tendon
M 12  diaphragm, crus, right
M 13  erector spinae
M 16  intercostal
M 20  intertransversarius, thoracis
M 21  latissimus dorsi
M 31  pectoralis, major
M 33  platysma
M 35  rhomboid, major
M 36  rhomboid, minor
M 38  scalene, anterior
M 39  scalene, middle
M 50  splenius, capitis
M 51  splenius, cervicis
M 52  sternocleidomastoid
M 54  sternothyroid
M 61  transversospinalis
M 62  transversus thoracis
M 63  trapezius

### Neural Structures
n  1  brachial plexus
n  2  intercostal nerve

### Nodes
N  2  bronchopulmonary (hilar)

### Organs
O  3  liver

### Pleural Structures
P  2  costal (parietal)
P  3  costodiaphragmatic recess
P  6  visceral

### Skeletal Structures
S  1  clavicle
S  3  rib (1st)
S  7  rib, costal cartilage (1st)
S  8  rib, head and neck (10th)
S 29  vertebra, transverse process
      (T10)

### Veins
V  0  pulmonary, superior, RUL
V  4  pulmonary, inferior, RLL
V  9  brachiocephalic (innominate),
      right
V 19  intercostal, posterior, right
V 27  jugular, internal, right
V 38  vena cava, inferior

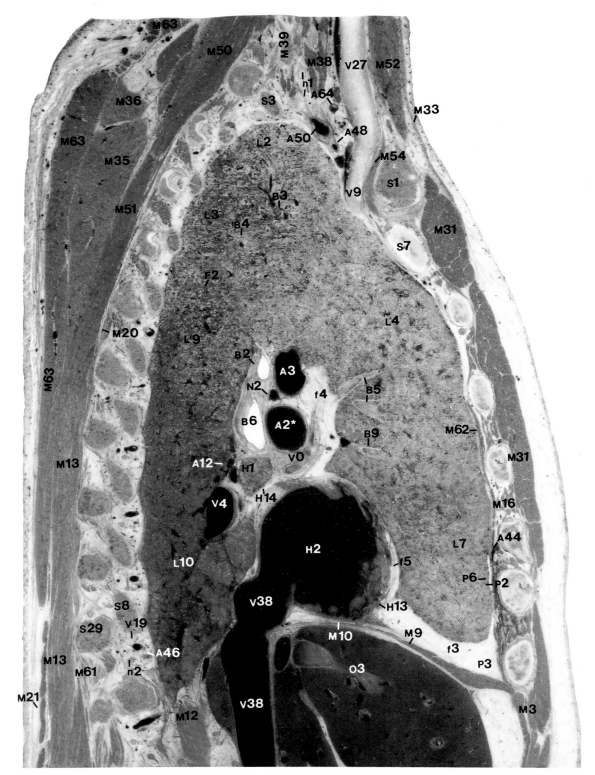

**Anatomic Specimen**

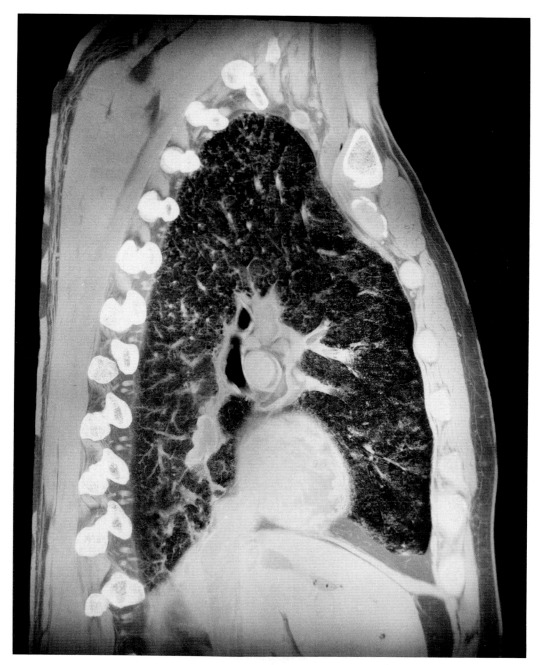

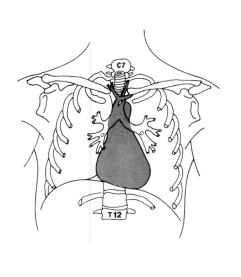

**Specimen Radiograph**

**Magnetic Resonance**

**Trispiral Tomogram**

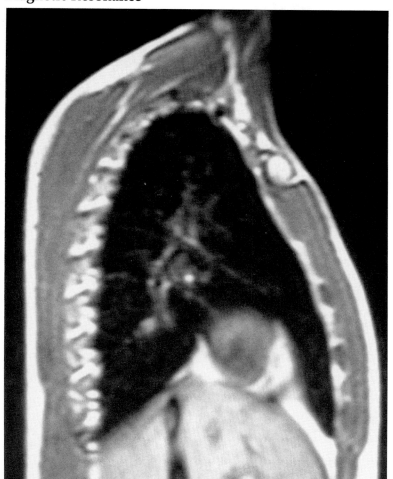

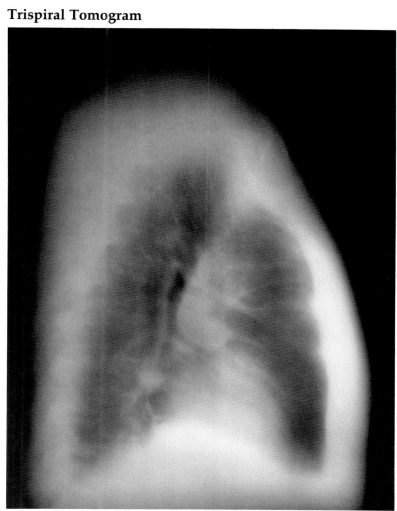

# Sagittal

## Section 15

## Anatomic Key

### Arteries
A 2* main pulmonary, right (interlobar part)
A 3 right upper lobe (RUL)
A 12 RLL, medial basal branch
A 46 intercostal, posterior, right
A 48 internal thoracic (mammary), right
A 50 subclavian, right
A 60 thyrocervical trunk, right
A 62 thyroid, inferior, right
A 66 vertebral, right

### Bronchi
B 1 main, right
B 6 intermediate, right
B 9 RML, medial segment

### Fat
f 2 epicardial
f 3 extrapleural
f 6 perirenal

### Fissures
F 2 right major (oblique)
F 3 right minor (horizontal)

### Heart
H 1 atrium, left
H 2 atrium, right
H 4 auricle, right
H 13 pericardium, parietal serous
H 14 pericardium, visceral serous

### Lobes
L 2 RUL, apical segment
L 3 RUL, posterior segment
L 4 RUL, anterior segment
L 7 RML, medial segment
L 9 RLL, superior segment
L 10 RLL, medial basal segment

### Muscles
M 3 abdominis, rectus
M 9 diaphragm
M 10 diaphragm, central tendon
M 12 diaphragm, crus, right
M 13 erector spinae
M 19 intertransversarius, cervicis
M 31 pectoralis, major
M 33 platysma
M 35 rhomboid, major
M 36 rhomboid, minor
M 50 splenius, capitis
M 51 splenius, cervicis
M 52 sternocleidomastoid
M 53 sternohyoid
M 54 sternothyroid
M 61 transversospinalis
M 62 transversus thoracis
M 63 trapezius

### Neural Structures
n 1 brachial plexus
n 2 intercostal nerve

### Nodes
N 2 bronchopulmonary (hilar)

### Organs
O 3 liver

### Pleural Structures
P 2 costal (parietal)
P 3 costodiaphragmatic recess
P 5 mediastinal (parietal)
P 6 visceral

### Skeletal Structures
S 1 clavicle
S 3 rib (1st)
S 7 rib, costal cartilage (1st)
S 8 rib, head and neck (10th)
S 23 vertebra, inferior articular process (C7)
S 29 vertebra, transverse process (T1)

### Veins
V 0 pulmonary, superior, RUL
V 4 pulmonary, inferior, RLL
V 7 azygos
V 9 brachiocephalic (innominate), right
V 19 intercostal, posterior, right
V 27 jugular, internal, right
V 38 vena cava, inferior
V 39 vena cava, superior

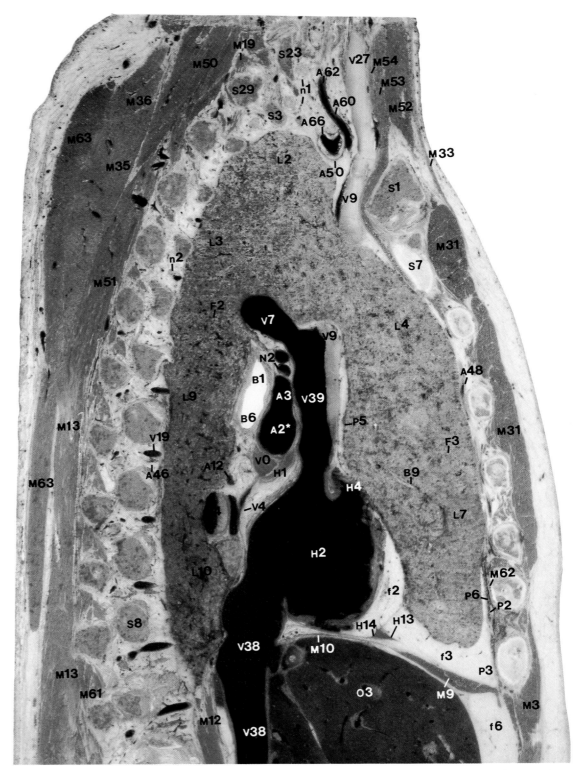

**Anatomic Specimen**

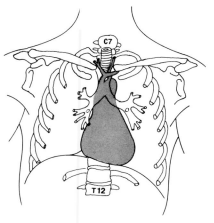

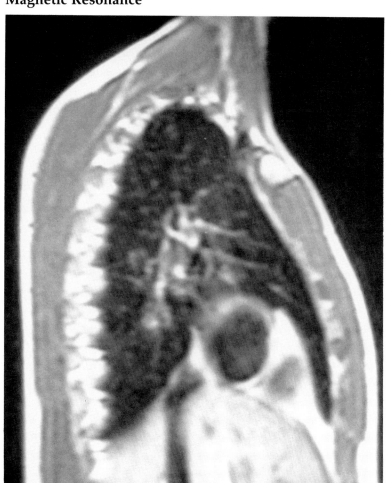

**Specimen Radiograph**

**Magnetic Resonance**

**Trispiral Tomogram**

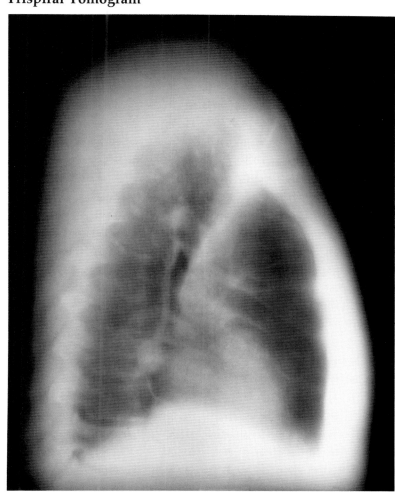

# Sagittal

## Section 16

### Anatomic Key

**Arteries**
A 2* main pulmonary, right (interlobar part)
A 38 carotid, common, right
A 46 intercostal, posterior, right
A 50 subclavian, right
A 66 vertebral, right

**Bronchi**
B 1 main, right

**Fat**
f 2 epicardial
f 3 extrapleural

**Fissures**
F 2 right major (oblique)
F 3 right minor (horizontal)

**Glands**
G 4 thyroid

**Heart**
H 1 atrium, left
H 2 atrium, right
H 4 auricle, right
H 11 coronary artery, right
H 13 pericardium, parietal serous
H 14 pericardium, visceral serous

**Lobes**
L 2 RUL, apical segment
L 3 RUL, posterior segment
L 4 RUL, anterior segment
L 7 RML, medial segment
L 9 RLL, superior segment

**Muscles**
M 3 abdominis, rectus
M 9 diaphragm
M 10 diaphragm, central tendon
M 12 diaphragm, crus, right
M 13 erector spinae
M 28 longus colli
M 31 pectoralis, major
M 33 platysma
M 35 rhomboid, major
M 36 rhomboid, minor
M 50 splenius, capitis
M 51 splenius, cervicis
M 52 sternocleidomastoid
M 53 sternohyoid
M 54 sternothyroid
M 62 transversus thoracis
M 63 trapezius

**Neural Structures**
n 2 intercostal nerve

**Nodes**
N 2 bronchopulmonary (hilar)

**Organs**
o 3 liver

**Pleural Structures**
P 5 mediastinal (parietal)

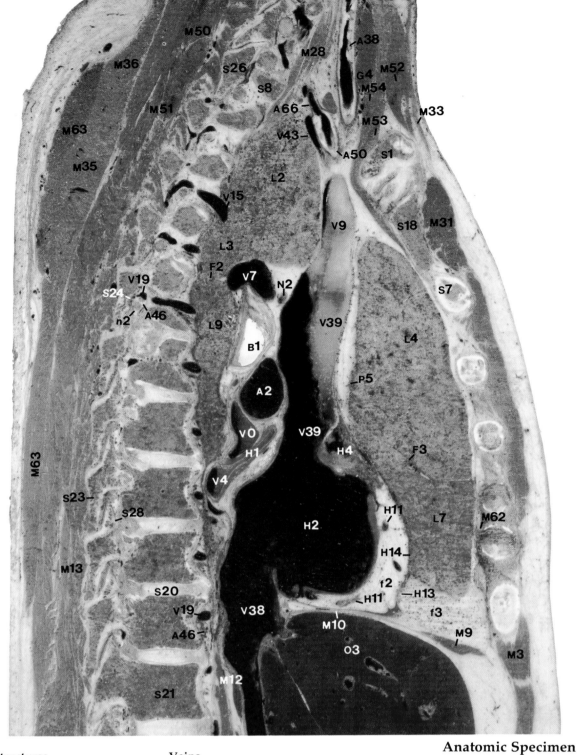

**Anatomic Specimen**

**Skeletal Structures**
s 1 clavicle
s 7 rib, costal cartilage (2nd)
s 8 rib, head and neck (1st)
s 18 sternum, manubrium
s 20 intervertebral disc (T9–T10)
s 21 vertebra, body (T11)
s 23 vertebra, inferior articular process (T8)
s 24 vertebra, intervertebral foramen (T4–T5)
s 26 vertebra, pedicle (T1)
s 28 vertebra, superior articular process (T9)

**Veins**
v 0 pulmonary, superior, RUL
v 4 pulmonary, inferior, RLL
v 7 azygos
v 9 brachiocephalic (innominate), right
v 15 intercostal, highest (superior), right
v 19 intercostal, posterior, right
v 38 vena cava, inferior
v 39 vena cava, superior
v 43 vertebral, right

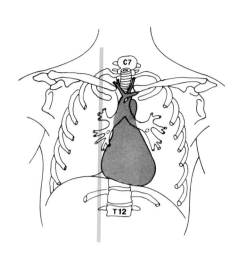

**Magnetic Resonance**

**Specimen Radiograph**

**Trispiral Tomogram**

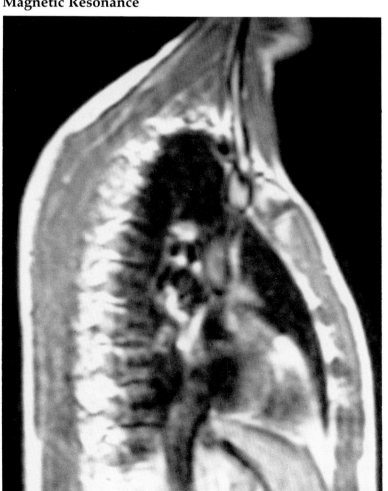

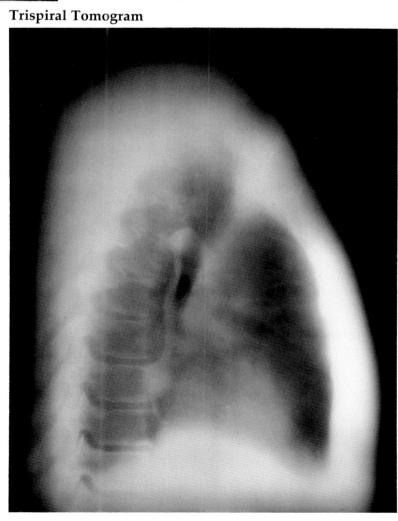

# Sagittal

## Section 17

## Anatomic Key

### Arteries

A  2  main pulmonary, right
A 34  brachiocephalic (innominate) trunk
A 38  carotid, common, right
A 46  intercostal, posterior, right
A 50  subclavian, right

### Bronchi

B  1  main, right

### Fat

f  2  epicardial
f  3  extrapleural
f  4  mediastinal

### Fissures

F  3  right minor (horizontal)

### Glands

G  4  thyroid

### Heart

H  1  atrium, left
H  2  atrium, right
H  4  auricle, right
H 11  coronary artery, right
H 13  pericardium, parietal serous
H 14  pericardium, visceral serous
H 22  ventricle, right

### Lobes

L  2  RUL, apical segment
L  4  RUL, anterior segment
L  7  RML, medial segment

### Muscles

M  3  abdominis, rectus
M  9  diaphragm
M 10  diaphragm, central tendon
M 12  diaphragm, crus, right
M 13  erector spinae
M 28  longus colli
M 31  pectoralis, major
M 35  rhomboid, major
M 36  rhomboid, minor
M 41  semispinalis, capitis
M 50  splenius, capitis
M 51  splenius, cervicis
M 52  sternocleidomastoid
M 53  sternohyoid
M 54  sternothyroid
M 63  trapezius

### Neural Structures

n  2  intercostal nerve

### Nodes

N  6  mediastinal, anterior

### Organs

O  3  liver

### Pleural Structures

P  2  costal (parietal)
P  5  mediastinal (parietal)
P  6  visceral

### Skeletal Structures

S  1  clavicle
S  7  rib, costal cartilage (2nd)
S 18  sternum, manubrium
S 20  intervertebral disc (T5–T6)
S 21  vertebra, body (T11)
S 23  vertebra, inferior articular process (C7)
S 26  vertebra, pedicle (T10)
S 28  vertebra, superior articular process (T1)

### Veins

V  0  pulmonary, superior, RUL
V  7  azygos
V  9  brachiocephalic (innominate), right
V 15  intercostal, highest (superior), right
V 17  intercostal, anterior, right
V 19  intercostal, posterior, right
V 35  thyroid, inferior, right
V 38  vena cava, inferior
V 39  vena cava, superior

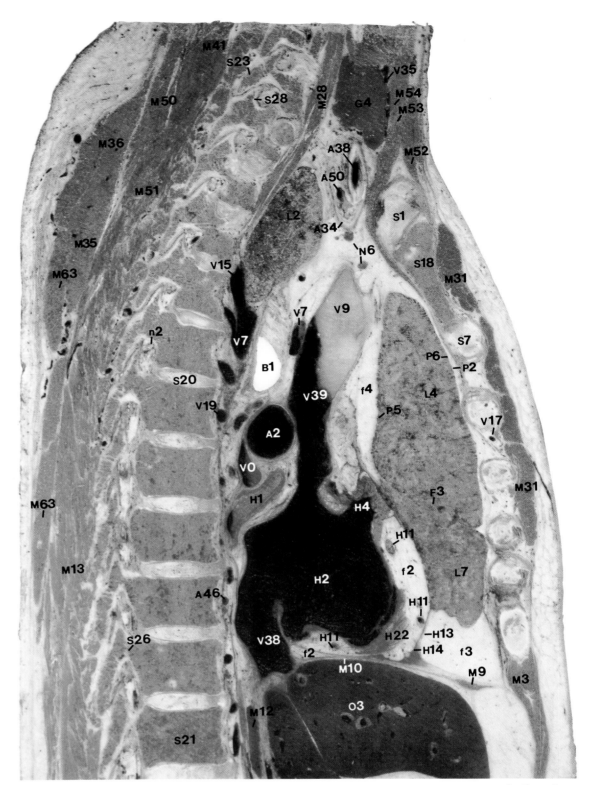

**Anatomic Specimen**

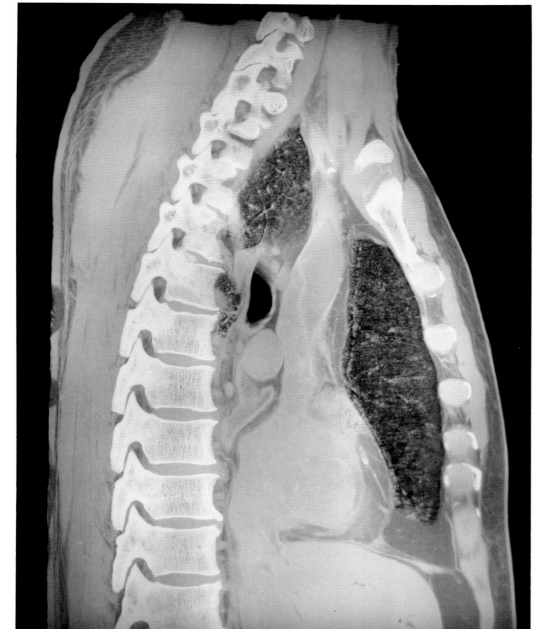

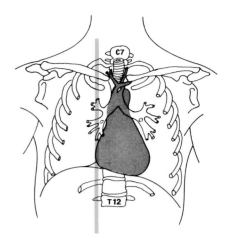

**Specimen Radiograph**

**Magnetic Resonance**

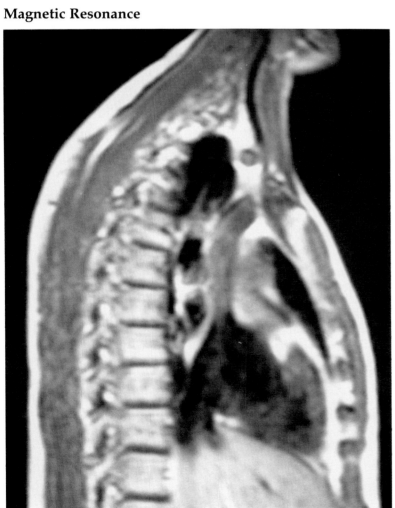

**Trispiral Tomogram**

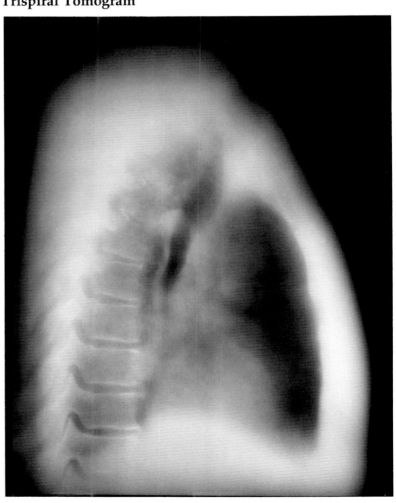

# Sagittal
## Section 18

## Anatomic Key

### Arteries
A 2 main pulmonary, right
A 30 aorta, ascending
A 34 brachiocephalic (innominate) trunk

### Bronchi
B 1 main, right

### Fat
f 2 epicardial
f 3 extrapleural
f 4 mediastinal

### Fissures
F 3 right minor (horizontal)

### Glands
G 4 thyroid

### Heart
H 1 atrium, left
H 2 atrium, right
H 4 auricle, right
H 11 coronary artery, right
H 13 pericardium, parietal serous
H 14 pericardium, visceral serous
H 22 ventricle, right

### Lobes
L 2 RUL, apical segment
L 4 RUL, anterior segment
L 7 RML, medial segment

### Muscles
M 3 abdominis, rectus
M 10 diaphragm, central tendon
M 12 diaphragm, crus, right
M 13 erector spinae
M 28 longus colli
M 31 pectoralis, major
M 41 semispinalis, capitis
M 52 sternocleidomastoid
M 53 sternohyoid
M 54 sternothyroid
M 62 transversus thoracis
M 63 trapezius

### Neural Structures
n 5 spinal cord

### Nodes
N 3 diaphragmatic
N 6 mediastinal, anterior
N 11 tracheobronchial, inferior (subcarinal)

### Organs
O 3 liver

### Pleural Structures
P 2 costal
P 5 mediastinal
P 6 visceral

### Skeletal Structures
S 7 rib, costal cartilage (7th)
S 17 sternum, body
S 18 sternum, manubrium
S 20 intervertebral disc (C7–T1)
S 21 vertebra, body (T11)
S 23 vertebra, inferior articular process (T1)
S 24 vertebra, intervertebral foramen (T2–T3)
S 26 vertebra, pedicle (T2)
S 27 vertebra, spinous process (T7)
S 28 vertebra, superior articular process (T2)

### Trachea and Larynx
T 1 trachea
T 3 larynx

### Veins
V 0 pulmonary, superior (RUL)
V 7 azygos
V 8 brachiocephalic (innominate), left
V 9 brachiocephalic (innominate), right
V 17 intercostal, anterior, right
V 19 intercostal, posterior, right
V 38 vena cava, inferior

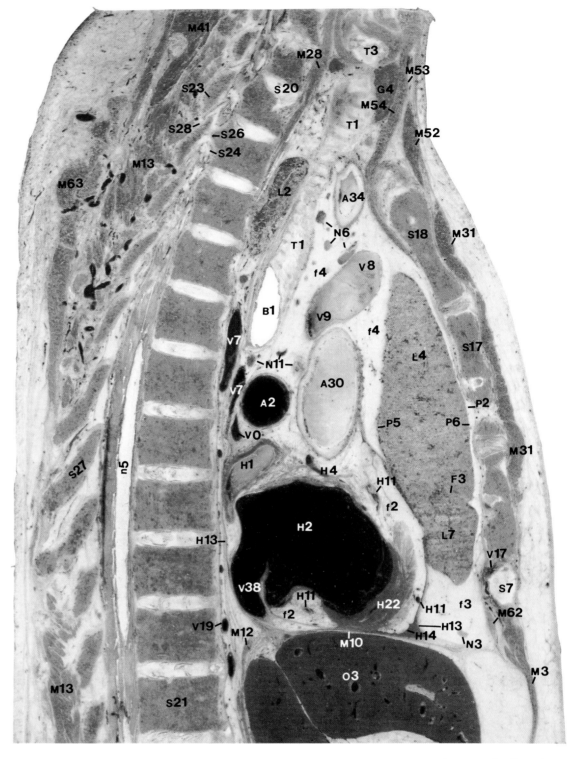

**Anatomic Specimen**

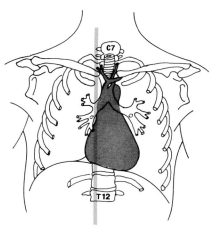

**Specimen Radiograph**

**Magnetic Resonance**

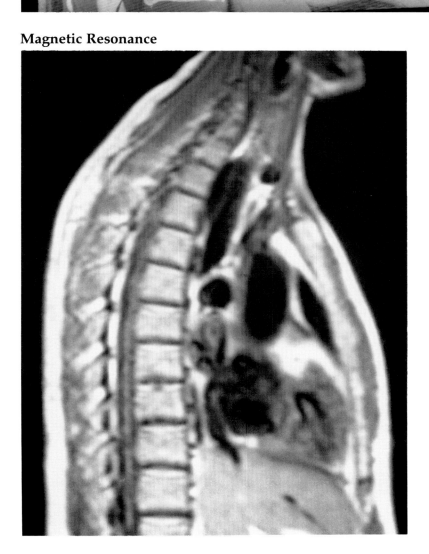

**Trispiral Tomogram**

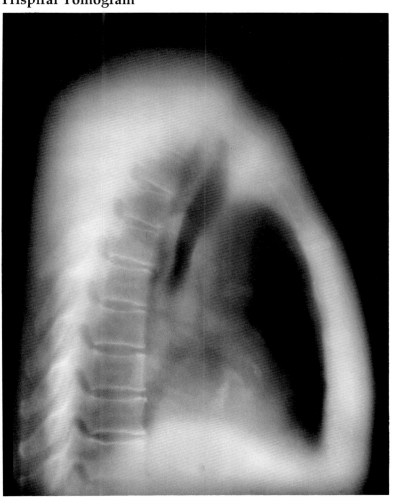

# Sagittal

## Section 19

## Anatomic Key

### Arteries
A 2 main pulmonary, right
A 30 aorta, ascending
A 34 brachiocephalic (innominate) trunk

### Esophagus
E 1 esophagus

### Fascia
F 6 perivisceral
F 7 prevertebral

### Fat
f 2 epicardial
f 3 extrapleural
f 4 mediastinal

### Fissures
F 3 right minor (horizontal)

### Glands
G 4 thyroid

### Heart
H 1 atrium, left
H 2 atrium, right
H 7 cardiac vein, small
H 11 coronary artery, right
H 12 coronary sinus
H 13 pericardium, parietal serous
H 14 pericardium, visceral serous
H 20 valve, aortic
H 22 ventricle, right

### Lobes
L 4 RUL, anterior segment
L 7 RML, medial segment

### Muscles
M 9 diaphragm
M 10 diaphragm, central tendon
M 12 diaphragm, crus, right
M 13 erector spinae
M 17 interspinalis, cervicis
M 18 interspinalis, thoracis
M 28 longus colli
M 31 pectoralis, major
M 33 platysma
M 52 sternocleidomastoid
M 54 sternothyroid
M 61 transversospinalis
M 63 trapezius

### Neural Structures
n 5 spinal cord

### Organs
O 1 colon
O 3 liver
O 4 small bowel

### Pleural Structures
P 5 mediastinal (parietal)
P 6 visceral

### Skeletal Structures
S 17 sternum, body
S 18 sternum, manubrium
S 19 sternum, xiphoid
S 20 intervertebral disc (C6–C7)
S 21 vertebra, body (T11)
S 25 vertebra, lamina (T3)
S 27 vertebra, spinous process (T3)

### Trachea and Larynx
T 1 trachea
T 2 carina (bifurcation of trachea)
T 3 larynx

### Veins
V 7 azygos
V 8 brachiocephalic (innominate), left
V 19 intercostal, posterior, right
V 38 vena cava, inferior

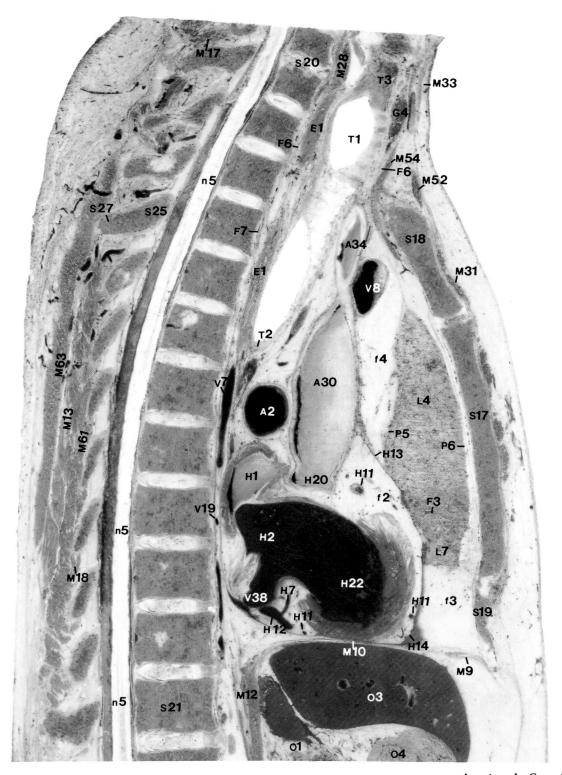

**Anatomic Specimen**

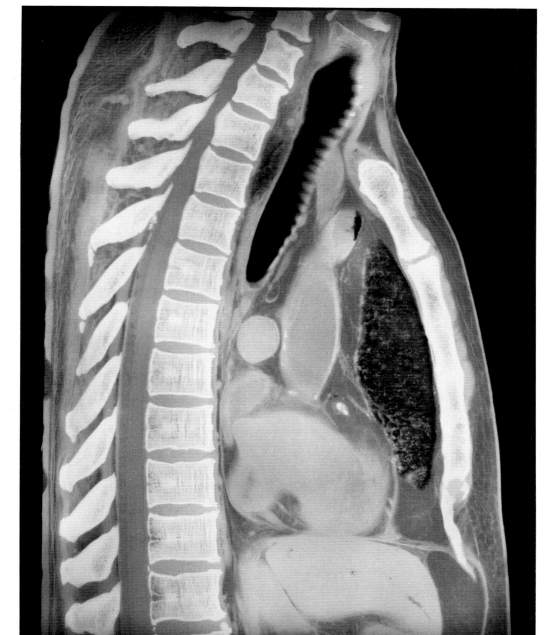

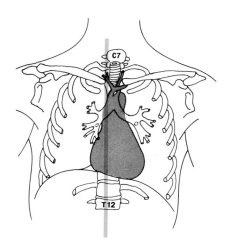

**Specimen Radiograph**

**Magnetic Resonance**

**Trispiral Tomogram**

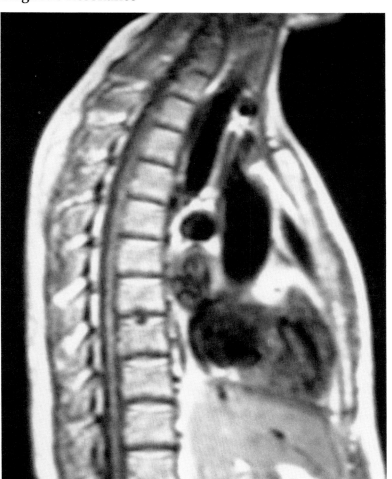

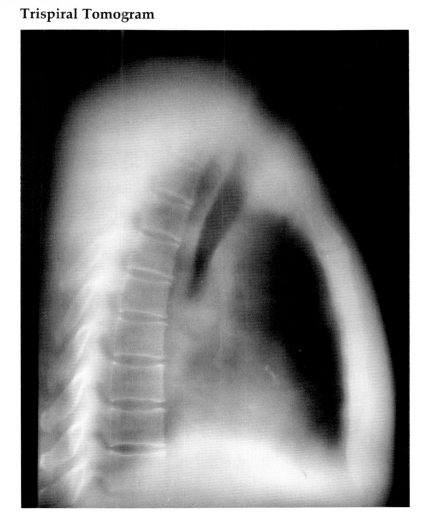

## Anatomic Key

### Arteries
A 2 main pulmonary, right
A 29 aortic arch
A 30 aorta, ascending
A 34 brachiocephalic (innominate) trunk

### Esophagus
E 1 esophagus

### Fascia
F 6 perivisceral
F 7 prevertebral
F 9 superficial (investing)

### Fat
f 2 epicardial
f 3 extrapleural
f 4 mediastinal

### Fissures
F 3 right minor (horizontal)

### Glands
G 4 thyroid

### Heart
H 1 atrium, left
H 2 atrium, right
H 6 cardiac vein, middle
H 9 coronary artery, posterior descending
H 11 coronary artery, right
H 13 pericardium, parietal serous
H 14 pericardium, visceral serous
H 15 septum, interatrial
H 17 valve, tricuspid
H 20 valve, aortic
H 21 ventricle, left
H 21 ventricle, right

### Lobes
L 4 RUL, anterior segment
L 7 RML, medial segment
L 16 LUL, anterior segment

### Muscles
M 9 diaphragm
M 10 diaphragm, central tendon
M 12 diaphragm, crus, right
M 13 erector spinae
M 28 longus colli
M 31 pectoralis, major
M 54 sternothyroid
M 61 transversospinalis
M 63 trapezius

### Neural Structures
n 2 intercostal nerve
n 5 spinal cord

### Organs
O 3 liver

### Pleural Structures
P 5 mediastinal (parietal)
P 6 visceral

### Skeletal Structures
S 17 sternum, body
S 18 sternum, manubrium
S 19 sternum, xiphoid
S 20 intervertebral disc (C7–T1)
S 21 vertebra, body (T11)
S 23 vertebra, inferior articular process (T6)
S 24 vertebra, intervertebral foramen (T7–T8)
S 26 vertebra, pedicle (T7)
S 27 vertebra, spinous process (T1)
S 28 vertebra, superior articular process (T7)

### Trachea and Larynx
T 1 trachea
T 2 carina (bifurcation of trachea)
T 3 larynx

### Veins
V 7 azygos
V 8 brachiocephalic (innominate), left
V 19 intercostal, posterior, right
V 38 vena cava, inferior

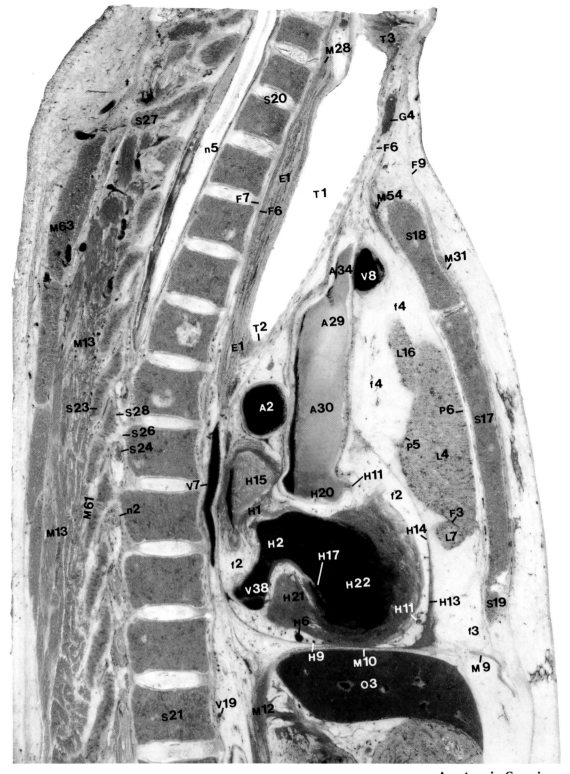

**Anatomic Specimen**

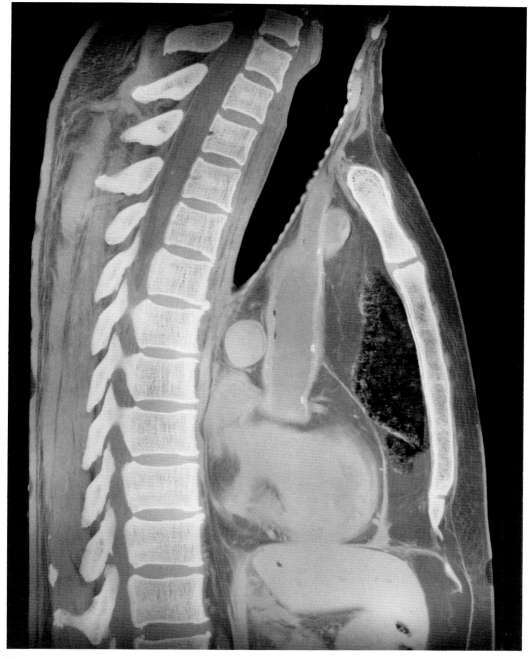

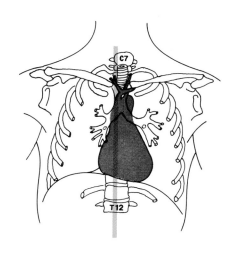

**Magnetic Resonance**

**Specimen Radiograph**

**Trispiral Tomogram**

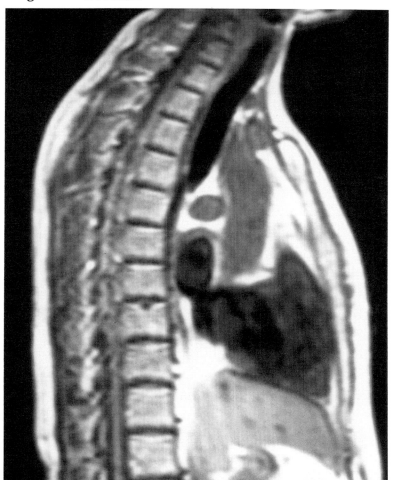

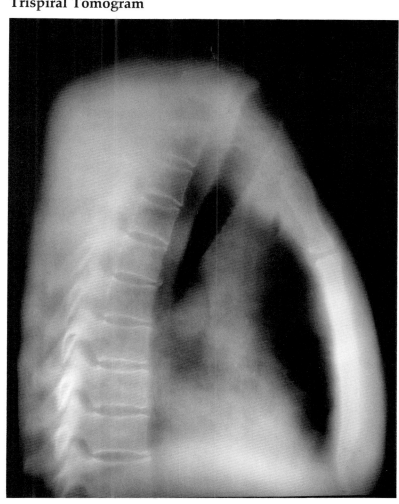

# Sagittal

## Section 21

### Anatomic Key

**Arteries**
A 2 main pulmonary, right
A 29 aortic arch
A 30 aorta, ascending
A 31 aorta, descending
A 37 carotid, common, left
A 43 intercostal, anterior, left

**Bronchi**
B 16 main, left

**Esophagus**
E 1 esophagus

**Fat**
f 2 epicardial
f 4 mediastinal
f 7 subcutaneous

**Glands**
G 4 thyroid

**Heart**
H 1 atrium, left
H 2 atrium, right
H 6 cardiac vein, middle
H 9 coronary artery, posterior
　　descending
H 15 septum, interatrial
H 16 septum, interventricular
H 20 valve, aortic
H 21 ventricle, left
H 22 ventricle, right

**Lobes**
L 4 RUL, anterior segment
L 16 LUL, anterior segment

**Muscles**
M 3 abdominis, rectus
M 10 diaphragm, central tendon
M 12 diaphragm, crus, right
M 13 erector spinae
M 31 pectoralis, major
M 33 platysma
M 35 rhomboid, major
M 36 rhomboid, minor
M 50 splenius, capitis
M 51 splenius, cervicis
M 52 sternocleidomastoid
M 54 sternothyroid
M 61 transversospinalis
M 63 trapezius

**Neural Structures**
n 2 intercostal nerve
n 5 spinal cord

**Organs**
O 3 liver
O 6 stomach

**Pleural Structures**
P 2 costal (parietal)
P 5 mediastinal (parietal)
P 6 visceral

**Skeletal Structures**
s 17 sternum, body
s 18 sternum, manubrium
s 19 sternum, xiphoid
s 20 intervertebral disc (C6–C7)
s 21 vertebra, body (T11)
s 23 vertebra, inferior articular
　　process (T3)
s 24 vertebra, intervertebral foramen
　　(T4–T5)
s 26 vertebra, pedicle (T5)
s 28 vertebra, superior articular
　　process (T6)

**Trachea and Larynx**
T 1 trachea
T 2 carina (bifurcation of trachea)
T 3 larynx

**Veins**
V 8 brachiocephalic (innominate),
　　left
V 12 hemiazygos
V 16 intercostal, anterior, left

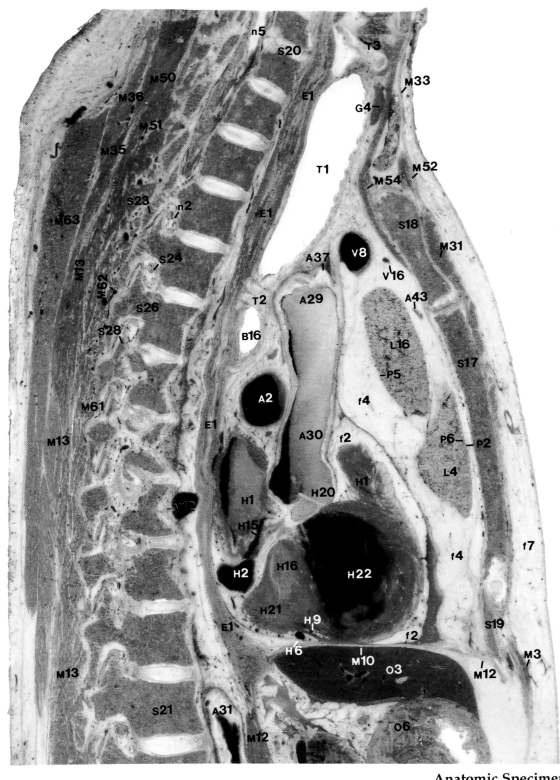

**Anatomic Specimen**

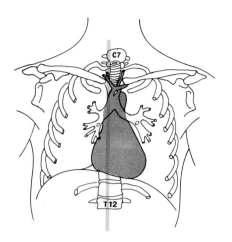

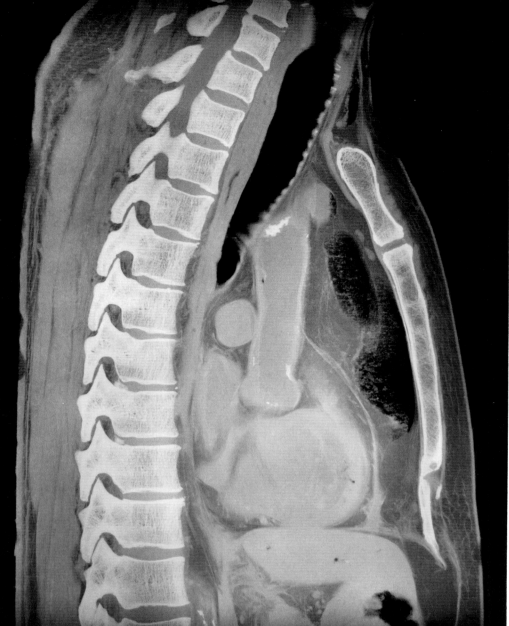

**Magnetic Resonance**

**Specimen Radiograph**

**Trispiral Tomogram**

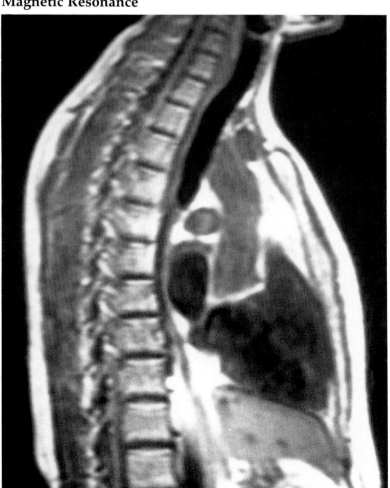

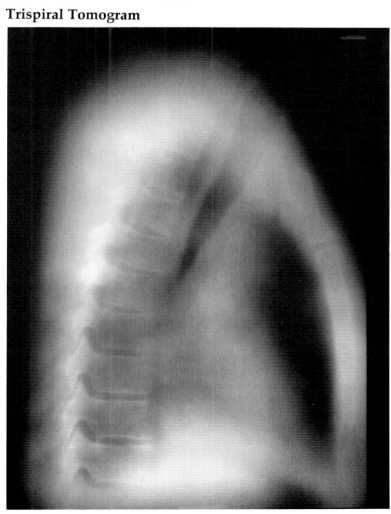

177

# Sagittal

## Section 22

### Anatomic Key

**Arteries**
A 1 pulmonary trunk
A 2 main pulmonary, right
A 29 aortic arch
A 30 aorta, ascending
A 31 aorta, descending
A 37 carotid, common, left
A 43 intercostal, anterior, left

**Bronchi**
B 16 main, left

**Esophagus**
E 1 esophagus
E 2 esophagogastric junction

**Fat**
f 2 epicardial
f 4 mediastinal

**Glands**
G 4 thyroid

**Heart**
H 1 atrium, left
H 2 atrium, right
H 6 cardiac vein, middle
H 9 coronary artery, posterior descending
H 13 pericardium, parietal serous
H 14 pericardium, visceral serous
H 16 septum, interventricular
H 20 valve, aortic
H 21 ventricle, left
H 22 ventricle, right

**Lobes**
L 15 LUL, apicoposterior segment
L 16 LUL, anterior segment
L 18 LUL, superior lingular segment

**Muscles**
M 3 abdominis, rectus
M 9 diaphragm
M 10 diaphragm, central tendon
M 12 diaphragm, crus, right
M 13 erector spinae
M 21 latissimus dorsi
M 28 longus colli
M 31 pectoralis, major
M 33 platysma
M 35 rhomboid, major
M 36 rhomboid, minor
M 41 semispinalis, capitis
M 50 splenius, capitis
M 51 splenius, cervicis
M 52 sternocleidomastoid
M 53 sternohyoid
M 54 sternothyroid
M 61 transversospinalis
M 63 trapezius

**Neural Structures**
n 2 intercostal nerve

**Organs**
o 1 colon
o 3 liver
o 6 stomach

**Pleural Structures**
P 5 mediastinal (parietal)
P 6 visceral

**Skeletal Structures**
s 1 clavicle
s 7 rib, costal cartilage (2nd)
s 8 rib, head and neck (10th)
s 17 sternum, body
s 18 sternum, manubrium
s 20 intervertebral disc (T11–T12)
s 21 vertebra, body (T2)
s 23 vertebra, inferior articular process (T1)
s 24 vertebra, intervertebral foramen (T1–T2)
s 26 vertebra, pedicle (T2)
s 28 vertebra, superior articular process (T2)
s 29 vertebra, transverse process (T10)

**Trachea and Larynx**
T 1 trachea
T 3 larynx

**Veins**
v 8 brachiocephalic (innominate), left
v 12 hemiazygos
v 13 hemiazygos, accessory
v 16 intercostal, anterior, left
v 18 intercostal, posterior, left
v 22 jugular, anterior, left

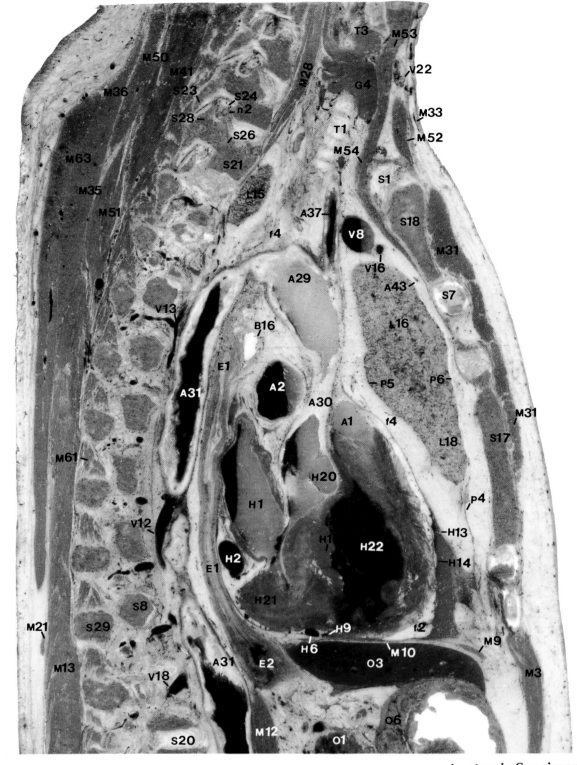

**Anatomic Specimen**

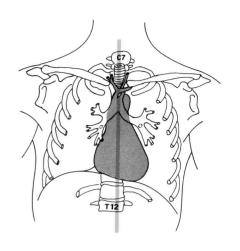

**Specimen Radiograph**

**Magnetic Resonance**

**Trispiral Tomogram**

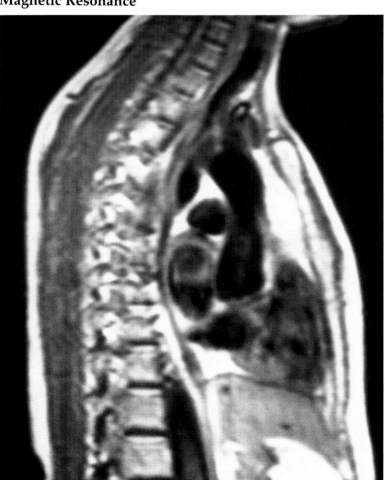

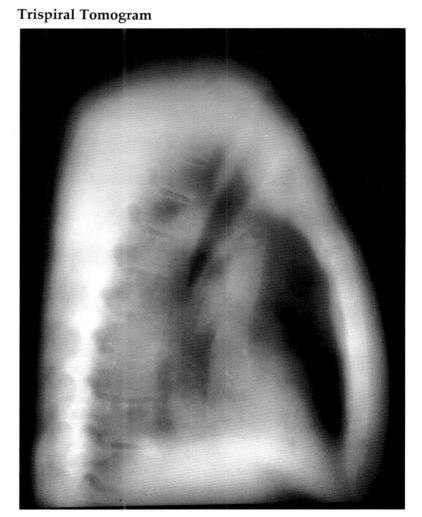

# Sagittal

## Section 23

## Anatomic Key

### Arteries
A 1 pulmonary trunk
A 29 aortic arch
A 30 aorta, ascending
A 31 aorta, descending
A 37 carotid, common, left
A 49 subclavian, left
A 65 vertebral, left

### Bronchi
B 16 main, left

### Esophagus
E 1 esophagus
E 2 esophagogastric junction

### Fat
f 2 epicardial
f 4 mediastinal

### Glands
G 4 thyroid

### Heart
H 1 atrium, left
H 6 cardiac vein, middle
H 9 coronary artery, posterior descending
H 11 coronary artery, right
H 12 coronary sinus
H 13 pericardium, parietal serous
H 14 pericardium, visceral serous
H 16 septum, interventricular
H 18 valve, pulmonary
H 19 valve, mitral
H 20 valve, aortic
H 21 ventricle, left
H 22 ventricle, right

### Lobes
L 15 LUL, apicoposterior segment
L 16 LUL, anterior segment
L 18 LUL, superior lingular segment

### Muscles
M 3 abdominis, rectus
M 9 diaphragm
M 10 diaphragm, central tendon
M 12 diaphragm, crus, right
M 13 erector spinae
M 21 latissimus dorsi
M 28 longus colli
M 31 pectoralis, major
M 33 platysma
M 35 rhomboid, major
M 36 rhomboid, minor
M 41 semispinalis, capitis
M 50 splenius, capitis
M 51 splenius, cervicis
M 52 sternocleidomastoid
M 53 sternohyoid
M 54 sternothyroid
M 63 trapezius

### Nodes
N 6 mediastinal, anterior

### Organs
O 3 liver
O 6 stomach

### Pleural Structures
P 2 costal (parietal)
P 5 mediastinal (parietal)
P 6 visceral

### Skeletal Structures
S 1 clavicle
S 7 rib, costal cartilage (2nd)
S 8 rib, head and neck (1st)
S 9 rib, tubercle (11th)
S 17 sternum, body
S 18 sternum, manubrium
S 23 vertebra, inferior articular process (T1)
S 28 vertebra, superior articular process (T2)
S 29 vertebra, transverse process (T11)

### Veins
V 2 pulmonary, superior, LUL
V 5 pulmonary, inferior, LLL
V 8 brachiocephalic (innominate), left
V 12 hemiazygos
V 14 intercostal, highest (superior), left
V 16 intercostal, anterior, left
V 18 intercostal, posterior, left
V 22 jugular, anterior, left
V 35 thyroid, inferior, left

**Anatomic Specimen**

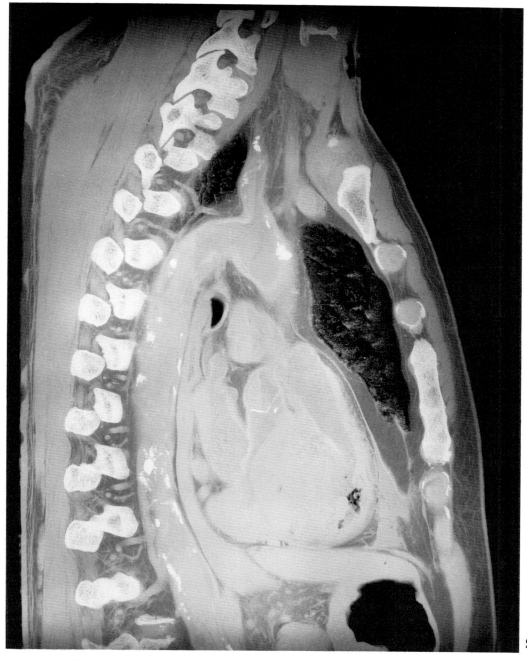

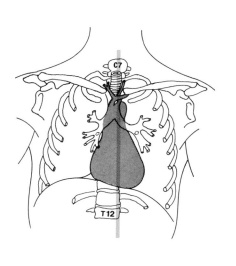

**Specimen Radiograph**

**Magnetic Resonance**

**Trispiral Tomogram**

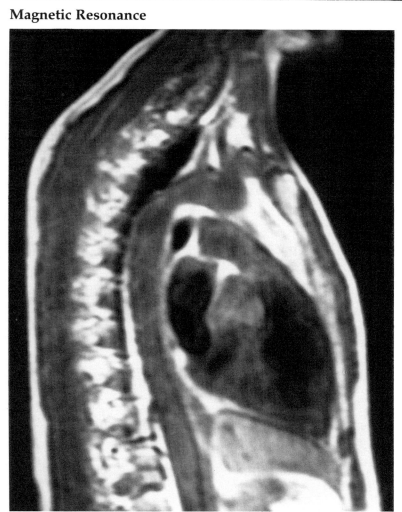

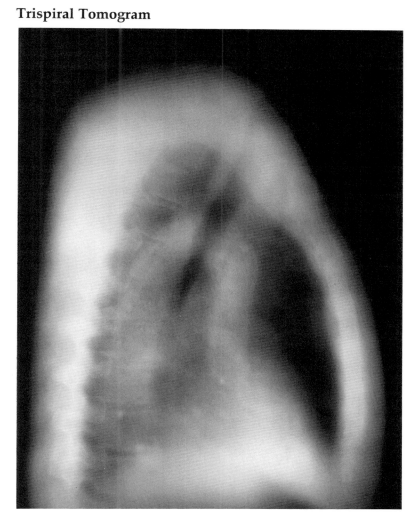

# Sagittal

## Section 24

## Anatomic Key

### Arteries
A 1 pulmonary trunk
A 29 aortic arch
A 30 aorta, ascending
A 31 aorta, descending
A 37 carotid, common, left
A 45 intercostal, posterior, left
A 49 subclavian, left
A 59 thyrocervical trunk, left

### Bronchi
B 16 main, left

### Esophagus
E 2 esophagogastric junction

### Fat
f 2 epicardial
f 3 extrapleural
f 4 mediastinal

### Fissures
F 1 left major (oblique)

### Glands
G 4 thyroid

### Heart
H 0 arteriosum, ligamentum
H 1 atrium, left
H 6 cardiac vein, middle
H 9 coronary artery, posterior descending
H 10 coronary artery, left
H 11 coronary artery, right
H 12 coronary sinus
H 16 septum, interventricular
H 19 valve, mitral
H 21 ventricle, left
H 22 ventricle, right

### Lobes
L 15 LUL, apicoposterior segment
L 16 LUL, anterior segment
L 18 LUL, superior lingular segment
L 21 LLL, superior segment
L 24 LLL, posterior basal segment

### Muscles
M 3 abdominis, rectus
M 4 abdominis, transversus
M 9 diaphragm
M 10 diaphragm, central tendon
M 12 diaphragm, crus, right
M 13 erector spinae
M 16 intercostal
M 21 latissimus dorsi
M 31 pectoralis, major
M 33 platysma
M 35 rhomboid, major
M 36 rhomboid, minor
M 38 scalene, anterior
M 41 semispinalis, capitis
M 50 splenius, capitis
M 51 splenius, cervicis
M 52 sternocleidomastoid
M 53 sternohyoid
M 54 sternothyroid
M 62 transversus thoracis
M 63 trapezius

### Neural Structures
n 2 intercostal nerve

### Nodes
N 6 mediastinal, anterior
N 10 supraclavicular

### Organs
O 3 liver
O 6 stomach

### Pleural Structures
P 2 costal (parietal)
P 6 visceral

### Skeletal Structures
S 1 clavicle
S 3 rib (8th)
S 6 rib, body, posterior (11th)
S 7 rib, costal cartilage (1st)
S 8 rib, head and neck (1st)
S 9 rib, tubercle (5th)
S 18 sternum, manubrium
S 29 vertebra, transverse process (T5)

### Veins
v 2 pulmonary, superior, LUL
v 5 pulmonary, inferior, LLL
v 8 brachiocephalic (innominate), left
v 12 hemiazygos
v 16 intercostal, anterior, left
v 18 intercostal, posterior, left
v 22 jugular, anterior, left
v 35 thyroid, inferior, left
v 42 vertebral, left

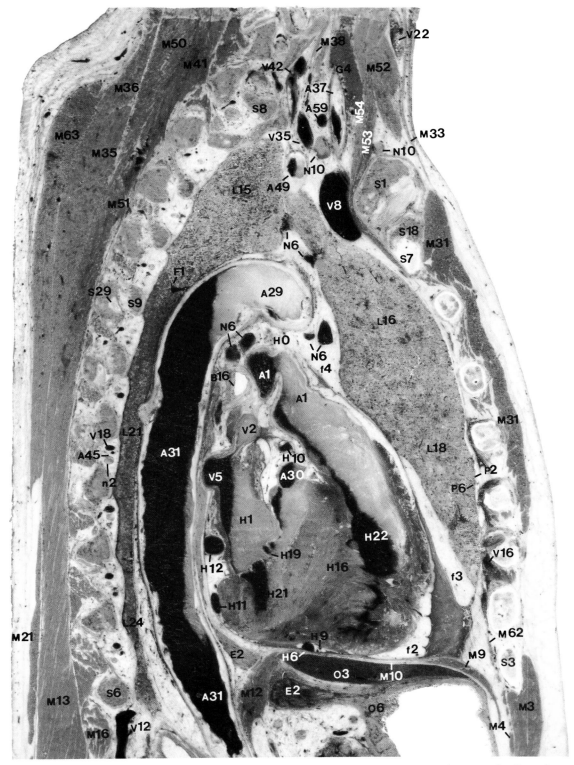

**Anatomic Specimen**

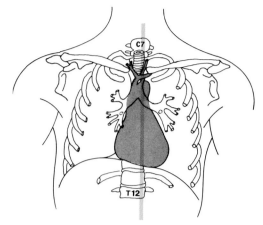

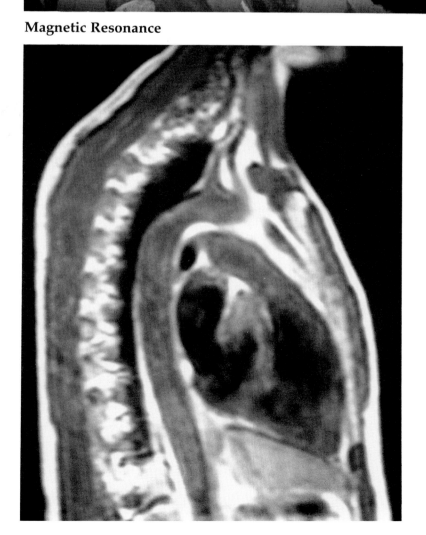

**Magnetic Resonance**

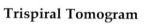

**Specimen Radiograph**

**Trispiral Tomogram**

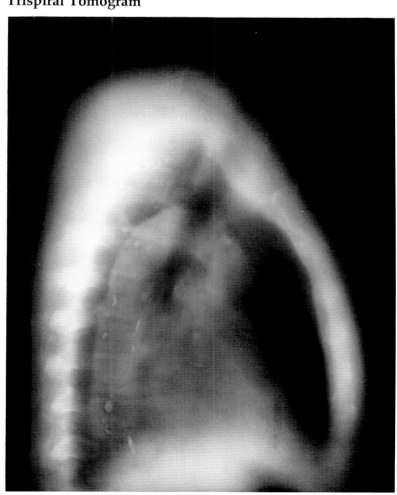

# Sagittal

## Section 25

## Anatomic Key

### Arteries
A  1  pulmonary trunk
A 16  main pulmonary, left
A 29  aortic arch
A 31  aorta, descending
A 45  intercostal, posterior, left
A 49  subclavian, left
A 59  thyrocervical trunk, left

### Bronchi
B 16  main, left

### Esophagus
E  2  esophagogastric junction

### Fat
f  2  epicardial
f  3  extrapleural
f  4  mediastinal
f  6  perirenal

### Fissures
F  1  left major (oblique)

### Heart
H  0  arteriosum, ligamentum
H  1  atrium, left
H  3  auricle, left
H  6  cardiac vein, middle
H  9  coronary artery, posterior
       descending
H 10  coronary artery, left
H 11  coronary artery, right
H 12  coronary sinus
H 13  pericardium, parietal serous
H 14  pericardium, visceral serous
H 16  septum, interventricular
H 19  valve, mitral
H 21  ventricle, left
H 22  ventricle, right

### Lobes
L 15  LUL, apicoposterior segment
L 16  LUL, anterior segment
L 18  LUL, superior lingular segment
L 21  LLL, superior segment
L 24  LLL, posterior basal segment

### Muscles
M  3  abdominis, rectus
M  4  abdominis, transversus
M  9  diaphragm
M 10  diaphragm, central tendon
M 11  diaphragm, crus, left
M 12  diaphragm, crus, right
M 13  erector spinae
M 16  intercostal
M 21  latissimus dorsi
M 31  pectoralis, major
M 33  platysma
M 35  rhomboid, major
M 36  rhomboid, minor
M 38  scalene, anterior
M 39  scalene, middle
M 41  semispinalis, capitis
M 50  splenius, capitis
M 51  splenius, cervicis
M 52  sternocleidomastoid
M 53  sternohyoid
M 54  sternothyroid
M 62  transversus thoracis
M 63  trapezius

### Neural Structures
n  1  brachial plexus
n  2  intercostal nerve

### Nodes
N  6  mediastinal, anterior
N 10  supraclavicular

### Organs
O  3  liver
O  6  stomach

### Pleural Structures
P  2  costal (parietal)
P  5  mediastinal (parietal)
P  6  visceral

### Skeletal Structures
S  1  clavicle
S  3  rib (1st)
S  6  rib, body, posterior (6th)
S  7  rib, costal cartilage (1st)

### Veins
v  2  pulmonary, superior, LUL
v  5  pulmonary, inferior, LLL
v  8  brachiocephalic (innominate),
       left
v 18  intercostal, posterior, left
v 26  jugular, internal, left
v 35  thyroid, inferior, left
v 42  vertebral, left

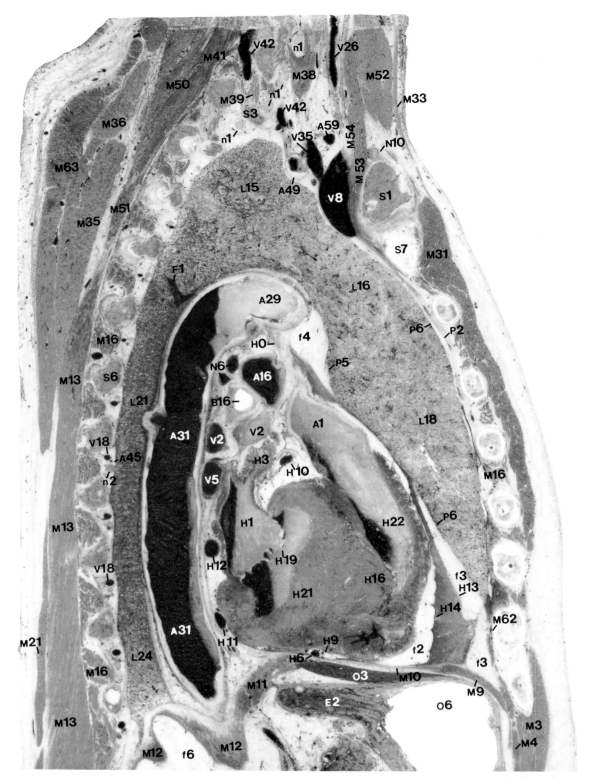

**Anatomic Specimen**

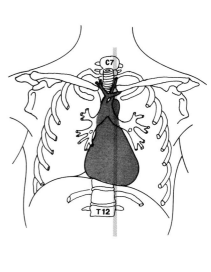

Specimen Radiograph

**Magnetic Resonance**

**Trispiral Tomogram**

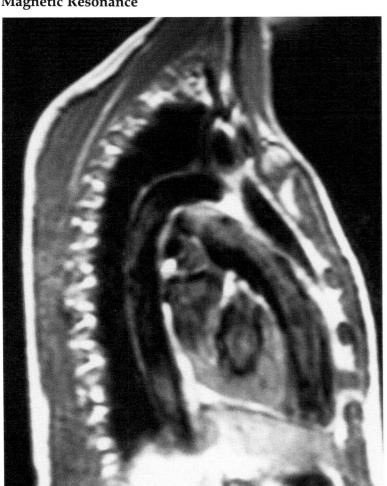

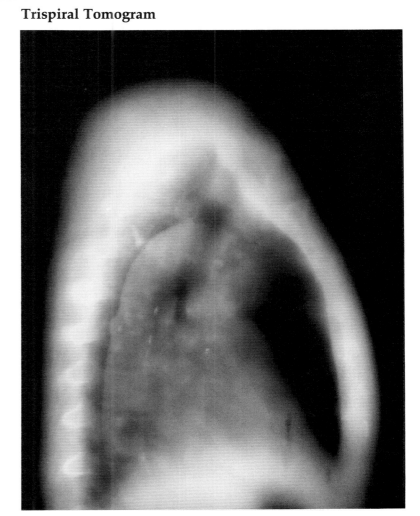

# Sagittal

## Section 26

### Anatomic Key

**Arteries**
A 1 pulmonary trunk
A 16 main pulmonary, left
A 29 aortic arch
A 31 aorta, descending
A 45 intercostal, posterior, left
A 49 subclavian, left
A 59 thyrocervical trunk, left

**Bronchi**
B 16 main, left

**Esophagus**
E 2 esophagogastric junction

**Fat**
f 2 epicardial
f 3 extrapleural
f 6 perirenal

**Fissures**
F 1 left major (oblique)

**Heart**
H 1 atrium, left
H 3 auricle, left
H 6 cardiac vein, middle
H 8 coronary artery, circumflex
H 9 coronary artery, anterior descending
H 11 coronary artery, right
H 12 coronary sinus
H 13 pericardium, parietal serous
H 14 pericardium, visceral serous
H 16 septum, interventricular
H 19 valve, mitral
H 21 ventricle, left
H 22 ventricle, right

**Lobes**
L 15 LUL, apicoposterior segment
L 16 LUL, anterior segment
L 18 LUL, superior lingular segment
L 21 LLL, superior segment
L 24 LLL, posterior basal segment

**Muscles**
M 3 abdominis, rectus
M 4 abdominis, transversus
M 9 diaphragm
M 10 diaphragm, central tendon
M 11 diaphragm, crus, left
M 13 erector spinae
M 16 intercostal
M 21 latissimus dorsi
M 31 pectoralis, major
M 33 platysma
M 35 rhomboid, major
M 36 rhomboid, minor
M 38 scalene, anterior
M 41 semispinalis, capitis
M 46 serratus, posterior superior
M 50 splenius, capitis
M 52 sternocleidomastoid
M 53 sternohyoid
M 62 transversus thoracis
M 63 trapezius

**Neural Structures**
n 1 brachial plexus
n 2 intercostal nerve

**Nodes**
N 6 mediastinal, anterior
N 10 supraclavicular

**Organs**
O 3 liver
O 6 stomach

**Pleural Structures**
P 2 costal (parietal)
P 5 mediastinal (parietal)
P 6 visceral

**Skeletal Structures**
S 1 clavicle
S 3 rib (1st)
S 6 rib, body, posterior (7th)
S 7 rib, costal cartilage (1st)

**Veins**
V 2 pulmonary, superior, LUL
V 5 pulmonary, inferior, LLL
V 8 brachiocephalic (innominate), left
V 16 intercostal, anterior, left
V 18 intercostal, posterior, left
V 26 jugular, internal, left
V 35 thyroid, inferior, left

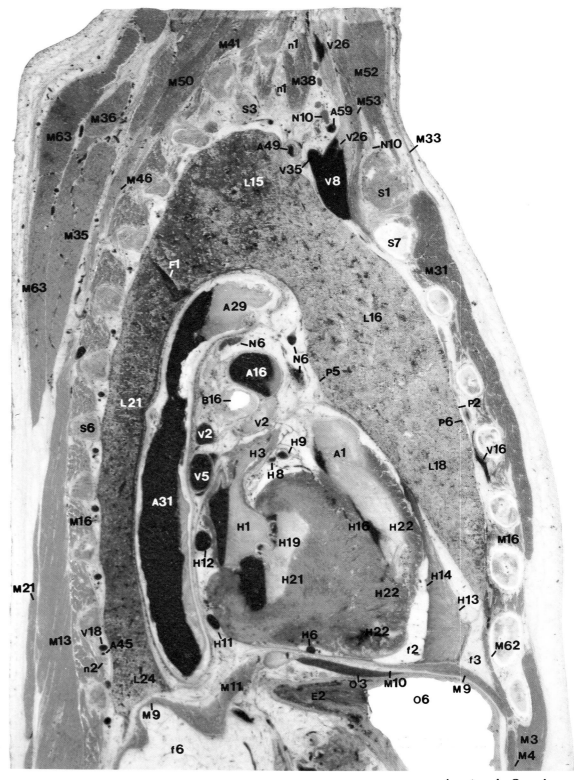

**Anatomic Specimen**

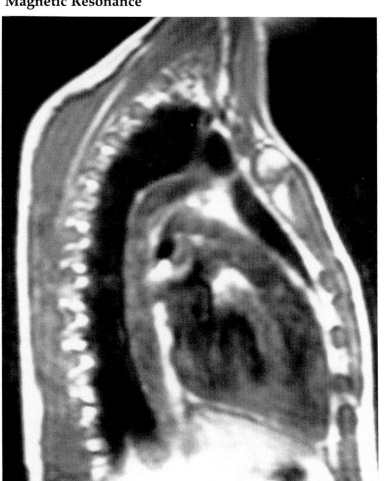

**Magnetic Resonance**

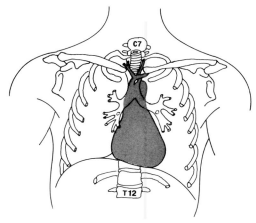

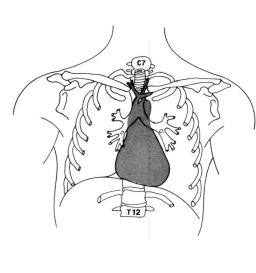

**Specimen Radiograph**

**Trispiral Tomogram**

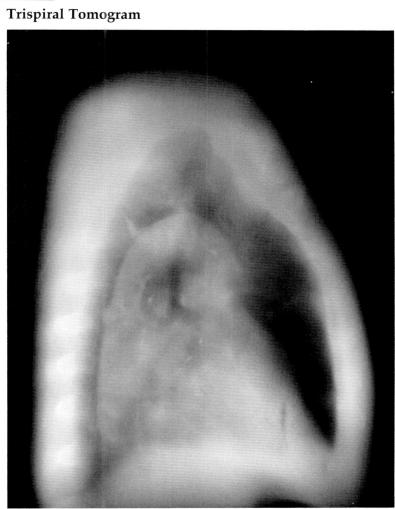

# Sagittal

## Section 27

### Anatomic Key

**Arteries**
A **16** main pulmonary, left
A **29** aortic arch
A **31** aorta, descending
A **35** bronchial, left
A **45** intercostal, posterior, left
A **49** subclavian, left
A **59** thyrocervical trunk, left

**Bronchi**
B **16** main, left
B **21** LUL, superior lingular segment

**Esophagus**
E **2** esophagogastric junction

**Fascia**
F **8** Sibson's

**Fat**
f **2** epicardial
f **3** extrapleural
f **6** perirenal

**Fissures**
F **1** left major (oblique)

**Heart**
H **1** atrium, left
H **3** auricle, left
H **5** cardiac vein, great
H **6** cardiac vein, middle
H **8** coronary artery, circumflex
H **9** coronary artery, anterior
       descending
H **12** coronary sinus
H **13** pericardium, parietal serous
H **14** pericardium, visceral serous
H **16** septum, interventricular
H **19** valve, mitral
H **21** ventricle, left
H **22** ventricle, right

**Lobes**
L **15** LUL, apicoposterior segment
L **16** LUL, anterior segment
L **18** LUL, superior lingular segment
L **21** LLL, superior segment
L **24** LLL, posterior basal segment

**Muscles**
M **3** abdominis, rectus
M **9** diaphragm
M **10** diaphragm, central tendon
M **11** diaphragm, crus, left
M **13** erector spinae
M **14** iliocostalis, thoracis
M **16** intercostal
M **21** latissimus dorsi
M **23** levator scapulae
M **31** pectoralis, major
M **33** platysma
M **35** rhomboid, major
M **36** rhomboid, minor
M **38** scalene, anterior
M **39** scalene, middle
M **40** scalene, posterior
M **46** serratus, posterior superior
M **50** splenius, capitis
M **52** sternocleidomastoid
M **54** sternothyroid
M **62** transversus thoracis
M **63** trapezius

**Neural Structures**
n **1** brachial plexus
n **2** intercostal nerve

**Nodes**
N **6** mediastinal, anterior
N **7** mediastinal, posterior
N **10** supraclavicular

**Organs**
O **2** kidney, left
O **6** stomach

**Pleural Structures**
P **2** costal (parietal)
P **5** mediastinal (parietal)
P **6** visceral

**Skeletal Structures**
S **1** clavicle
S **3** rib (1st)
S **6** rib, body, posterior (8th)
S **7** rib, costal cartilage (1st)

**Veins**
V **2** pulmonary, superior, LUL
V **5** pulmonary, inferior, LLL
V **8** brachiocephalic (innominate),
       left
V **16** intercostal, anterior, left
V **18** intercostal, posterior, left
V **26** jugular, internal, left

**Anatomic Specimen**

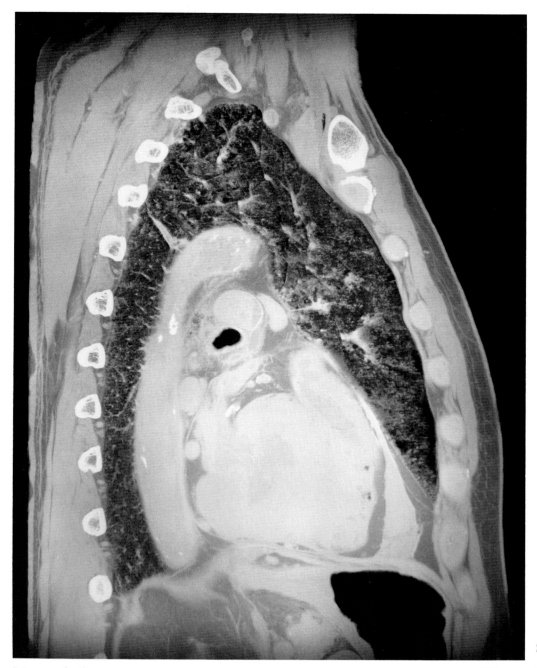

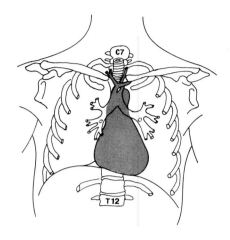

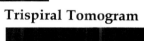

**Specimen Radiograph**

**Magnetic Resonance**

**Trispiral Tomogram**

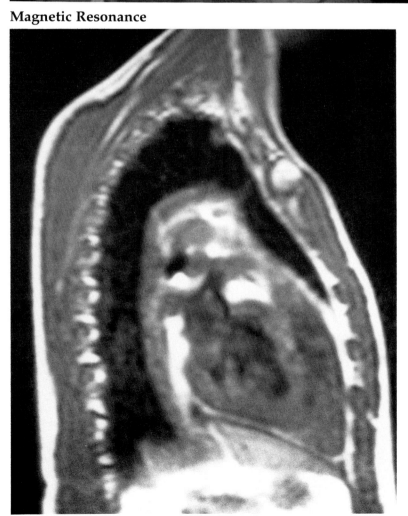

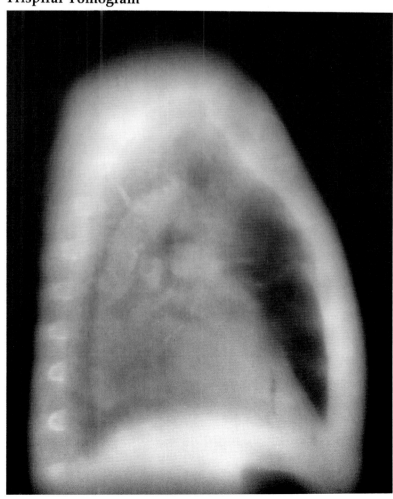

# Sagittal

## Section 28

## Anatomic Key

### Arteries
A 16 main pulmonary, left
A 23 left lower lobe (LLL)
A 29 aortic arch
A 31 aorta, descending
A 45 intercostal, posterior, left
A 49 subclavian, left
A 59 thyrocervical trunk, left

### Bronchi
B 17 left upper lobe (LUL)
B 21 LUL, superior lingular segment
B 23 left lower lobe (LLL)

### Fat
f 2 epicardial
f 3 extrapleural
f 6 perirenal

### Fissures
F 1 left major (oblique)

### Heart
H 1 atrium, left
H 3 auricle, left
H 6 cardiac vein, middle
H 7 cardiac vein, small
H 8 coronary artery, circumflex
H 9 coronary artery, anterior descending
H 12 coronary sinus
H 13 pericardium, parietal serous
H 14 pericardium, visceral serous
H 16 septum, interventricular
H 21 ventricle, left
H 22 ventricle, right

### Lobes
L 15 LUL, apicoposterior segment
L 16 LUL, anterior segment
L 18 LUL, superior lingular segment
L 19 LUL, inferior lingular segment
L 21 LLL, superior segment
L 22 LLL, anteromedial basal segment
L 24 LLL, posterior basal segment

### Muscles
M 3 abdominis, rectus
M 9 diaphragm
M 10 diaphragm, central tendon
M 13 erector spinae
M 14 iliocostalis, thoracis
M 16 intercostal
M 21 latissimus dorsi
M 23 levator scapulae
M 31 pectoralis, major
M 33 platysma
M 35 rhomboid, major
M 36 rhomboid, minor
M 38 scalene, anterior
M 39 scalene, middle
M 40 scalene, posterior
M 46 serratus, posterior superior
M 50 splenius, capitis
M 52 sternocleidomastoid
M 55 subclavius
M 62 transversus thoracis
M 63 trapezius

### Neural Structures
n 1 brachial plexus
n 2 intercostal nerve

### Organs
O 2 kidney, left
O 6 stomach

### Pleural Structures
P 2 costal (parietal)
P 3 costodiaphragmatic recess
P 5 mediastinal (parietal)
P 6 visceral

### Skeletal Structures
S 1 clavicle
S 3 rib (8th)
S 5 rib, body, lateral (1st)
S 6 rib, body, posterior (11th)
S 7 rib, costal cartilage (1st)

### Veins
V 2 pulmonary, superior, LUL
V 5 pulmonary, inferior, LLL
V 16 intercostal, anterior, left
V 18 intercostal, posterior, left
V 24 jugular, external, left
V 26 jugular, internal, left
V 30 subclavian, left

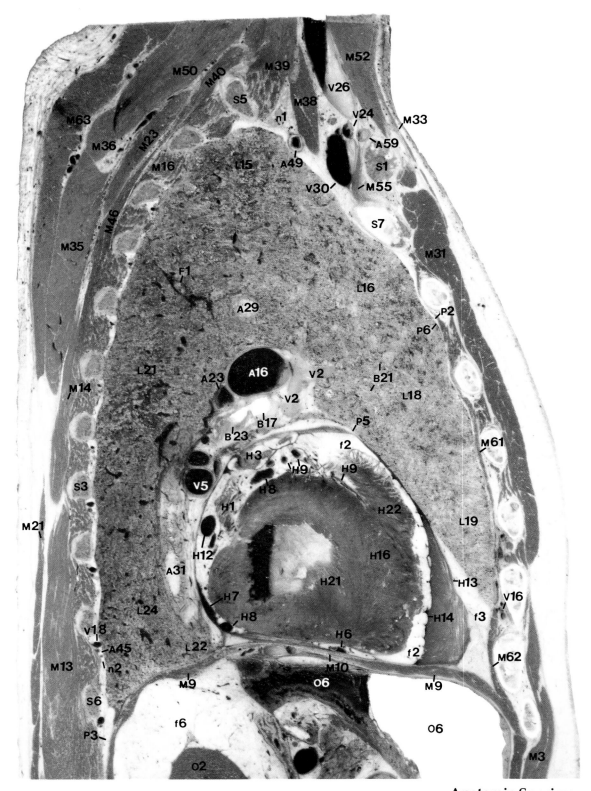

**Anatomic Specimen**

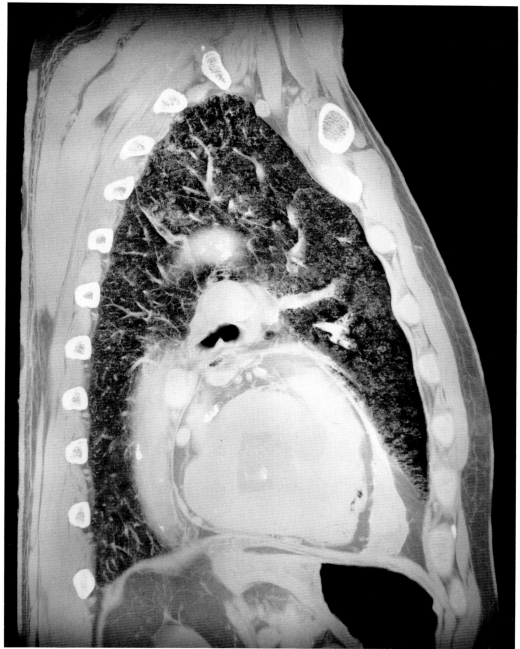

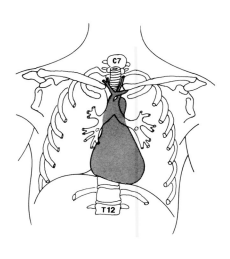

**Magnetic Resonance**

**Specimen Radiograph**

**Trispiral Tomogram**

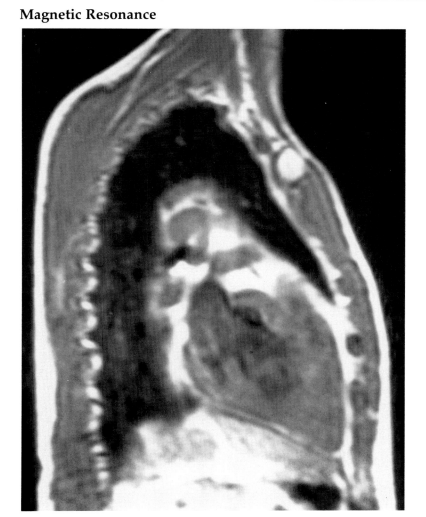

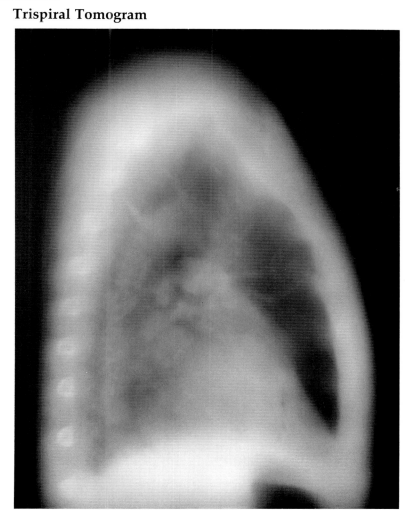

# Sagittal

## Section 29

## Anatomic Key

### Arteries
A 16 main pulmonary, left
A 17 left upper lobe (LUL)
A 22 LUL, inferior lingular branch
A 23 left lower lobe (LLL)
A 24 LLL, superior branch
A 27 LLL, posterior basal branch
A 45 intercostal, posterior, left
A 49 subclavian, left

### Bronchi
B 17 left upper lobe (LUL)
B 18 LUL, apicoposterior segment
B 19 LUL, anterior segment
B 21 LUL, superior lingular segment
B 22 LUL, inferior lingular segment
B 23 left lower lobe (LLL)
B 24 LLL, superior segment

### Fat
f 2 epicardial
f 3 extrapleural
f 6 perirenal

### Fissures
F 1 left major (oblique)

### Glands
G 1 adrenal, left
G 2 pancreas

### Heart
H 3 auricle, left
H 5 cardiac vein, great
H 8 coronary artery, circumflex
H 9 coronary artery, anterior descending
H 12 coronary sinus
H 13 pericardium, parietal serous
H 14 pericardium, visceral serous
H 21 ventricle, left
H 22 ventricle, right

### Lobes
L 15 LUL, apicoposterior segment
L 16 LUL, anterior segment
L 18 LUL, superior lingular segment
L 19 LUL, inferior lingular segment
L 21 LLL, superior segment
L 22 LLL, anteromedial basal segment
L 24 LLL, posterior basal segment

### Muscles
M 3 abdominis, rectus
M 9 diaphragm
M 13 erector spinae
M 14 iliocostalis, thoracis
M 16 intercostal
M 21 latissimus dorsi
M 30 omohyoid
M 31 pectoralis, major
M 33 platysma
M 35 rhomboid, major
M 36 rhomboid, minor
M 38 scalene, anterior
M 39 scalene, middle
M 40 scalene, posterior
M 46 serratus, posterior superior
M 50 splenius, capitis
M 52 sternocleidomastoid
M 62 transversus thoracis
M 63 trapezius

### Neural Structures
n 1 brachial plexus
n 2 intercostal nerve

### Nodes
N 10 supraclavicular

### Organs
O 5 spleen
O 6 stomach

### Pleural Structures
P 2 costal (parietal)
P 3 costodiaphragmatic recess
P 5 mediastinal (parietal)
P 6 visceral

### Skeletal Structures
S 1 clavicle
S 3 rib (1st)
S 6 rib, body, posterior (8th)
S 7 rib, costal cartilage (1st)

### Veins
v 2 pulmonary, superior, LUL
v 5 pulmonary, inferior, LLL
v 18 intercostal, posterior, left
v 24 jugular, external, left
v 26 jugular, internal, left
v 30 subclavian, left

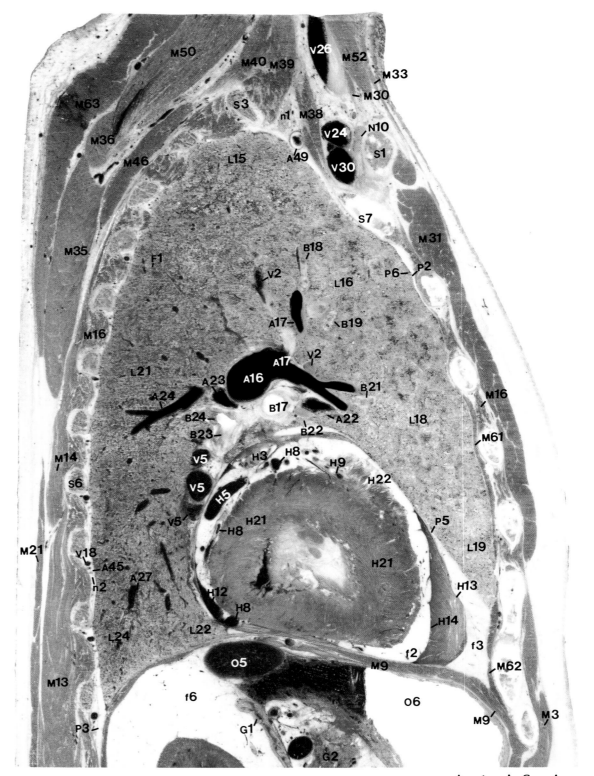

**Anatomic Specimen**

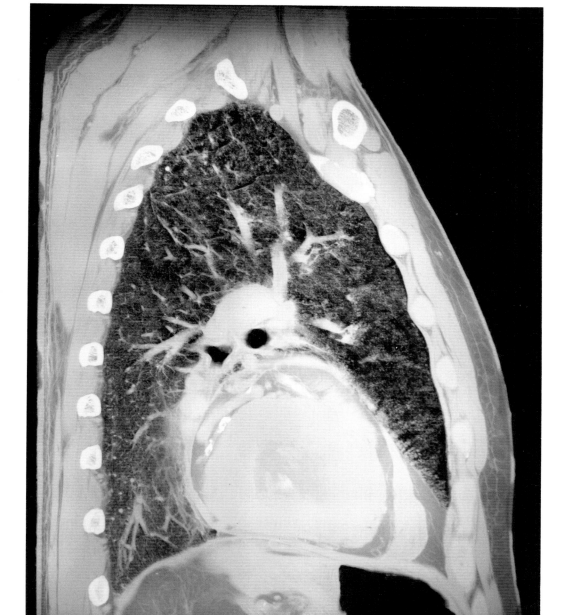

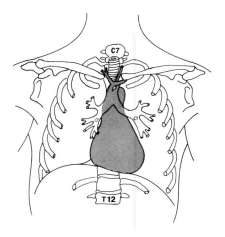

**Specimen Radiograph**

**Magnetic Resonance**

**Trispiral Tomogram**

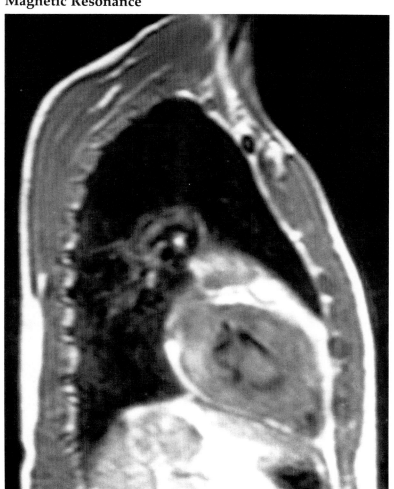

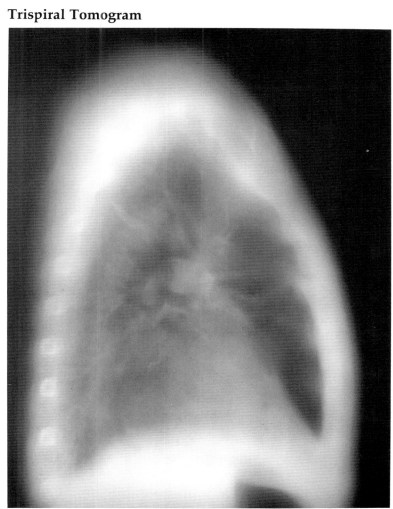

# Sagittal

## Section 30

### Anatomic Key

**Arteries**
A 16 main pulmonary, left
A 17 left upper lobe (LUL)
A 20 LUL, lingula
A 21 LUL, superior lingular branch
A 23 left lower lobe (LLL)
A 24 LLL, superior branch
A 27 LLL, posterior basal branch
A 49 subclavian, left

**Bronchi**
B 17 left upper lobe (LUL)
B 18 LUL, apicoposterior segment
B 21 LUL, superior lingular segment
B 22 LUL, inferior lingular segment
B 23 left lower lobe (LLL)
B 24 LLL, superior segment

**Fat**
f 2 epicardial
f 3 extrapleural
f 6 perirenal

**Fissures**
F 1 left major (oblique)

**Heart**
H 8 coronary artery, circumflex
H 9 coronary artery, anterior descending
H 12 coronary sinus
H 21 ventricle, left

**Lobes**
L 15 LUL, apicoposterior segment
L 16 LUL, anterior segment
L 18 LUL, superior lingular segment
L 19 LUL, inferior lingular segment
L 21 LLL, superior segment
L 22 LLL, anteromedial basal segment
L 24 LLL, posterior basal segment

**Muscles**
M 3 abdominis, rectus
M 9 diaphragm
M 13 erector spinae
M 16 intercostal
M 21 latissimus dorsi
M 30 omohyoid
M 31 pectoralis, major
M 33 platysma
M 35 rhomboid, major
M 36 rhomboid, minor
M 39 scalene, middle
M 40 scalene, posterior
M 44 serratus, anterior
M 46 serratus, posterior superior
M 50 splenius, capitis
M 52 sternocleidomastoid
M 58 supraspinatus
M 62 transversus thoracis
M 63 trapezius

**Neural Structures**
n 1 brachial plexus

**Organs**
o 5 spleen
o 6 stomach

**Pleural Structures**
P 2 costal (parietal)
P 3 costodiaphragmatic recess
P 5 mediastinal (parietal)
P 6 visceral

**Skeletal Structures**
s 1 clavicle
s 3 rib (10th)
s 6 rib, body, posterior (1st)
s 7 rib, costal cartilage (1st)
s 10 scapula (medial margin)

**Veins**
v 2 pulmonary, superior, LUL
v 5 pulmonary, inferior, LLL
v 18 intercostal, posterior, left
v 24 jugular, external, left
v 26 jugular, internal, left
v 30 subclavian, left

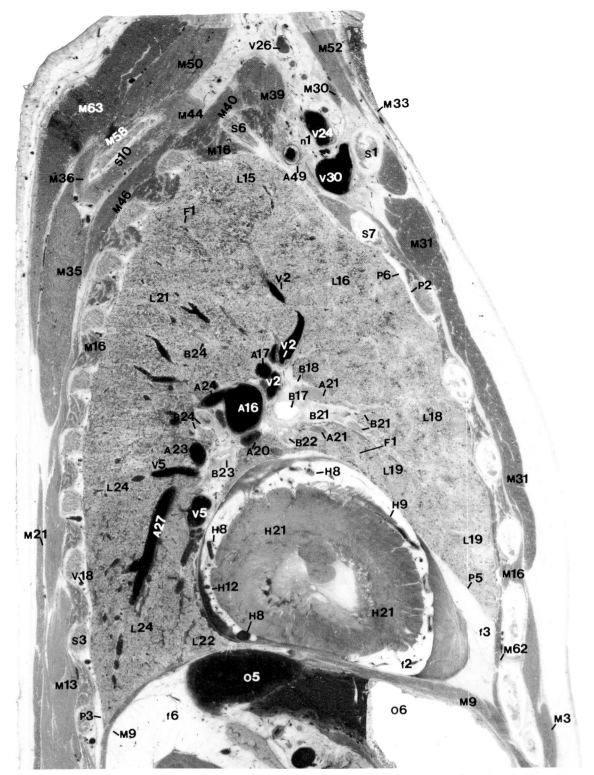

**Anatomic Specimen**

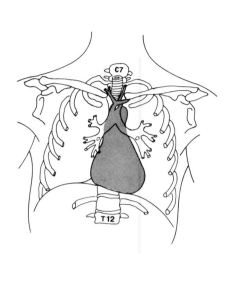

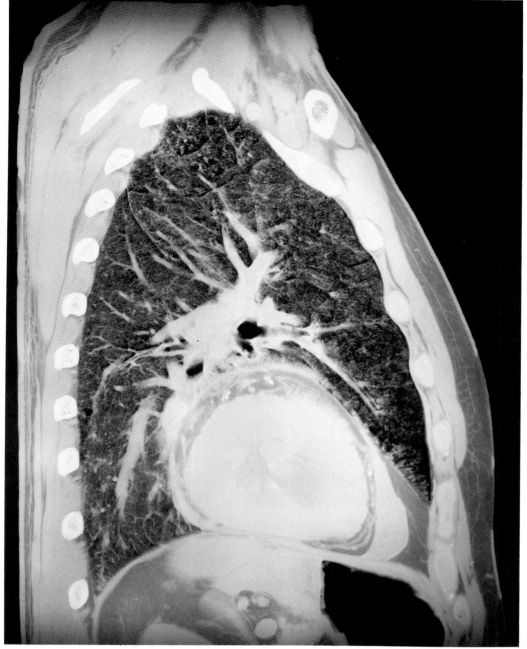

**Magnetic Resonance**

**Specimen Radiograph**

**Trispiral Tomogram**

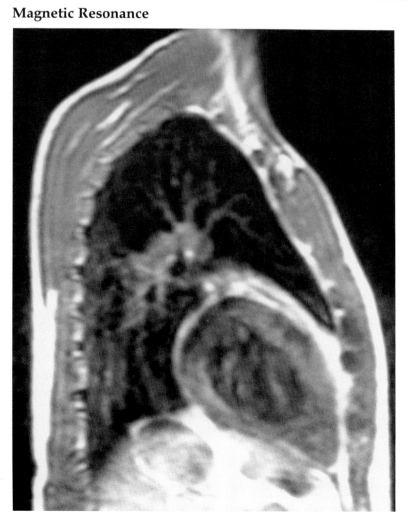

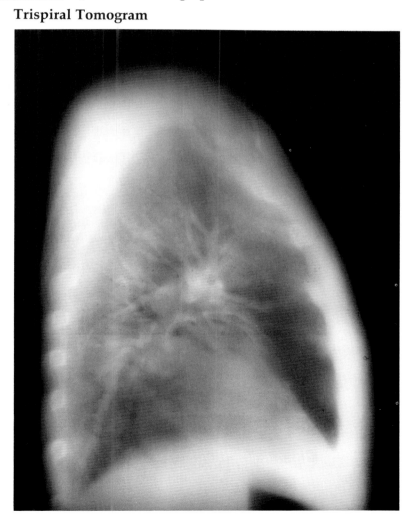

# Sagittal

## Section 31

### Anatomic Key

#### Arteries
A **16** main pulmonary, left
A **20** LUL, lingula
A **23** left lower lobe (LLL)
A **24** LLL, superior branch
A **27** LLL, posterior basal branch
A **45** intercostal, posterior, left
A **49** subclavian, left
A **57** thoracoacromial, left

#### Bronchi
B **17** left upper lobe (LUL)
B **18** LUL, apicoposterior segment
B **19** LUL, anterior segment
B **21** LUL, superior lingular segment
B **22** LUL, inferior lingular segment
B **23** left lower lobe (LLL)
B **24** LLL, superior segment
B **27** LLL, posterior basal segment

#### Fat
f **2** epicardial
f **3** extrapleural
f **6** perirenal

#### Fissures
F **1** left major (oblique)

#### Heart
H **8** coronary artery, circumflex
H **13** pericardium, parietal serous
H **14** pericardium, visceral serous
H **21** ventricle, left

#### Lobes
L **15** LUL, apicoposterior segment
L **16** LUL, anterior segment
L **18** LUL, superior lingular segment
L **19** LUL, inferior lingular segment
L **21** LLL, superior segment
L **22** LLL, anteromedial basal segment
L **24** LLL, posterior basal segment

#### Muscles
M **3** abdominis, rectus
M **9** diaphragm
M **13** erector spinae
M **16** intercostal
M **21** latissimus dorsi
M **30** omohyoid
M **31** pectoralis, major
M **33** platysma
M **35** rhomboid, major
M **39** scalene, middle
M **40** scalene, posterior
M **44** serratus, anterior
M **46** serratus, posterior superior
M **50** splenius, capitis
M **52** sternocleidomastoid
M **58** supraspinatus
M **62** transversus thoracis
M **63** trapezius

#### Neural Structures
n **1** brachial plexus

#### Organs
O **5** spleen
O **6** stomach

#### Pleural Structures
P **2** costal (parietal)
P **3** costodiaphragmatic recess
P **5** mediastinal (parietal)
P **6** visceral

#### Skeletal Structures
S **1** clavicle
S **3** rib (10th)
S **4** rib, body, anterior (1st)
S **6** rib, body, posterior (1st)
S **7** rib, costal cartilage (8th)
S **10** scapula

#### Veins
v **2** pulmonary, superior, LUL
v **5** pulmonary, inferior, LLL
v **18** intercostal, posterior, left
v **24** jugular, external, left
v **30** subclavian, left
v **34** thoracoacromial, left

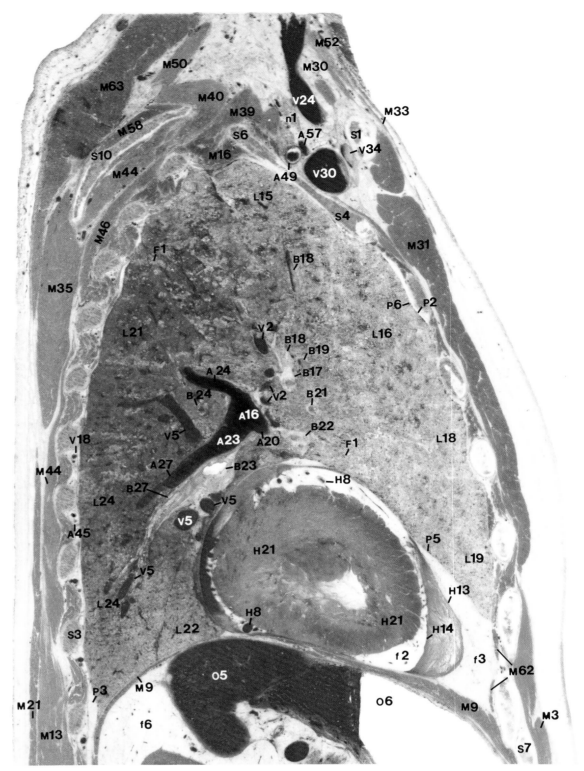

**Anatomic Specimen**

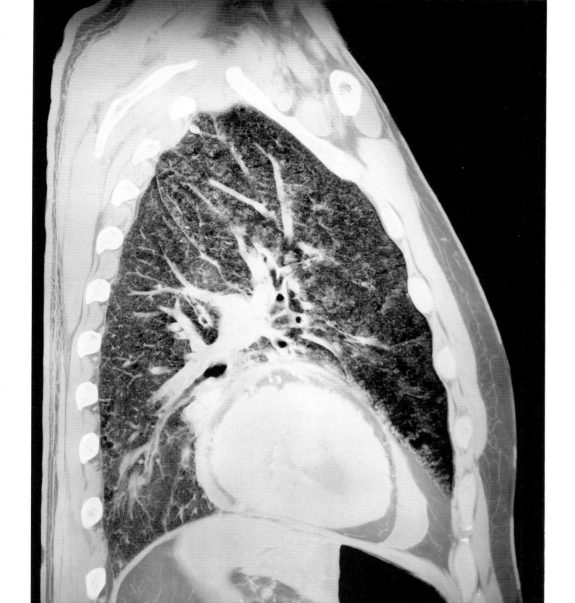

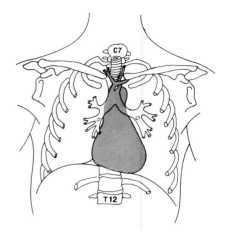

**Specimen Radiograph**

**Magnetic Resonance**

**Trispiral Tomogram**

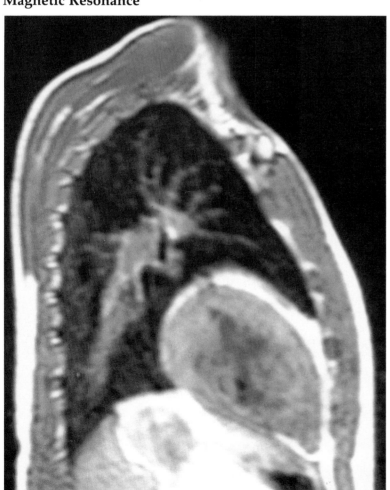

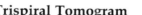

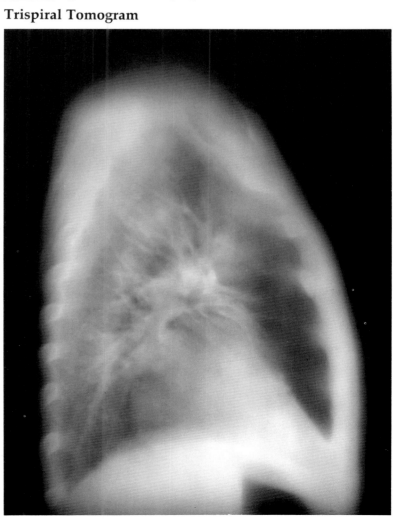

# Sagittal

## Section 32

### Anatomic Key

**Arteries**
A 23 left lower lobe (LLL)
A 27 LLL, posterior basal branch
A 49 subclavian, left
A 53 suprascapular, left

**Bronchi**
B 18 LUL, apicoposterior segment
B 19 LUL, anterior segment
B 21 LUL, superior lingular segment
B 22 LUL, inferior lingular segment
B 23 left lower lobe (LLL)
B 24 LLL, superior segment
B 25 LLL, anteromedial basal
    segment
B 26 LLL, lateral basal segment
B 27 LLL, posterior basal segment

**Fat**
f 2 epicardial
f 3 extrapleural
f 6 perirenal

**Fissures**
F 1 left major (oblique)

**Heart**
H 5 cardiac vein, great
H 8 coronary artery, circumflex
H 13 pericardium, parietal serous
H 14 pericardium, visceral serous
H 21 ventricle, left

**Lobes**
L 15 LUL, apicoposterior segment
L 16 LUL, anterior segment
L 18 LUL, superior lingular segment
L 19 LUL, inferior lingular segment
L 21 LLL, superior segment
L 22 LLL, anteromedial basal
    segment
L 24 LLL, posterior basal segment

**Muscles**
M 3 abdominis, rectus
M 9 diaphragm
M 13 erector spinae
M 15 infraspinatus
M 16 intercostal
M 21 latissimus dorsi
M 30 omohyoid
M 31 pectoralis, major
M 32 pectoralis, minor
M 33 platysma
M 35 rhomboid, major
M 44 serratus, anterior
M 46 serratus, posterior superior
M 57 subscapularis
M 58 supraspinatus
M 62 transversus thoracis
M 63 trapezius

**Neural Structures**
n 1 brachial plexus

**Nodes**
N 2 bronchopulmonary (hilar)

**Organs**
O 5 spleen
O 6 stomach

**Pleural Structures**
P 2 costal (parietal)
P 3 costodiaphragmatic recess
P 5 mediastinal (parietal)
P 6 visceral

**Skeletal Structures**
S 1 clavicle
S 3 rib (1st)
S 4 rib, body, anterior (3rd)
S 6 rib, body, posterior (10th)
S 7 rib, costal cartilage (7th)
S 15 scapula, spine

**Veins**
V 3 pulmonary, superior, LUL,
    lingula
V 5 pulmonary, inferior, LLL
V 18 intercostal, posterior, left
V 24 jugular, external, left
V 30 subclavian, left
V 34 thoracoacromial, left

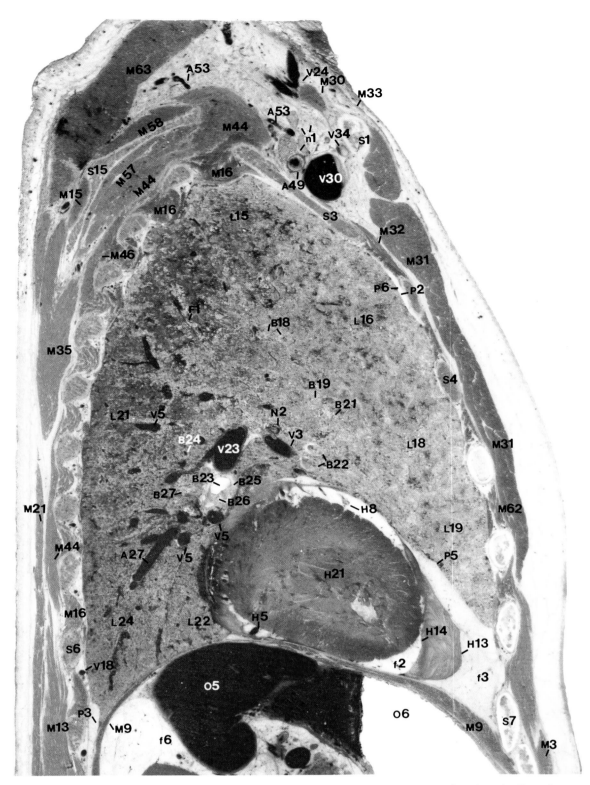

**Anatomic Specimen**

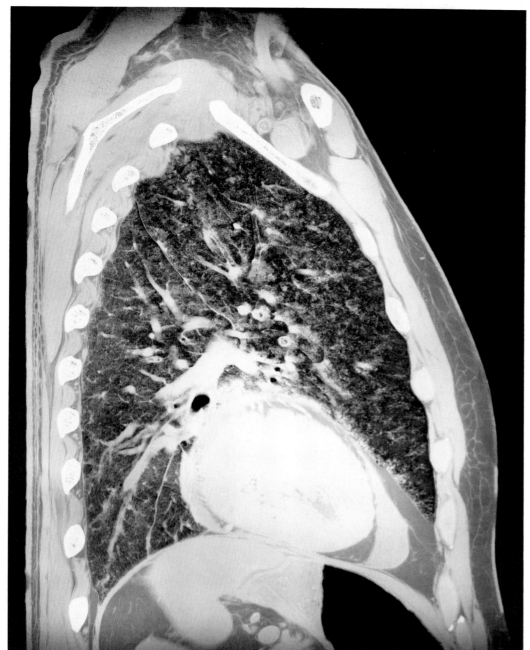

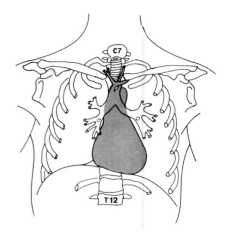

**Magnetic Resonance**

**Specimen Radiograph**

**Trispiral Tomogram**

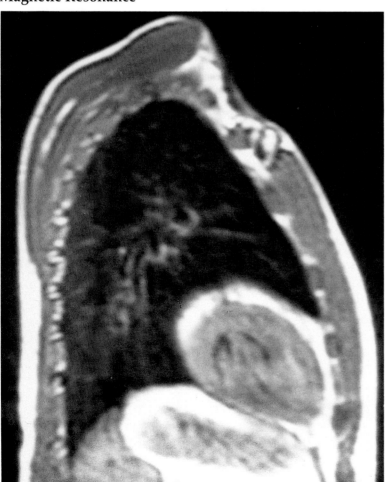

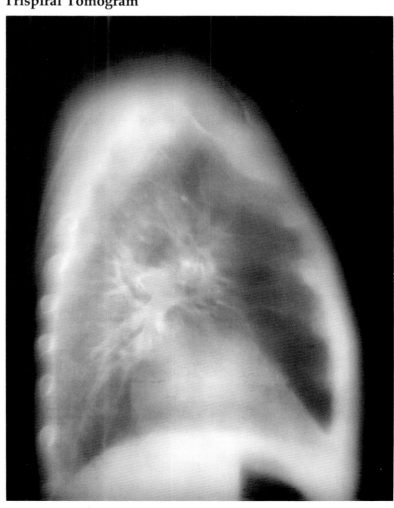

# Sagittal

## Section 33

### Anatomic Key

**Arteries**

A 18 LUL, apicoposterior branch
A 23 left lower lobe (LLL)
A 24 LLL, superior branch
A 49 subclavian, left

**Bronchi**

B 18 LUL, apicoposterior segment
B 19 LUL, anterior segment
B 21 LUL, superior lingular segment
B 22 LUL, inferior lingular segment
B 24 LLL, superior segment
B 25 LLL, anteromedial basal
        segment
B 26 LLL, lateral basal segment
B 27 LLL, posterior basal segment

**Fat**

f  2 epicardial
f  3 extrapleural
f  6 perirenal

**Fissures**

F  1 left major (oblique)

**Heart**

H 13 pericardium, parietal serous
H 14 pericardium, visceral serous
H 21 ventricle, left

**Lobes**

L 15 LUL, apicoposterior segment
L 16 LUL, anterior segment
L 18 LUL, superior lingular segment
L 19 LUL, inferior lingular segment
L 21 LLL, superior segment
L 22 LLL, anteromedial basal
        segment
L 24 LLL, posterior basal segment

**Muscles**

M  9 diaphragm
M 13 erector spinae
M 15 infraspinatus
M 16 intercostal
M 21 latissimus dorsi
M 30 omohyoid
M 31 pectoralis, major
M 32 pectoralis, minor
M 33 platysma
M 35 rhomboid, major
M 44 serratus, anterior
M 56 subcostal
M 57 subscapularis
M 58 supraspinatus
M 62 transversus thoracis
M 63 trapezius

**Neural Structures**

n  1 brachial plexus

**Organs**

o  5 spleen
o  6 stomach

**Pleural Structures**

P  3 costodiaphragmatic recess
P  5 mediastinal (parietal)

**Skeletal Structures**

s  1 clavicle
s  3 rib (2nd)
s  4 rib, body, anterior (2nd)
s  5 rib, body, lateral (1st)
s  6 rib, body, posterior (8th)
s  7 rib, costal cartilage (7th)
s 10 scapula
s 15 scapula, spine

**Veins**

v  3 pulmonary, superior, LUL,
        lingula
v  5 pulmonary, inferior, LLL
v 18 intercostal, posterior, left
v 30 subclavian, left
v 34 thoracoacromial, left

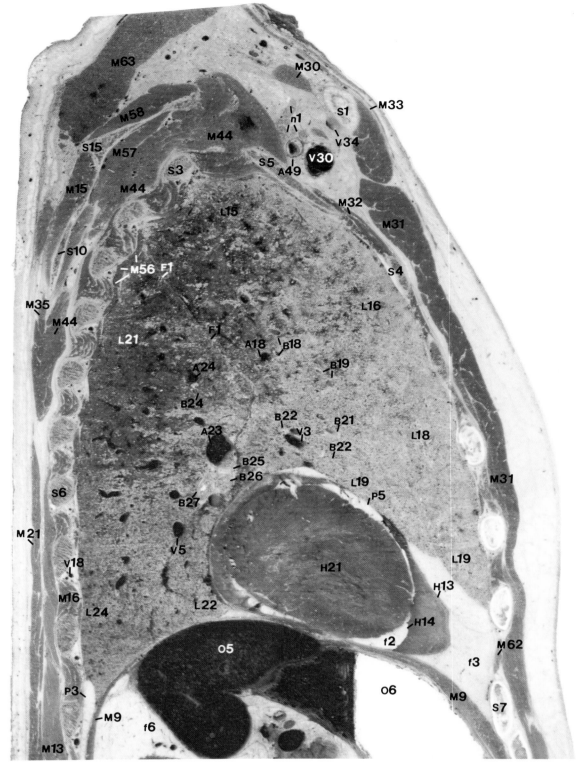

**Anatomic Specimen**

200

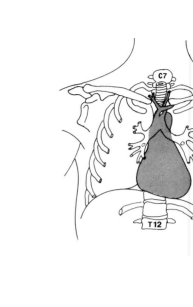

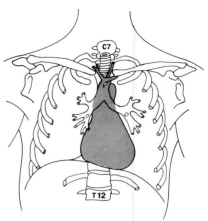

**Specimen Radiograph**

**Magnetic Resonance**

**Trispiral Tomogram**

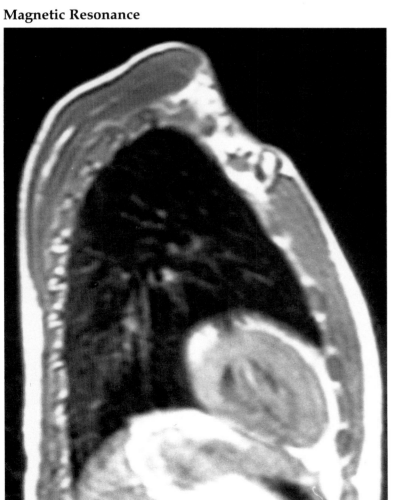

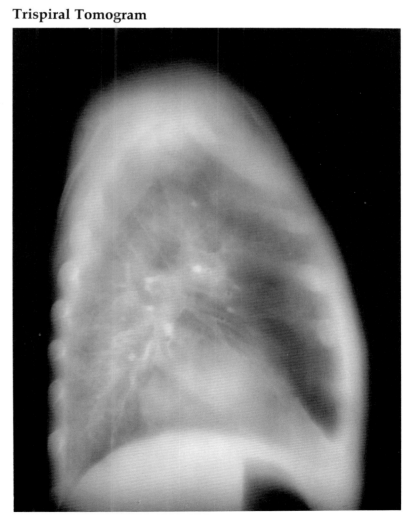

# Sagittal

## Section 34

### Anatomic Key

**Arteries**
A 24 LLL, superior branch
A 25 LLL, anteromedial basal branch
A 26 LLL, lateral basal branch
A 27 LLL, posterior basal branch
A 49 subclavian, left
A 53 suprascapular, left

**Bronchi**
B 18 LUL, apicoposterior segment
B 21 LUL, superior lingular segment
B 22 LUL, inferior lingular segment
B 25 LLL, anteromedial basal segment
B 26 LLL, lateral basal segment
B 27 LLL, posterior basal segment

**Fat**
f 2 epicardial
f 3 extrapleural
f 6 perirenal

**Fissures**
F 1 left major (oblique)

**Heart**
H 13 pericardium, parietal serous
H 14 pericardium, visceral serous
H 21 ventricle, left

**Lobes**
L 15 LUL, apicoposterior segment
L 16 LUL, anterior segment
L 18 LUL, superior lingular segment
L 19 LUL, inferior lingular segment
L 21 LLL, superior segment
L 22 LLL, anteromedial basal segment
L 24 LLL, posterior basal segment

**Muscles**
M 3 abdominis, rectus
M 9 diaphragm
M 15 infraspinatus
M 16 intercostal
M 21 latissimus dorsi
M 30 omohyoid
M 31 pectoralis, major
M 32 pectoralis, minor
M 33 platysma
M 44 serratus, anterior
M 55 subclavius
M 57 subscapularis
M 58 supraspinatus
M 62 transversus thoracis
M 63 trapezius

**Neural Structures**
n 1 brachial plexus

**Organs**
O 5 spleen
O 6 stomach

**Pleural Structures**
P 2 costal (parietal)
P 3 costodiaphragmatic recess
P 5 mediastinal (parietal)
P 6 visceral

**Skeletal Structures**
s 1 clavicle
s 3 rib (2nd)
s 4 rib, body, anterior (2nd)
s 5 rib, body, lateral (1st)
s 6 rib, body, posterior (8th)
s 7 rib, costal cartilage (5th)
s 10 scapula
s 15 scapula, spine

**Veins**
v 5 pulmonary, inferior, LLL
v 18 intercostal, posterior, left
v 30 subclavian, left
v 33 suprascapular, right
v 34 thoracoacromial, left

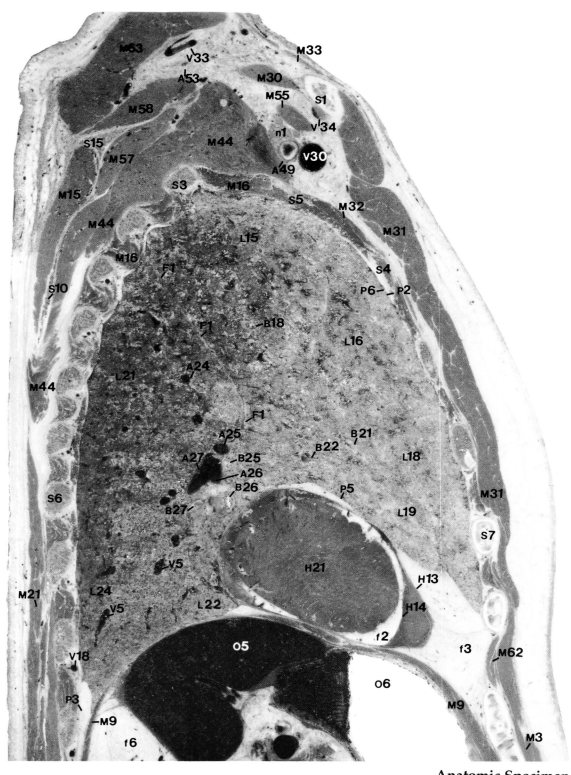

**Anatomic Specimen**

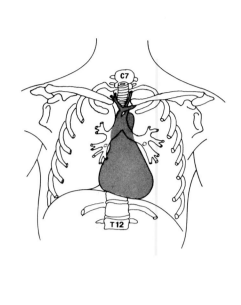

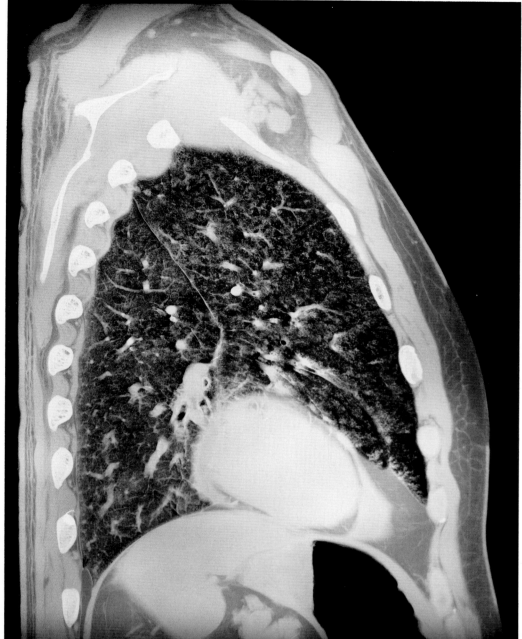

**Specimen Radiograph**

**Magnetic Resonance**

**Trispiral Tomogram**

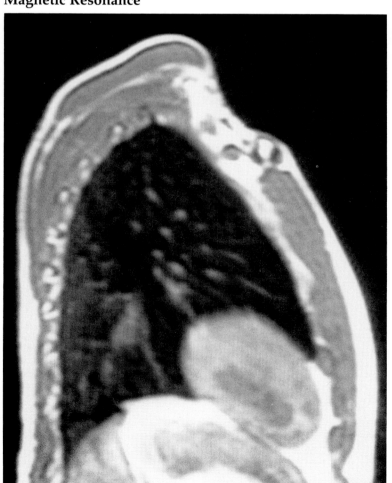

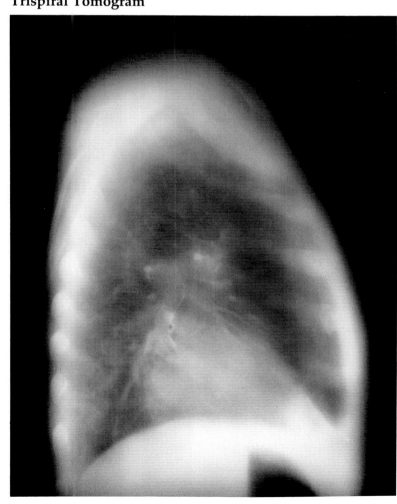

# Sagittal

## Section 35

## Anatomic Key

### Arteries
A 24 LLL, superior branch
A 25 LLL, anteromedial basal branch
A 26 LLL, lateral basal branch
A 27 LLL, posterior basal branch
A 32 axillary, left
A 53 suprascapular, left

### Bronchi
B 18 LUL, apicoposterior segment
B 21 LUL, superior lingular segment
B 22 LUL, inferior lingular segment
B 25 LLL, anteromedial basal
  segment
B 26 LLL, lateral basal segment
B 27 LLL, posterior basal segment

### Fat
f 2 epicardial
f 3 extrapleural
f 6 perirenal

### Fissures
F 1 left major (oblique)

### Glands
G 2 pancreas

### Heart
H 13 pericardium, parietal serous
H 14 pericardium, visceral serous
H 21 ventricle, left

### Lobes
L 15 LUL, apicoposterior segment
L 16 LUL, anterior segment
L 18 LUL, superior lingular segment
L 19 LUL, inferior lingular segment
L 21 LLL, superior segment
L 22 LLL, anteromedial basal
  segment
L 24 LLL, posterior basal segment

### Muscles
M 1 abdominal, external oblique
M 9 diaphragm
M 13 erector spinae
M 15 infraspinatus
M 16 intercostal
M 21 latissimus dorsi
M 31 pectoralis, major
M 32 pectoralis, minor
M 44 serratus, anterior
M 50 splenius, capitis
M 55 subclavius
M 57 subscapularis
M 58 supraspinatus
M 63 trapezius

### Neural Structures
n 1 brachial plexus

### Organs
O 5 spleen
O 6 stomach

### Pleural Structures
P 2 costal (parietal)
P 3 costodiaphragmatic recess
P 5 mediastinal (parietal)
P 6 visceral

### Skeletal Structures
S 1 clavicle
S 3 rib (11th)
S 4 rib, body, anterior (2nd)
S 6 rib, body, posterior (2nd)
S 7 rib, costal cartilage (5th)
S 10 scapula
S 15 scapula, spine

### Veins
V 3 pulmonary, superior, LUL,
  lingula
V 5 pulmonary, inferior, LLL
V 6 axillary, left
V 10 cephalic, left
V 18 intercostal, posterior, left
V 33 suprascapular, left
V 34 thoracoacromial, left

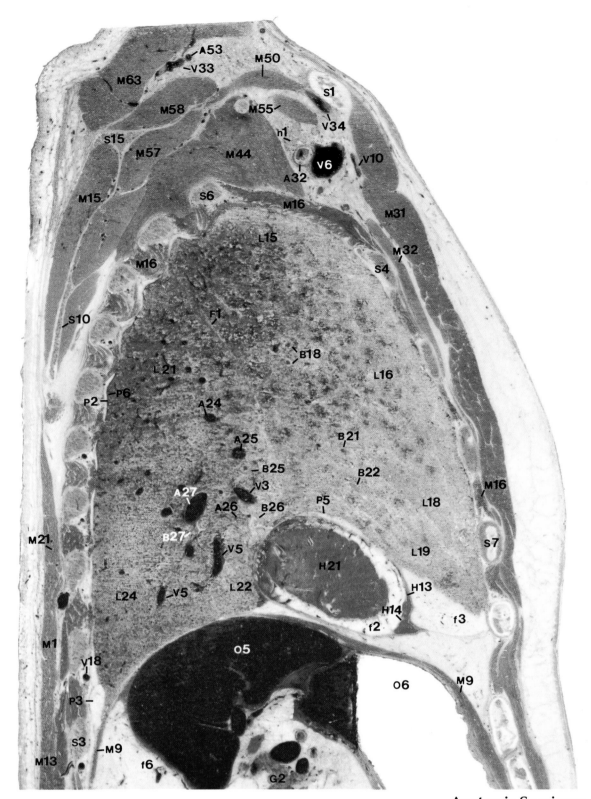

**Anatomic Specimen**

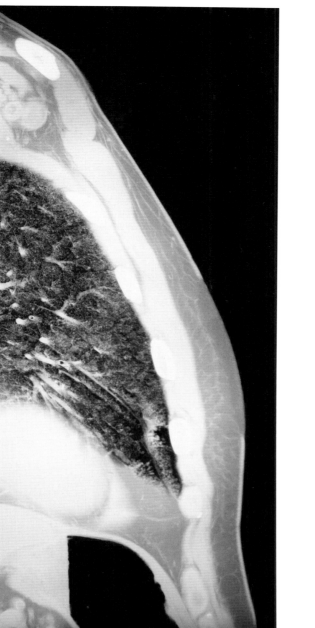

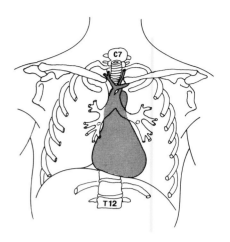

**Specimen Radiograph**

**Magnetic Resonance**

**Trispiral Tomogram**

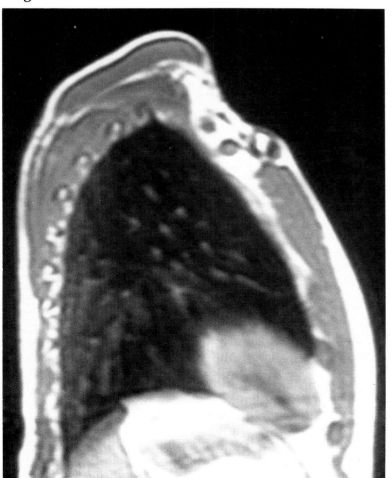

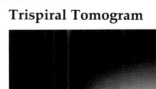

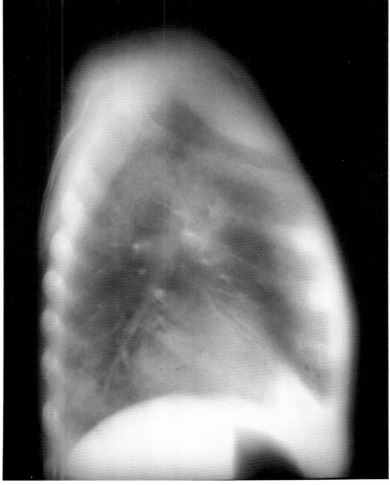

# Sagittal

## Section 36

### Anatomic Key

**Arteries**
A 25 LLL, anteromedial basal branch
A 26 LLL, lateral basal branch
A 27 LLL, posterior basal branch
A 32 axillary, left
A 53 suprascapular, left
A 57 thoracoacromial, left

**Bronchi**
B 18 LUL, apicoposterior segment
B 19 LUL, anterior segment
B 21 LUL, superior lingular segment
B 22 LUL, inferior lingular segment
B 24 LLL, superior segment
B 25 LLL, anteromedial basal segment
B 26 LLL, lateral basal segment
B 27 LLL, posterior basal segment

**Fat**
f 1 axillary
f 3 extrapleural

**Fissures**
F 1 left major (oblique)

**Lobes**
L 15 LUL, apicoposterior segment
L 16 LUL, anterior segment
L 18 LUL, superior lingular segment
L 19 LUL, inferior lingular segment
L 21 LLL, superior segment
L 22 LLL, anteromedial basal segment
L 23 LLL, lateral basal segment
L 24 LLL, posterior basal segment

**Muscles**
M 9 diaphragm
M 15 infraspinatus
M 16 intercostal
M 21 latissimus dorsi
M 31 pectoralis, major
M 32 pectoralis, minor
M 44 serratus, anterior
M 57 subscapularis
M 58 supraspinatus
M 63 trapezius

**Neural Structures**
n 1 brachial plexus

**Nodes**
N 1 axillary

**Organs**
O 5 spleen
O 6 stomach

**Pleural Structures**
P 3 costodiaphragmatic recess

**Skeletal Structures**
S 1 clavicle
S 3 rib (3rd)
S 4 rib, body, anterior (5th)
S 5 rib, body, lateral (2nd)
S 6 rib, body, posterior (10th)
S 7 rib, costal cartilage (7th)
S 10 scapula
S 15 scapula, spine

**Veins**
V 5 pulmonary, inferior, LLL
V 6 axillary, left
V 18 intercostal, posterior, left
V 33 suprascapular, left
V 34 thoracoacromial, left

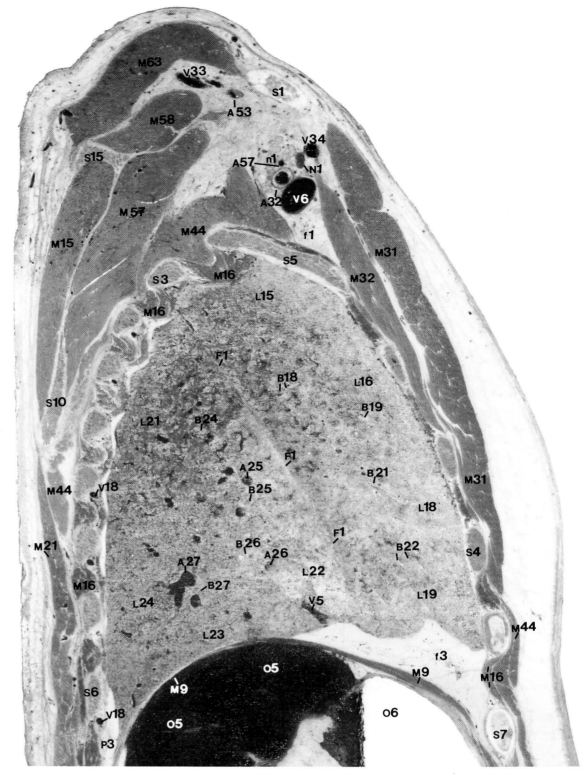

**Anatomic Specimen**

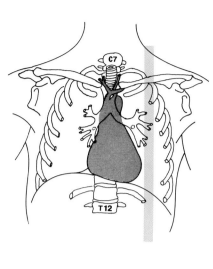

Specimen Radiograph

**Magnetic Resonance**

**Trispiral Tomogram**

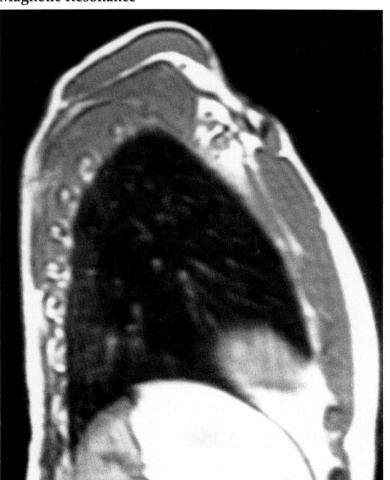

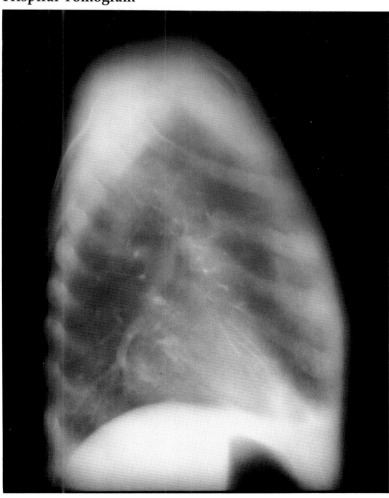

# Sagittal

## Section 37

## Anatomic Key

### Arteries
A 32 axillary, left
A 57 thoracoacromial, left

### Bronchi
B 24 LLL, superior segment
B 25 LLL, anteromedial basal
    segment

### Fat
f 1 axillary
f 3 extrapleural

### Fissures
F 1 left major (oblique)

### Lobes
L 15 LUL, apicoposterior segment
L 16 LUL, anterior segment
L 18 LUL, superior lingular segment
L 19 LUL, inferior lingular segment
L 21 LLL, superior segment
L 22 LLL, anteromedial basal
    segment
L 23 LLL, lateral basal segment
L 24 LLL, posterior basal segment

### Muscles
M 9 diaphragm
M 15 infraspinatus
M 16 intercostal
M 21 latissimus dorsi
M 31 pectoralis, major
M 32 pectoralis, minor
M 44 serratus, anterior
M 57 subscapularis
M 58 supraspinatus
M 59 teres major
M 63 trapezius

### Neural Structures
n 1 brachial plexus

### Nodes
N 1 axillary

### Organs
O 5 spleen
O 6 stomach

### Pleural Structures
P 2 costal (parietal)
P 3 costodiaphragmatic recess
P 6 visceral

### Skeletal Structures
s 1 clavicle
s 4 rib, body, anterior (3rd)
s 5 rib, body, lateral (3rd)
s 6 rib, body, posterior (10th)
s 7 rib, costal cartilage (7th)
s 10 scapula
s 15 scapula, spine

### Veins
v 6 axillary, left
v 10 cephalic, left
v 28 lateral thoracic, left
v 32 subscapular, left
v 33 suprascapular, left

### Nonanatomic
NB nylon bolt

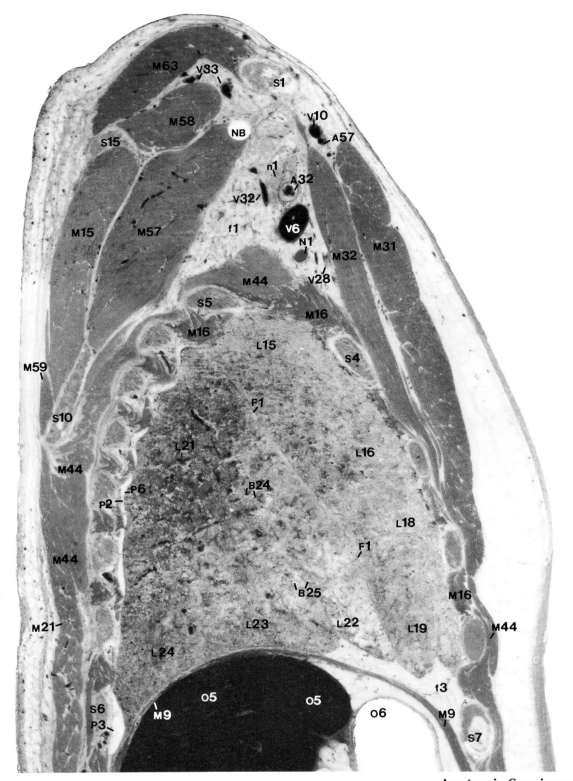

**Anatomic Specimen**

208

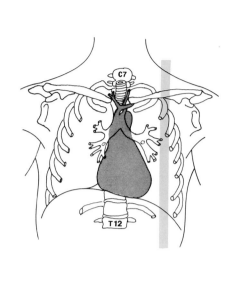

Specimen Radiograph

**Magnetic Resonance**

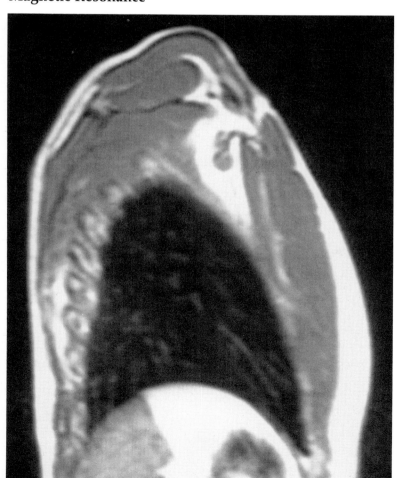

**Trispiral Tomogram**

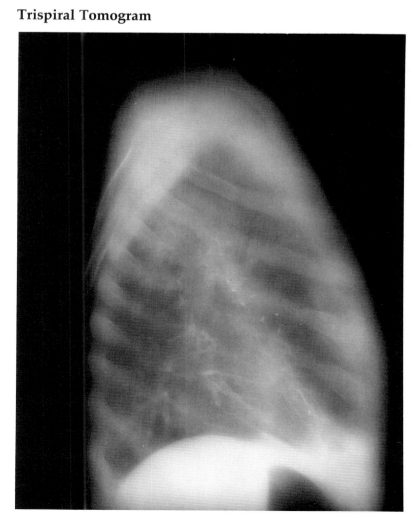

# Sagittal

## Section 38

### Anatomic Key

**Arteries**
A 32 axillary, left
A 45 intercostal, posterior, left
A 53 suprascapular, left
A 55 thoracic, lateral, left

**Bronchi**
B 25 LLL, anteromedial basal
        segment
B 26 LLL, lateral basal segment

**Fat**
f 1 axillary

**Fissures**
F 1 left major (oblique)

**Lobes**
L 15 LUL, apicoposterior segment
L 16 LUL, anterior segment
L 17 LUL, lingula
L 21 LLL, superior segment
L 22 LLL, anteromedial basal
        segment
L 23 LLL, lateral basal segment
L 24 LLL, posterior basal segment

**Muscles**
M 1 abdominal, external oblique
M 8 deltoid
M 9 diaphragm
M 15 infraspinatus
M 16 intercostal
M 21 latissimus dorsi
M 31 pectoralis, major
M 32 pectoralis, minor
M 44 serratus, anterior
M 57 subscapularis
M 58 supraspinatus
M 59 teres major
M 63 trapezius

**Neural Structures**
n 1 brachial plexus

**Organs**
o 5 spleen

**Skeletal Structures**
s 1 clavicle
s 4 rib, body, anterior (4th)
s 5 rib, body, lateral (3rd)
s 6 rib, body, posterior (10th)
s 7 rib, costal cartilage (7th)
s 10 scapula
s 15 scapula, spine

**Veins**
v 6 axillary, left
v 10 cephalic, left
v 18 intercostal, posterior, left
v 28 lateral thoracic, left
v 33 suprascapular, left

**Nonanatomic**
NB nylon bolt

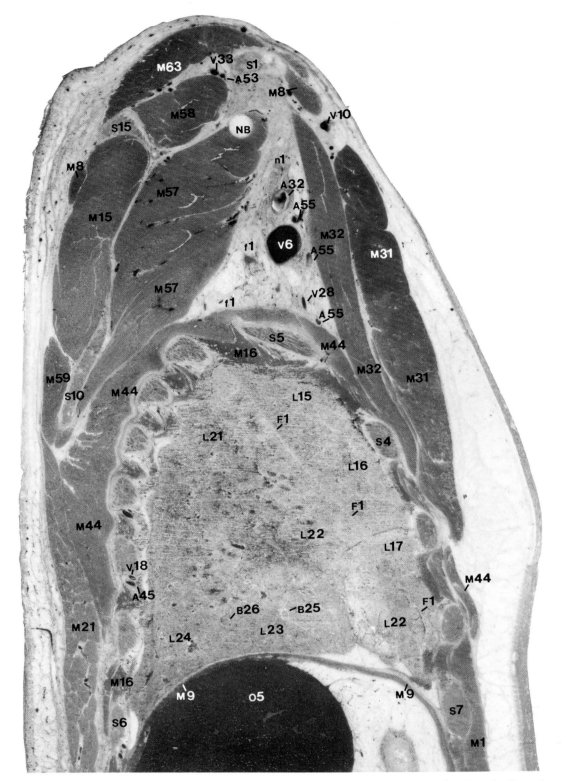

**Anatomic Specimen**

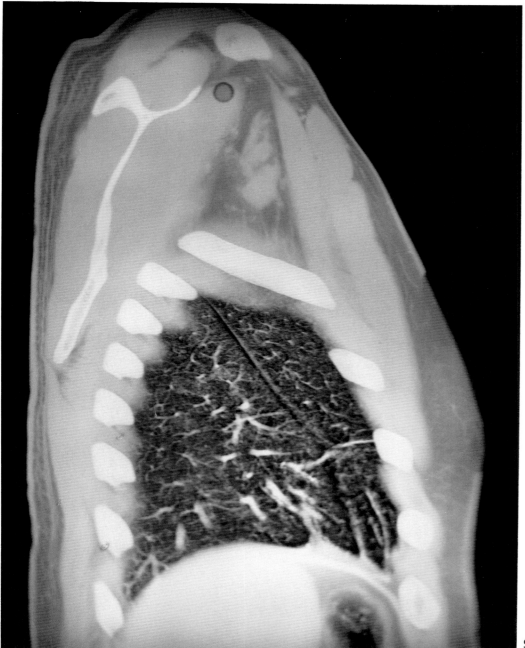

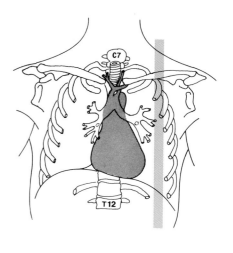

**Magnetic Resonance**

**Specimen Radiograph**

**Trispiral Tomogram**

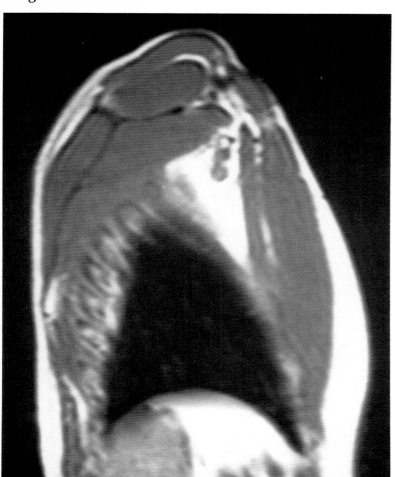

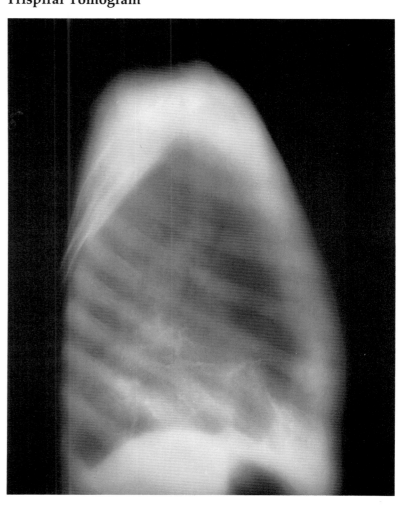

# Sagittal

## Section 39

### Anatomic Key

**Arteries**
A 32 axillary, left
A 55 thoracic, lateral, left

**Fat**
f 1 axillary

**Lobes**
L 21 LLL, superior segment
L 22 LLL, anteromedial basal segment
L 23 LLL, lateral basal segment
L 24 LLL, posterior basal segment

**Muscles**
M 1 abdominal, external oblique
M 8 deltoid
M 9 diaphragm
M 15 infraspinatus
M 16 intercostal
M 21 latissimus dorsi
M 31 pectoralis, major
M 32 pectoralis, minor
M 44 serratus, anterior
M 57 subscapularis
M 58 supraspinatus
M 59 teres major
M 63 trapezius

**Organs**
O 5 spleen

**Skeletal Structures**
S 4 rib, body, anterior (7th)
S 5 rib, body, lateral (4th)
S 6 rib, body, posterior (10th)
S 10 scapula
S 12 scapula, coracoid process
S 15 scapula, spine

**Veins**
V 6 axillary, left
V 10 cephalic, left
V 28 lateral thoracic, left
V 33 suprascapular, left

**Nonanatomic**
NB nylon bolt

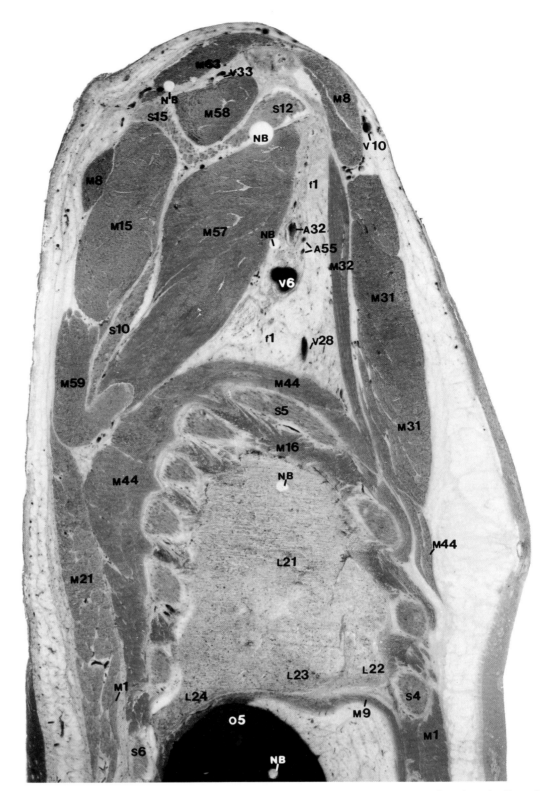

**Anatomic Specimen**

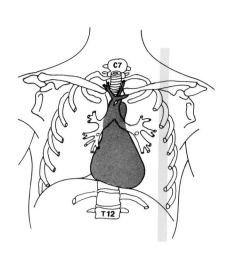

**Magnetic Resonance**

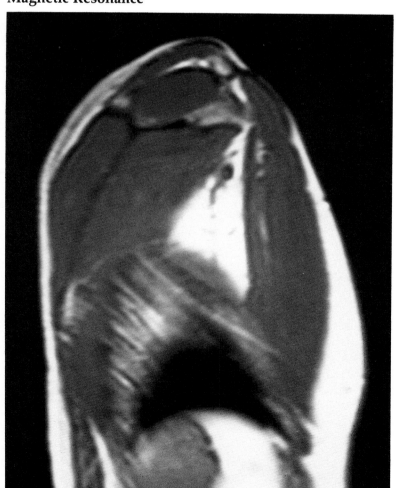

**Specimen Radiograph**

**Trispiral Tomogram**

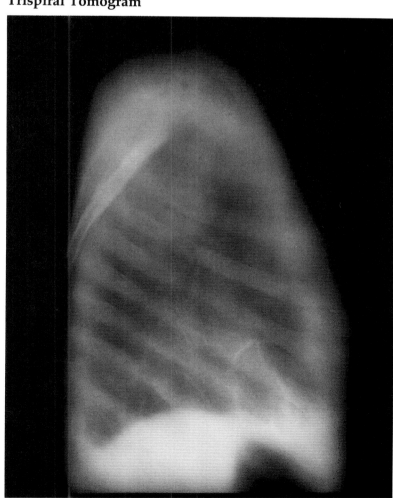

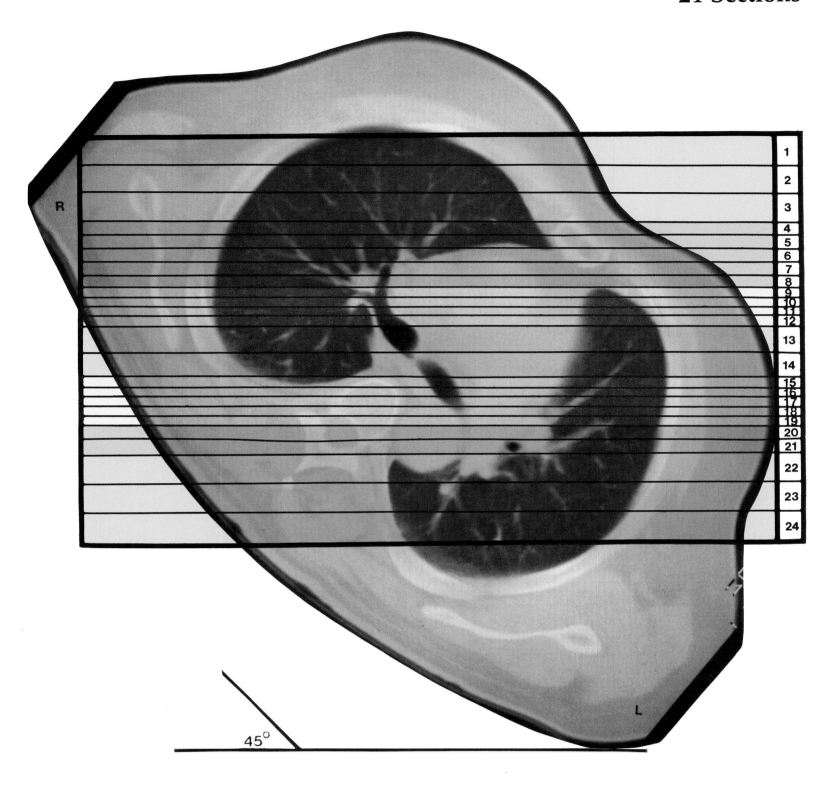

# Chapter 5
# Left Posterior Oblique Plane

## The Anatomic Specimen

The cadaver for the left posterior oblique sections was that of a 33-year-old white male who willed his body to the Alabama Anatomical Board prior to a judicial execution in the electric chair. The trachea was intubated a few minutes following death and the lungs fully expanded. The postmortem radiograph of the chest demonstrated normal structures with some early evidence of chronic obstructive pulmonary disease. By Alabama statute, an autopsy limited to the head was performed. The body was placed in the freezer in a neutral supine position where it remained for 49 months prior to sectioning.

During the making of the trispiral tomograms, it became apparent that postmortem air filled the major mediastinal vessels and cardiac chambers. At the time of the limited autopsy of the head, the major vessels entering the skull base were not ligated, and the air probably entered through the dissected cerebral vessels incident to relative negative pressure in the thorax with insufflation of the lungs.

After careful planning from the trispiral tomograms and CT sections, a plane was selected that would permit either direct entry or transcardiac catheterization of the air-filled structures. The frozen specimen was securely bolted to a 45° plastic frame, and a single bivalving cut was made through the thorax between Sections 13 and 14 (see Section Plan). Under fluoroscopic control, the cardiac chambers were easily filled with packed red cells to prevent subsequent sludging. Difficulty was experienced in entering the superior vena cava, but ultimately through a transbronchial needle puncture, the inferior portion of the vena cava and the azygos vein were successfully filled. In an effort to fill the superior portion of the superior vena cava, a second needle puncture was made and a catheter was threaded into the superior vena cava. Unfortunately, the tip of the catheter advanced though the vessel wall and a small amount of blood seeped out into the medial portion of the RUL, soiling this pulmonary segment. No anatomic structures were distorted, but the extravasated blood can be noted in the anatomic specimen photographs and the specimen radiographs (Sections 7–10).

Test films showed that normal radiographic density of the heart and great vessels had been restored, and the trispiral tomogram series was made. A small amount of air can still be seen in the superior vena cava in Tomograms 7–10.

Sections 11–13 were made 1 day following the bivalving procedure. An unavoidable delay of several months interrupted the cutting sequence of the reminder of the specimen. Despite careful wrapping of each hemithorax in Saran Wrap®, increased freezer dehydration of structures not covered by skin occurred. The dehydration will be noted as a color change in the anatomic specimens over the neck and supraclavicular areas and the posterior surfaces in all of the remaining sections. The dehydrated areas are especially evident as a decrease in radiographic density in the specimen radiographs. In addition, there is a color difference of the cut surface of the lung in all of the anatomic sections of this specimen when compared to the other four cadavers. We attribute this phenomenon to the fact that the subject was an excessive cigarette smoker, which also probably accounts for the multiple pulmonary cysts (PC) found in both lungs.

## Section Plan

The plan for the left posterior oblique sections (facing page), as shown in colored strips over an axial CT hard copy radiograph of the chest, represents a precise charting of the position of the gross sections. The sections are color-coded according to section thickness: **blue**, 10 mm; **pink**, 5 mm; **yellow**, 3 mm. The level and thickness of section is indicated throughout the chapter by the color bar overlying the miniature reference key at the top right of each double-page layout.

The most anterior sections (Sections 1–3) are 10 mm thick and extend through the lateral portion of the right lung to the level of the right lateral margin of the sternum anteriorly. Sections 4–8 are 5-mm sections which continue into the lateral portion of the right hilum and the midcardiac structures. Sections 9–12 are 3 mm thick and identify the critical structures in the right hilum and the midportion of the heart. Sections 13 and 14 occur on either side of the bivalving procedure and due to damage of lung tissue on some parts of the lung surface during filling of the heart and great vessels with blood, it was necessary to make thicker sections (approximately 10 mm) to reestablish the surface planes. Sections 15–19 are 3 mm thick and visualize the critical structures in the left hilum. Sections 20 and 21 are 5 mm thick and show the lateral segments of the left hilum. The remaining three sections of the left hemithorax are 10 mm thick and conclude near the lateral portion of the left lung, lateral to the spine.

## Orientation of Illustrations

Structures are viewed as seen from front to back with the thorax positioned obliquely at a 45° rotation from the coronal plane, with the left side tilted posteriorly. Throughout the atlas this projection is referred to as a *left posterior oblique*, which is the same as, and may also be called, a *right anterior oblique* projection. Structures on the right side of the chest appear on the reader's left, and structures on the left side of the chest appear on the reader's right. The anterior structures appear on the reader's right in all illustrations. The right lung appears in Sections 1–20; the left lung first appears in Section 4 and persists throughout the remaining sections.

## Anatomic–Radiographic Correlation

Several structures are seen to excellent advantage in this projection. There is no film from the trispiral tomogram series to match to Anatomic Section 1.

### Bronchial Structures
The bronchi that are especially well seen in the left posterior oblique projection are summarized in Table 5.1.
**Right Lung**: In Sections 5 and 6, the proximal portions of the lateral (B8) and medial (B9) divisions of the RML are visualized. The mainstem bronchus to the RML (B7) is seen in Section 7. Sections 9 and 10 visualize the apical (B3) and anterior (B5) segments of the RUL. Section 11 is an unusual profile of the right bronchopulmonary system. The distal trachea (T1) is seen extending inferiorly into the right mainstem bronchus (B1). Also shown are the origin of the RUL bronchus (B2), with its apical segment (B3); the intermediate bronchus (B6); and the RLL bronchus (B10), with its superior (B11), medial basal (B12), anterior basal (B13), and lateral basal (B14) segments. The superior segment of the RLL (B11) is best seen in Section 12. The carina (T2) and the posterior basal segment to the RLL (B15) are seen in Section 13.
**Left Lung**: The carina (T2) is clearly seen in Section 13. In Sections 14–16 the left mainstem bronchus (B16) is seen to good advantage. In the specimen radiograph of Section 16 and the trispiral tomograms of Sections 16 and 17, the lingular bronchus (B20) is very clearly seen dividing into the superior (B21) and inferior (B22) lingular segments. In Section 17, the mainstem bronchus

to the LLL (B23) is just beginning. Section 18 shows the LUL bronchus (B17) for the first time with its anterior segmental bronchus (B19). The LLL bronchus (B23), is seen to best advantage in the left posterior oblique projection in Sections 17 and 18, and particularly Section 19 where its entire extent is seen to end in the bifurcation into the anteromedial (B25) and lateral basal (B26) segments. The posterior basal bronchus to the LLL (B27) is quite well seen in Section 22.

### Fissures
The basic anatomic detail of this subject and the orientation of the left oblique projection have resulted in a clear display of the fissures in the anatomic specimen photographs, the specimen radiographs, and some of the trispiral tomograms. On the right side, the major (F2) and minor (F3) fissures are well seen in anatomic specimens (Sections 1–7) and specimen radiographs (Sections 3–5). Only the horizontal or minor fissure (F3) is well imaged in trispiral tomographic Sections 3–6.

On the left side, the major fissure (F1) is seen in anatomic specimens (Sections 11–24) and specimen radiographs (Sections 11, 14, and 17–24).

**Anatomic Variant**: In the left lung, an aberrant fissure (F4) was encountered as a complete fissure separating the LUL from the lingula, similar to two instances described by Medlar (30). In Medlar's experience, the lingular bronchus in both instances arose as a branch from the LLL bronchus. In this subject, however, there is a normal origin of the lingular bronchus (B20) from the left mainstem bronchus (B16). This anomalous fissure (F4) is well seen in anatomic specimen photographs (Sections 6–23), specimen radiographs (Sections 6–24), and trispiral tomograms (Sections 11–13).

### Neural Structures
The right brachial plexus (n1) can be well seen in the anatomic specimens (Sections 2–6) and the left brachial plexus (n1) in Sections 16–22.

### Vascular Structures
Many cardiac structures are best seen in oblique planes. Oblique projections can delineate the ventricular and atrial septa, more nearly optimally separate the four heart chambers, and demonstrate the relationship of the brachiocephalic vessels to the aortic arch.

**Table 5.1**
**Index to Optimal Bronchial Imaging LPO Projection**
Specimen Radiograph and Trispiral Tomogram
Right                    Bronchus                    Left

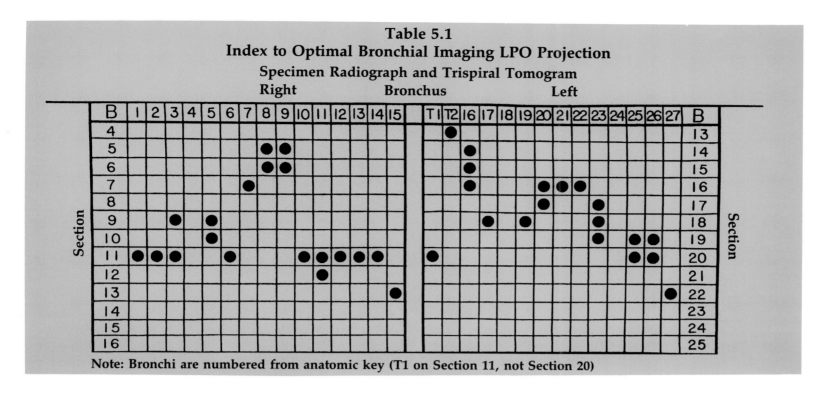

Note: Bronchi are numbered from anatomic key (T1 on Section 11, not Section 20)

Some vascular structures are of special interest. The right pulmonary artery (A2) and its interlobar part (A2*) are well seen in Sections 9 and 10, as is the left pulmonary artery (A16) in Section 16. The right inferior pulmonary vein (V4) as it enters the left atrium is well seen in Section 11. The superior pulmonary veins from the LUL (V2) and the lingula (V3) are graphically seen to fuse and enter the left atrium (H1) in Section 15. Right intercostal veins (V19) are clearly seen to enter the azygos vein (V7) in Section 16.

### Other Structures
The anterior pleural stripe (anterior junction line) (P1) is prominent in Sections 5 and 6.

### Magnetic Resonance Images
The MR vascular imaging of the heart and great vessels of the size-related volunteer correlates well with anatomic specimens and specimen radiographs. Absolute correlation was not possible as previously described. Note the right internal thoracic vessels in MR Sections 2–4. In Section 8 the superior vena cava (V39) dividing into the right (V9) and left (V8) brachiocephalic veins is clearly seen in both the anatomic specimen and its corresponding MR image. The brachiocephalic artery (A34) is seen crossing behind the left brachiocephalic vein (V8) in MR Section 9. Just posterior to this in Section 10, the image of the vein is completely obscured, while the artery (A34) is sharply seen arising from the aortic arch (A29). The aortic valve (H20) is also well seen in MR Section 10 and in Anatomic Section 12. The left common carotid artery (A37) is seen in MR Section 11, and the left subclavian artery (A49) coursing over the apex of the left lung is well seen in MR Sections 12 and 13.

# Left
# Posterior
# Oblique

## Section 1

### Anatomic Key

**Bronchi**
B 9 RML, medial segment

**Fissures**
F 2 right major (oblique)
F 3 right minor (horizontal)

**Lobes**
L 4 RUL, anterior segment
L 6 RML, lateral segment
L 7 RML, medial segment
L 11 RLL, anterior basal segment

**Muscles**
M 1 abdominal, external oblique
M 2 abdominal, internal oblique
M 3 abdominis, rectus
M 7 coracobrachialis
M 8 deltoid
M 9 diaphragm
M 16 intercostal
M 31 pectoralis, major
M 32 pectoralis, minor
M 44 serratus, anterior
M 62 transversus thoracis

**Organs**
O 3 liver

**Pleural Structures**
P 2 costal (parietal)
P 6 visceral

**Skeletal Structures**
S 1 clavicle, right
S 5 rib, body, lateral (3rd right)
S 6 rib, body, posterior (9th right)
S 7 rib, costal cartilage (8th right)
S 12 scapula, coracoid process

**Nonanatomic**
NB nylon bolt

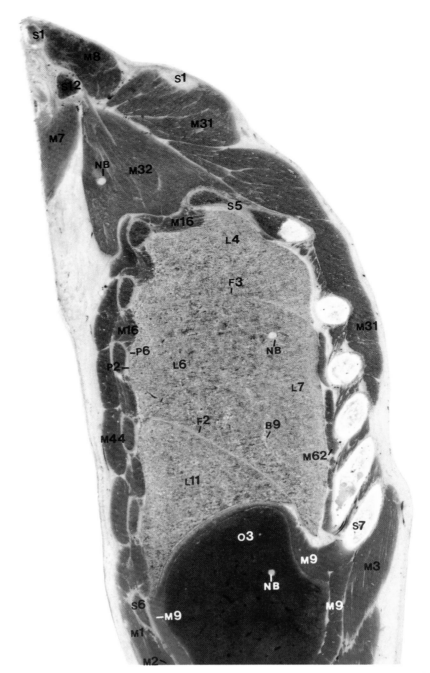

**Anatomic Specimen**

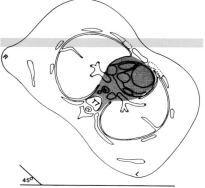

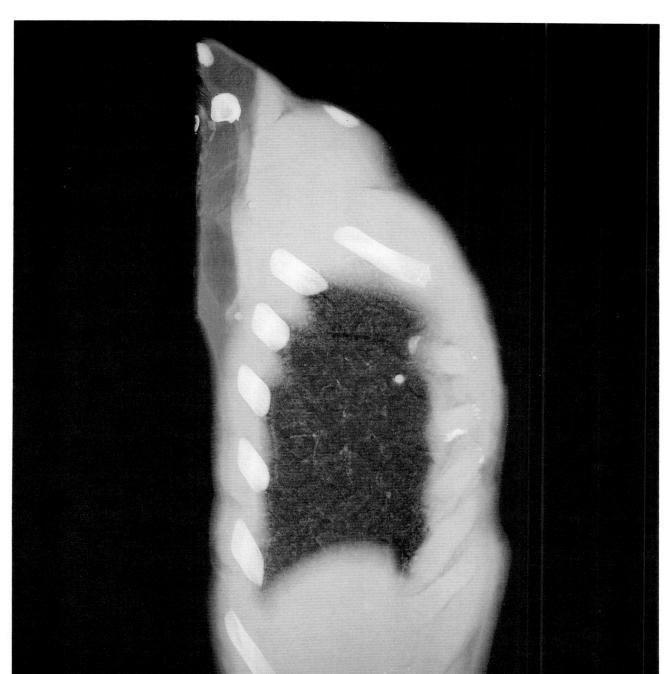

**Specimen Radiograph**

**Magnetic Resonance**

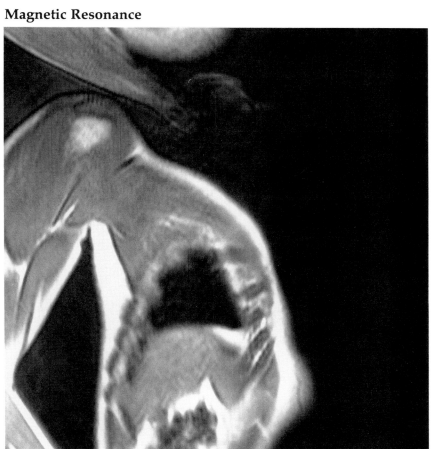

No
Trispiral Tomogram
made

# Left Posterior Oblique

## Section 2

### Anatomic Key

**Arteries**
A 33 axillary, right
A 44 intercostal, anterior right
A 46 intercostal, posterior, right

**Bronchi**
B 9 RML, medial segment

**Fat**
f 1 axillary
f 3 extrapleural

**Fissures**
F 2 right major (oblique)
F 3 right minor (horizontal)

**Lobes**
L 4 RUL, anterior segment
L 6 RML, lateral segment
L 7 RML, medial segment
L 11 RLL, anterior basal segment

**Muscles**
M 1 abdominal, external oblique
M 2 abdominal, internal oblique
M 3 abdominis, rectus
M 7 coracobrachialis
M 9 diaphragm
M 16 intercostal
M 31 pectoralis, major
M 32 pectoralis, minor
M 33 platysma
M 44 serratus, anterior
M 55 subclavius
M 57 subscapularis

**Neural Structures**
n 1 brachial plexus
n 2 intercostal nerve

**Nodes**
N 1 axillary

**Organs**
O 3 liver

**Pleural Structures**
P 2 costal (parietal)
P 6 visceral

**Skeletal Structures**
S 1 clavicle, right
S 3 rib (2nd right)
S 4 rib, body, anterior (9th right)
S 5 rib, body, lateral (5th right)
S 7 rib, costal cartilage (2nd right)
S 12 scapula, coracoid process
S 13 scapula, glenoid cavity
S 17 sternum, body

**Veins**
V 6 axillary, right
V 21 internal thoracic (mammary), right

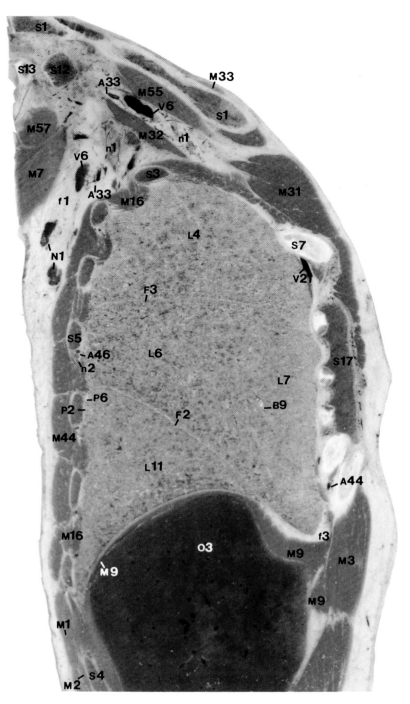

**Anatomic Specimen**

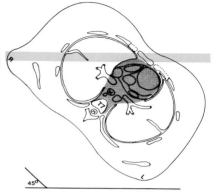

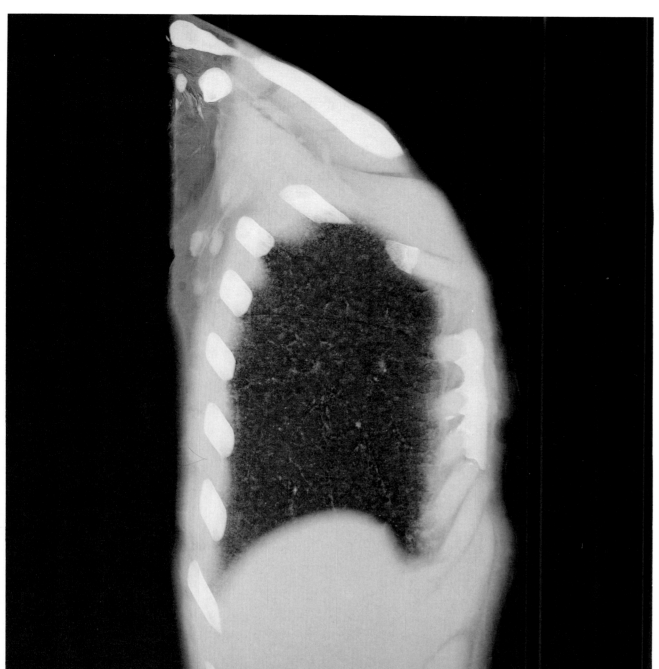

**Specimen Radiograph**

**Magnetic Resonance**

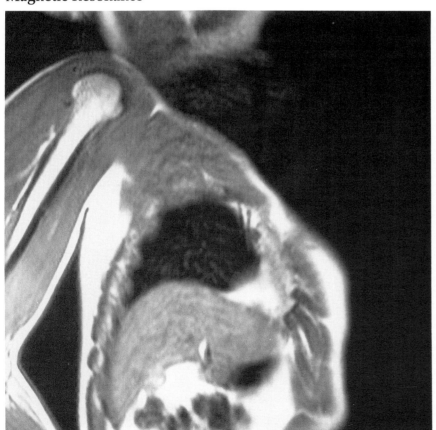

**Trispiral Tomogram**

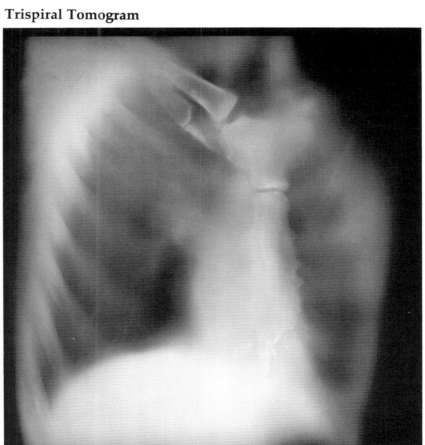

# Left Posterior Oblique

## Section 3

### Anatomic Key

**Arteries**
A 33 axillary, right
A 46 intercostal, posterior, right
A 56 thoracic, lateral, right
A 58 thoracoacromial, right

**Bronchi**
B 5 RUL, anterior segment
B 8 RML, lateral segment
B 9 RML, medial segment
B 13 RLL, anterior basal segment

**Fat**
f 1 axillary
f 3 extrapleural
f 7 subcutaneous

**Fissures**
F 2 right major (oblique)
F 3 right minor (horizontal)

**Lobes**
L 2 RUL, apical segment
L 4 RUL, anterior segment
L 6 RML, lateral segment
L 7 RML, medial segment
L 11 RLL, anterior basal segment

**Muscles**
M 1 abdominal, external oblique
M 2 abdominal, internal oblique
M 3 abdominis, rectus
M 9 diaphragm
M 16 intercostal
M 21 latissimus dorsi
M 31 pectoralis, major
M 32 pectoralis, minor
M 33 platysma
M 44 serratus, anterior
M 52 sternocleidomastoid
M 55 subclavius
M 57 subscapularis
M 58 supraspinatus

**Neural Structures**
n 1 brachial plexus

**Nodes**
N 1 axillary

**Organs**
O 3 liver

**Pleural Structures**
P 2 costal (parietal)
P 6 visceral

**Skeletal Structures**
S 1 clavicle, right
S 5 rib, body, lateral (5th right)
S 7 rib, costal cartilage (1st right)
S 12 scapula, coracoid process
S 13 scapula, glenoid cavity
S 17 sternum, body
S 18 sternum, manubrium
S 19 sternum, xiphoid

**Veins**
V 6 axillary, right
V 31 subclavian, right
V 33 suprascapular, right

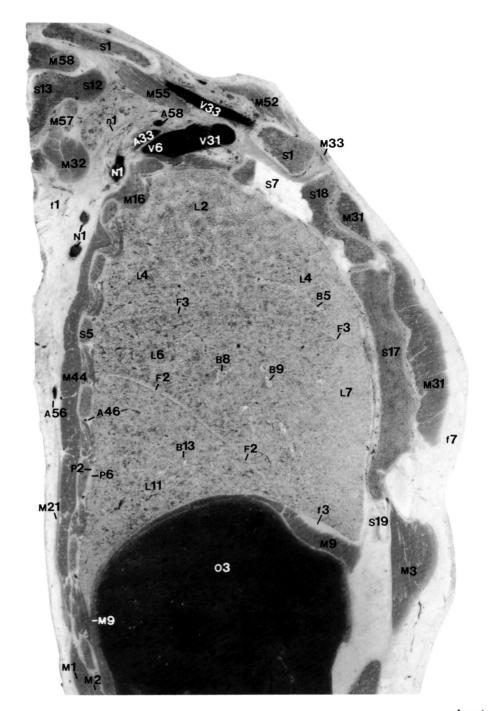

**Anatomic Specimen**

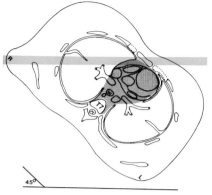

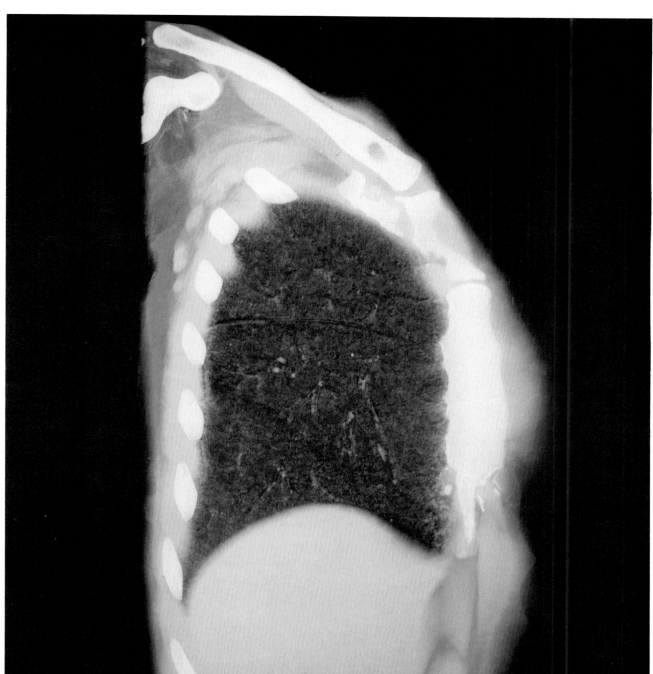

**Specimen Radiograph**

**Magnetic Resonance**

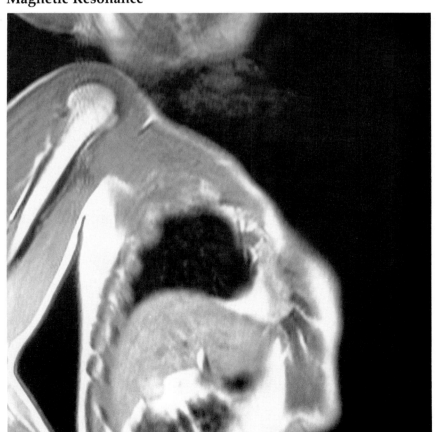

**Trispiral Tomogram**

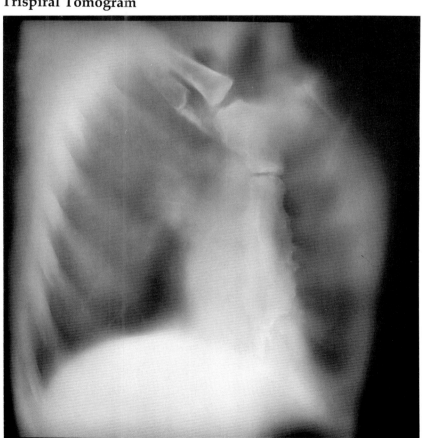

# Left Posterior Oblique

## Section 4

### Anatomic Key

**Arteries**
A 50 subclavian, right
A 56 thoracic, lateral, right

**Bronchi**
B 3 RUL, apical segment
B 5 RUL, anterior segment
B 8 RML, lateral segment
B 9 RML, medial segment
B 13 RLL, anterior basal segment
B 14 RLL, lateral basal segment

**Fat**
f 1 axillary
f 3 extrapleural

**Fissures**
F 2 right major (oblique)
F 3 right minor (horizontal)

**Lobes**
L 2 RUL, apical segment
L 4 RUL, anterior segment
L 6 RML, lateral segment
L 7 RML, medial segment
L 11 RLL, anterior basal segment
L 12 RLL, lateral basal segment
L 17 LUL, lingula

**Muscles**
M 1 abdominal, external oblique
M 3 abdominis, rectus
M 9 diaphragm
M 16 intercostal
M 21 latissimus dorsi
M 31 pectoralis, major
M 33 platysma
M 44 serratus, anterior
M 52 sternocleidomastoid
M 55 subclavius
M 57 subscapularis
M 58 supraspinatus

**Neural Structures**
n 1 brachial plexus

**Organs**
o 3 liver

**Pleural Structures**
P 1 anterior junction line
P 2 costal (parietal)
P 6 visceral

**Skeletal Structures**
S 1 clavicle, right
S 3 rib (2nd right)
S 4 rib, body, anterior (1st right)
S 5 rib, body, lateral (5th right)
S 7 rib, costal cartilage (8th right)
S 11 scapula, acromion
S 12 scapula, coracoid process
S 13 scapula, glenoid cavity
S 17 sternum, body
S 18 sternum, manubrium
S 19 sternum, xiphoid

**Veins**
V 6 axillary, right
V 31 subclavian, right
V 33 suprascapular, right

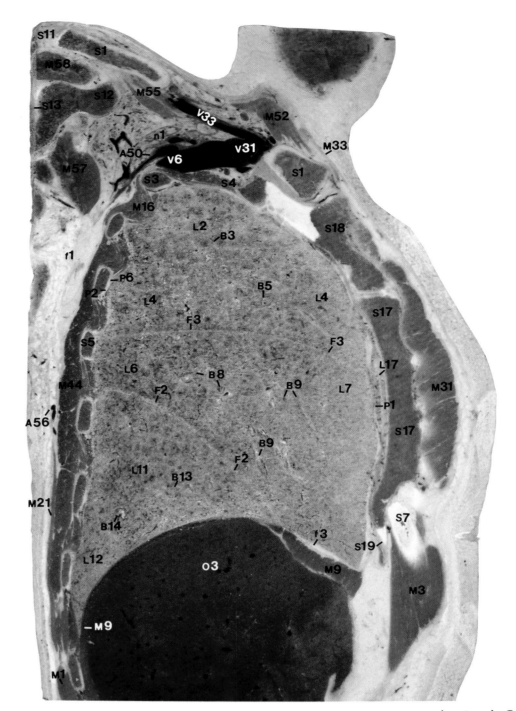

**Anatomic Specimen**

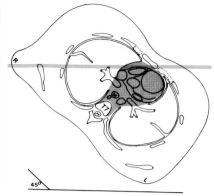

# Left Posterior Oblique

## Section 4

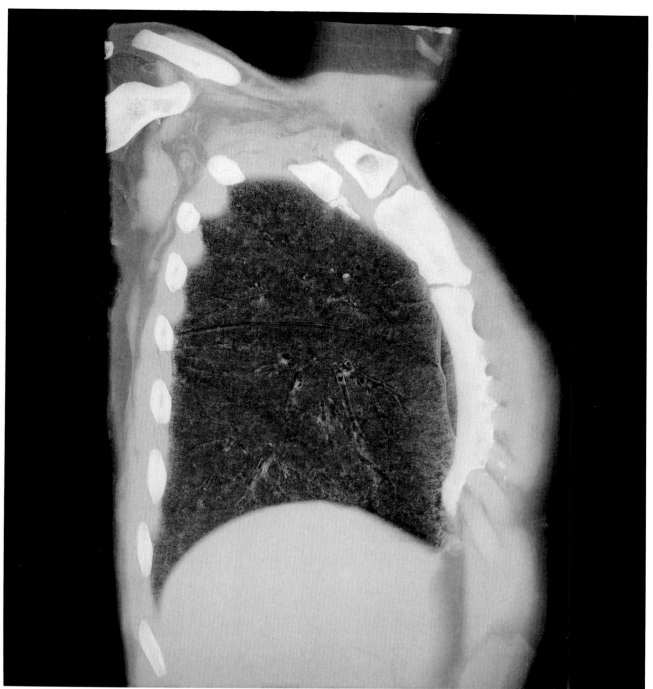

Specimen Radiograph

**Magnetic Resonance**

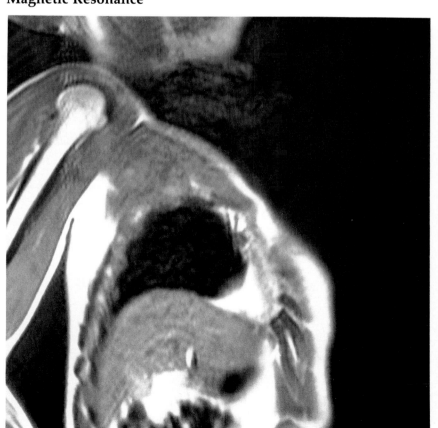

**Trispiral Tomogram**

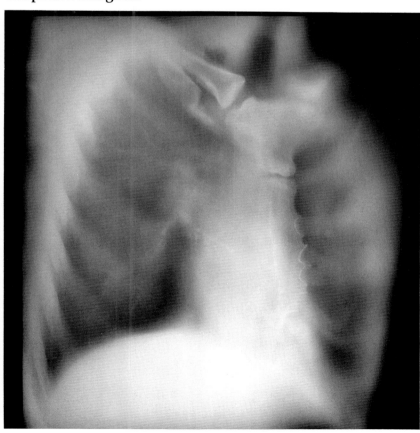

225

# Left Posterior Oblique

## Section 5

## Anatomic Key

### Arteries
A **8** RML, lateral branch
A **9** RML, medial branch
A **50** subclavian, right
A **52** subscapular, right
A **54** suprascapular, right

### Bronchi
B **3** RUL, apical segment
B **5** RUL, anterior segment
B **8** RML, lateral segment
B **9** RML, medial segment
B **13** RLL, anterior basal segment
B **14** RLL, lateral basal segment

### Fat
f **2** epicardial
f **3** extrapleural
f **7** subcutaneous

### Fissures
F **2** right major (oblique)
F **3** right minor (horizontal)

### Glands
G **4** thyroid

### Heart
H **7** cardic vein, small
H **11** coronary artery, right
H **14** pericardium, visceral serous
H **22** ventricle, right

### Lobes
L **2** RUL, apical segment
L **4** RUL, anterior segment
L **6** RML, lateral segment
L **7** RML, medial segment
L **11** RLL, anterior basal segment
L **12** RLL, lateral basal segment
L **16** LUL, anterior segment
L **18** LUL, superior lingular segment

### Muscles
M **3** abdominis, rectus
M **4** abdominis, transversus
M **9** diaphragm
M **10** diaphragm, central tendon
M **15** infraspinatus
M **16** intercostal
M **21** latissimus dorsi
M **31** pectoralis, major
M **38** scalene, anterior
M **44** serratus, anterior
M **52** sternocleidomastoid
M **53** sternohyoid
M **54** sternothyroid
M **57** subscapularis
M **58** supraspinatus
M **62** transversus thoracis
M **63** trapezius

### Neural Structures
n **1** brachial plexus

### Organs
O **3** liver

### Pleural Structures
P **1** anterior junction line
P **2** costal (parietal)
P **6** visceral

### Skeletal Structures
S **1** clavicle, right
S **3** rib (7th left)
S **4** rib, body, anterior (1st right)
S **5** rib, body, lateral (9th right)
S **7** rib, costal cartilage (2nd left)
S **14** scapula, neck
S **15** scapula, spine
S **18** sternum, manubrium

### Trachea and Larynx
T **1** trachea
T **3** larynx

### Veins
V **16** intercostal, anterior, left
V **20** internal thoracic (mammary), left
V **25** jugular, external, right
V **27** jugular, internal, right
V **31** subclavian, right
V **33** suprascapular, right

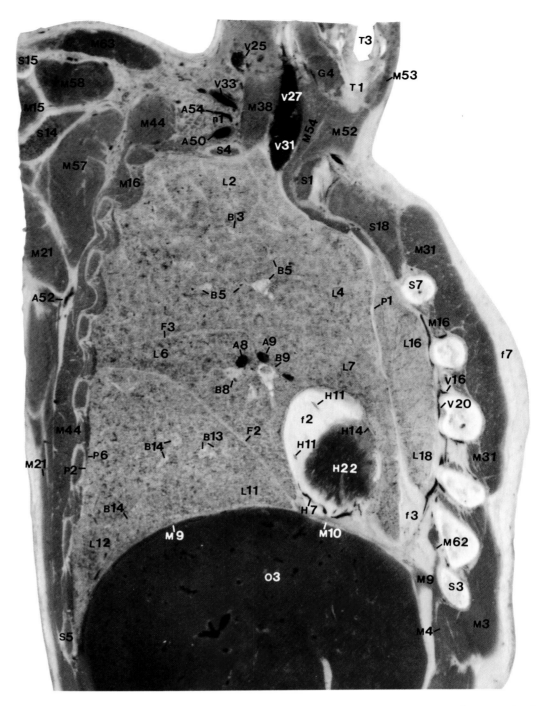

**Anatomic Specimen**

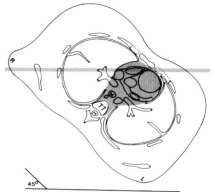

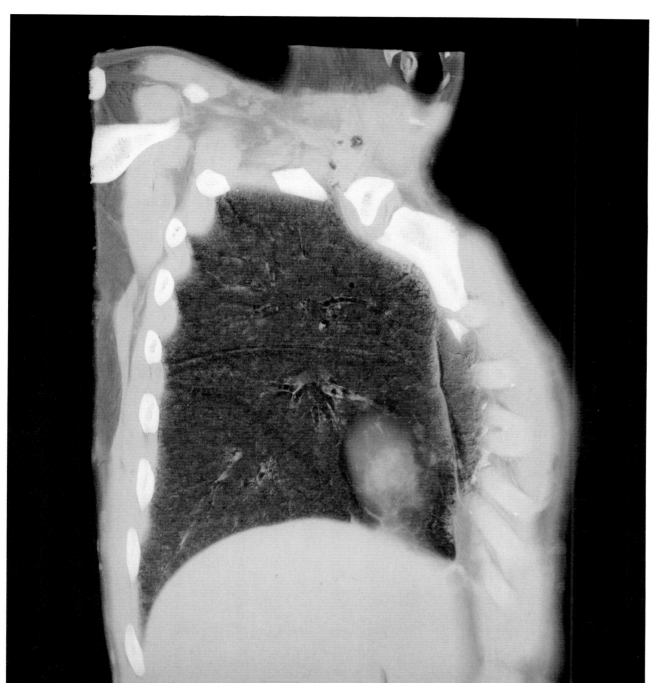

**Specimen Radiograph**

**Magnetic Resonance**

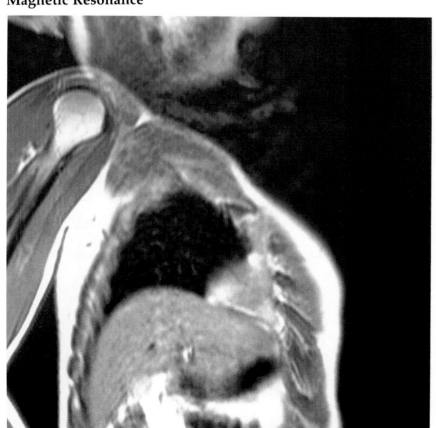

**Trispiral Tomogram**

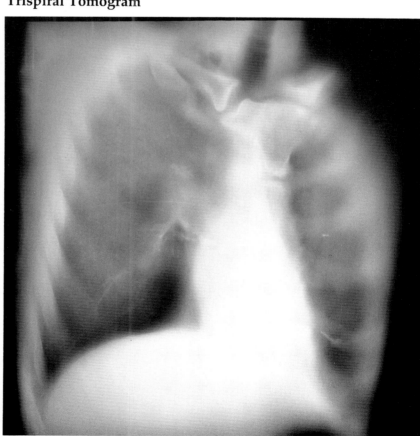

# Left Posterior Oblique

## Section 6

### Anatomic Key

**Arteries**
A 7 right middle lobe (RML)
A 52 subscapular, right

**Bronchi**
B 3 RUL, apical segment
B 5 RUL, anterior segment
B 8 RML, lateral segment
B 9 RML, medial segment
B 14 RLL, lateral basal segment
B 15 RLL, posterior basal segment

**Fat**
f 2 epicardial
f 3 extrapleural
f 4 mediastinal

**Fissures**
F 2 right major (oblique)
F 3 right minor (horizontal)
F 4 aberrant (LUL)

**Glands**
G 4 thyroid

**Heart**
H 2 atrium, right
H 11 coronary artery, right
H 14 pericardium, visceral serous
H 22 ventricle, right

**Lobes**
L 2 RUL, apical segment
L 4 RUL, anterior segment
L 6 RML, lateral segment
L 7 RML, medial segment
L 10 RLL, medial basal segment
L 11 RLL, anterior basal segment
L 12 RLL, lateral basal segment
L 15 LUL, apicoposterior segment
L 16 LUL, anterior segment
L 18 LUL, superior lingular segment
L 19 LUL, inferior lingular segment

**Muscles**
M 3 abdominis, rectus
M 9 diaphragm
M 10 diaphragm, central tendon
M 15 infraspinatus
M 16 intercostal
M 21 latissimus dorsi
M 31 pectoralis, major
M 38 scalene, anterior
M 39 scalene, middle
M 40 scalene, posterior
M 44 serratus, anterior
M 53 sternohyoid
M 54 sternothyroid
M 57 subscapularis
M 58 supraspinatus
M 62 transversus thoracis
M 63 trapezius

**Neural Structures**
n 1 brachial plexus

**Nodes**
N 10 supraclavicular

**Organs**
O 3 liver

**Pleural Structures**
P 1 anterior junction line
P 2 costal (parietal)
P 5 mediastinal (parietal)
P 6 visceral

**Skeletal Structures**
S 1 clavicle, right
S 3 rib (2nd left)
S 4 rib, body, anterior (1st right)
S 5 rib, body, lateral (9th right)
S 7 rib, costal cartilage (5th left)
S 10 scapula
S 15 scapula, spine
S 18 sternum, manubrium

**Trachea**
T 1 trachea

**Veins**
V 9 brachiocephalic (innominate), right
V 16 intercostal, anterior, left
V 35 thyroid, inferior, right
V 43 vertebral, right

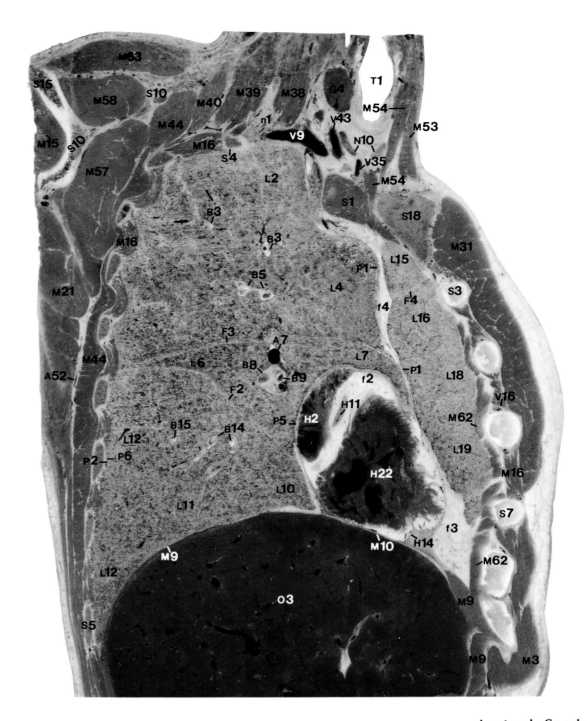

**Anatomic Specimen**

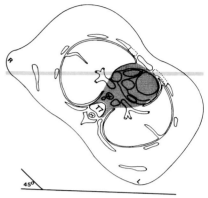

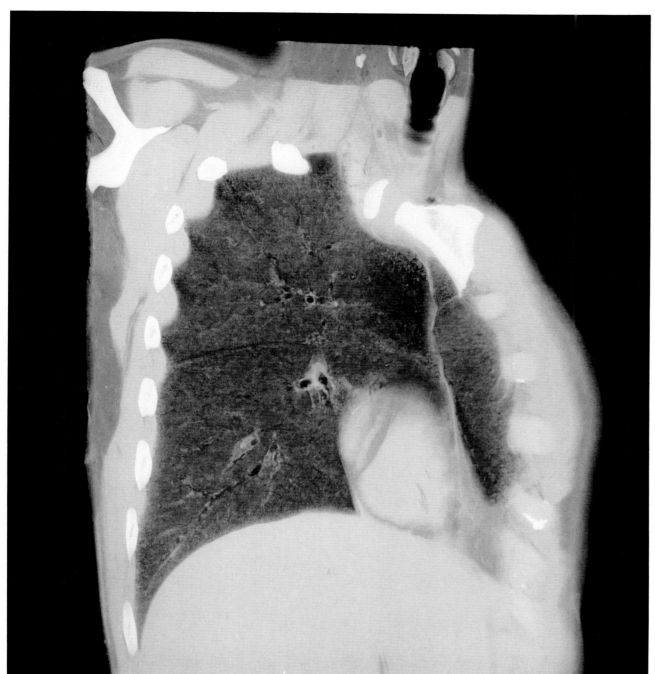

**Specimen Radiograph**

**Magnetic Resonance**

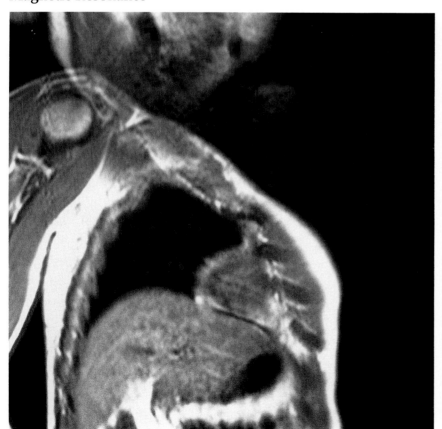

**Trispiral Tomogram**

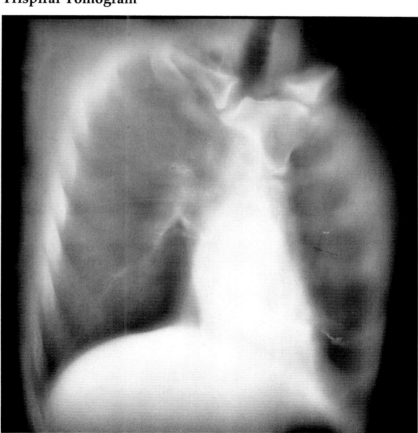

# Left Posterior Oblique

## Section 7

## Anatomic Key

**Arteries**
A 7 right middle lobe (RML)
A 34 brachiocephalic (innominate) trunk
A 38 carotid, common, right
A 40 costocervical trunk, right
A 50 subclavian, right
A 52 subscapular, right

**Bronchi**
B 3 RUL, apical segment
B 5 RUL, anterior segment
B 7 right middle lobe (RML)
B 14 RLL, lateral basal segment
B 15 RLL, posterior basal segment

**Fat**
f 2 epicardial
f 3 extrapleural
f 4 mediastinal

**Fissures**
F 2 right major (oblique)
F 3 right minor (horizontal)
F 4 aberrant (LUL)

**Glands**
G 4 thyroid

**Heart**
H 2 atrium, right
H 4 auricle, right
H 11 coronary artery, right
H 17 valve, tricuspid
H 22 ventricle, right

**Lobes**
L 2 RUL, apical segment
L 3 RUL, posterior segment
L 4 RUL, anterior segment
L 6 RML, lateral segment
L 7 RML, medial segment
L 10 RLL, medial basal segment
L 11 RLL, anterior basal segment
L 12 RLL, lateral basal segment
L 15 LUL, apicoposterior segment
L 16 LUL, anterior segment
L 18 LUL, superior lingular segment
L 19 LUL, inferior lingular segment

**Muscles**
M 3 abdominis, rectus
M 9 diaphragm
M 10 diaphragm, central tendon
M 15 infraspinatus
M 16 intercostal
M 21 latissimus dorsi
M 28 longus colli
M 31 pectoralis, major
M 33 platysma
M 38 scalene, anterior
M 39 scalene, middle
M 40 scalene, posterior
M 44 serratus, anterior
M 53 sternohyoid
M 54 sternothyroid
M 57 subscapularis
M 58 supraspinatus
M 62 transversus thoracis
M 63 trapezius

**Nodes**
N 6 mediastinal, anterior
N 9 paratracheal

**Organs**
O 3 liver

**Pleural Structures**
P 2 costal (parietal)
P 5 mediastinal (parietal)
P 6 visceral

**Skeletal Structures**
S 3 rib, (9th right)
S 5 rib, body, lateral (5th right)
S 6 rib, body, posterior (1st right)
S 7 rib, costal cartilage (2nd left)
S 10 scapula
S 15 scapula, spine
S 18 sternum, manubrium

**Trachea**
T 1 trachea

**Veins**
v 0 pulmonary, superior, RUL
v 1 pulmonary, superior, RML
v 4 pulmonary, inferior, RLL
v 9 brachiocephalic (innominate), right
v 16 intercostal, anterior, left
v 19 intercostal, posterior, right
v 20 internal thoracic (mammary), left

**Nonanatomic**
‡ postmortem trauma
PC pulmonary cyst

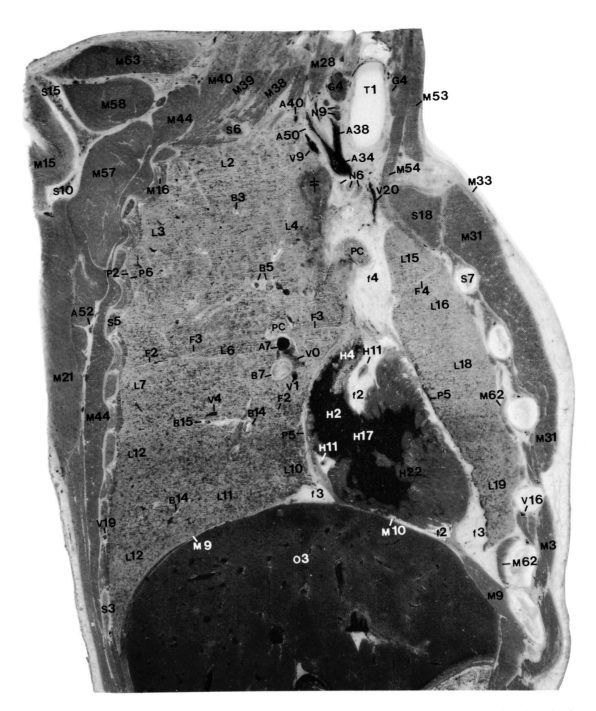

**Anatomic Specimen**

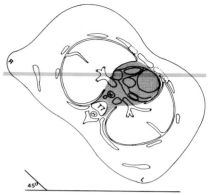

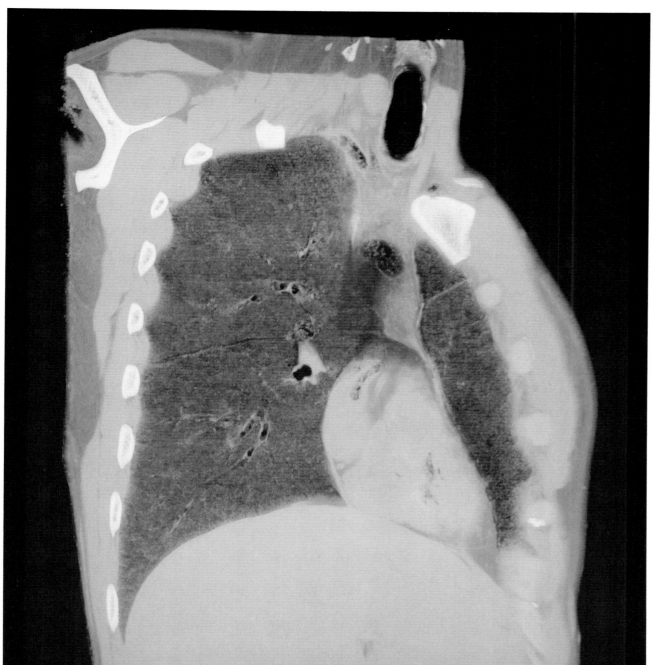

**Specimen Radiograph**

**Magnetic Resonance**

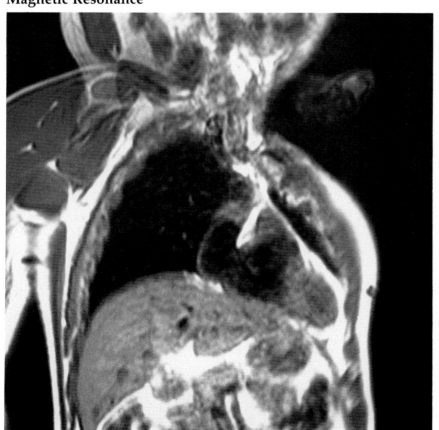

**Trispiral Tomogram**

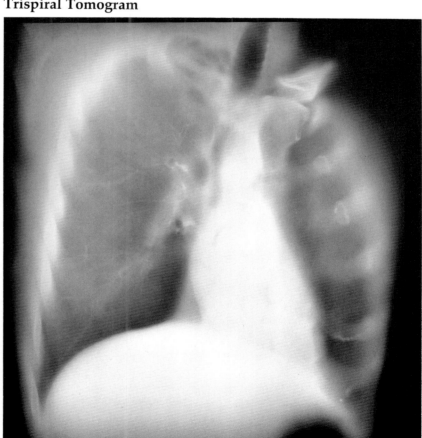

# Left Posterior Oblique

## Section 8

## Anatomic Key

### Arteries
A 1 pulmonary trunk
A 2* main pulmonary, right (interlobar part)
A 4 RUL, apical branch
A 30 aorta, ascending
A 66 vertebral, right

### Bronchi
B 2 right upper lobe (RUL)
B 3 RUL, apical segment
B 5 RUL, anterior segment
B 6 intermediate, right
B 11 RLL, superior segment
B 12 RLL, medial basal segment
B 15 RLL, posterior basal segment
B 19 LUL, anterior segment

### Esophagus
E 1 esophagus

### Fat
f 2 epicardial
f 3 extrapleural
f 4 mediastinal

### Fissures
F 2 right major (oblique)
F 4 aberrant (LUL)

### Glands
G 4 thyroid

### Heart
H 2 atrium, right
H 4 auricle, right
H 11 coronary artery, right
H 13 pericardium, parietal serous
H 14 pericardium, visceral serous
H 16 septum, interventricular
H 17 valve, tricuspid
H 18 valve, pulmonary
H 22 ventricle, right

### Lobes
L 2 RUL, apical segment
L 3 RUL, posterior segment
L 4 RUL, anterior segment
L 9 RLL, superior segment
L 10 RLL, medial basal segment
L 13 RLL, posterior basal segment
L 15 LUL, apicoposterior segment
L 16 LUL, anterior segment
L 18 LUL, superior lingular segment
L 19 LUL, inferior lingular segment

### Muscles
M 1 abdominal, external oblique
M 9 diaphragm
M 10 diaphragm, central tendon
M 15 infraspinatus
M 16 intercostal
M 21 latissimus dorsi
M 28 longus colli
M 31 pectoralis, major
M 33 platysma
M 38 scalene, anterior
M 39 scalene, middle
M 44 serratus, anterior
M 52 sternocleidomastoid
M 53 sternohyoid
M 54 sternothyroid
M 57 subscapularis
M 58 supraspinatus
M 63 trapezius

### Nodes
N 2 bronchopulmonary (hilar)
N 6 mediastinal, anterior
N 9 paratracheal

### Organs
O 3 liver
O 6 stomach

### Pleural Structures
P 2 costal (parietal)
P 5 mediastinal (parietal)
P 6 visceral

### Skeletal Structures
S 1 clavicle, left
S 3 rib (7th left)
S 5 rib, body, lateral (9th right)
S 6 rib, body, posterior (1st right)
S 7 rib, costal cartilage (2nd left)
S 10 scapula
S 15 scapula, spine
S 18 sternum, manubrium
S 21 vertebra, body (C6)
S 29 vertebra, transverse process (C7)

### Trachea
T 1 trachea

### Veins
V 0 pulmonary, superior, RUL
V 4 pulmonary, inferior, RLL
V 8 brachiocephalic (innominate), left
V 9 brachiocephalic (innominate), right
V 39 vena cava, superior

### Nonanatomic
PC pulmonary cyst

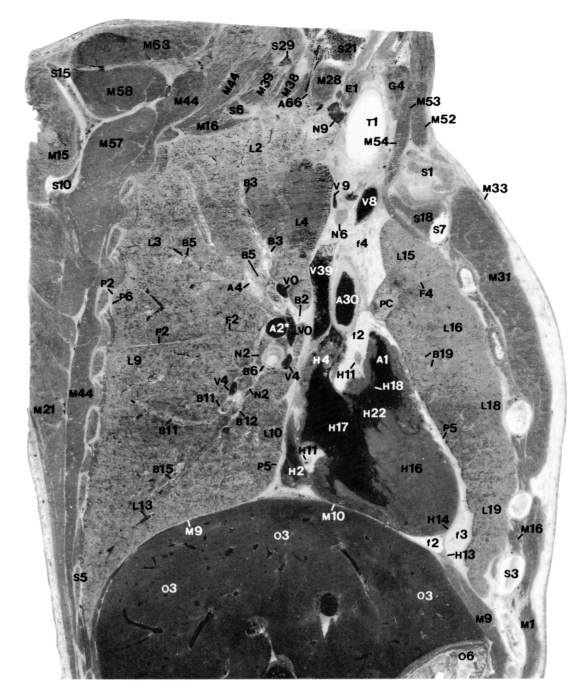

**Anatomic Specimen**

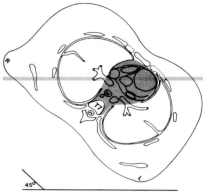

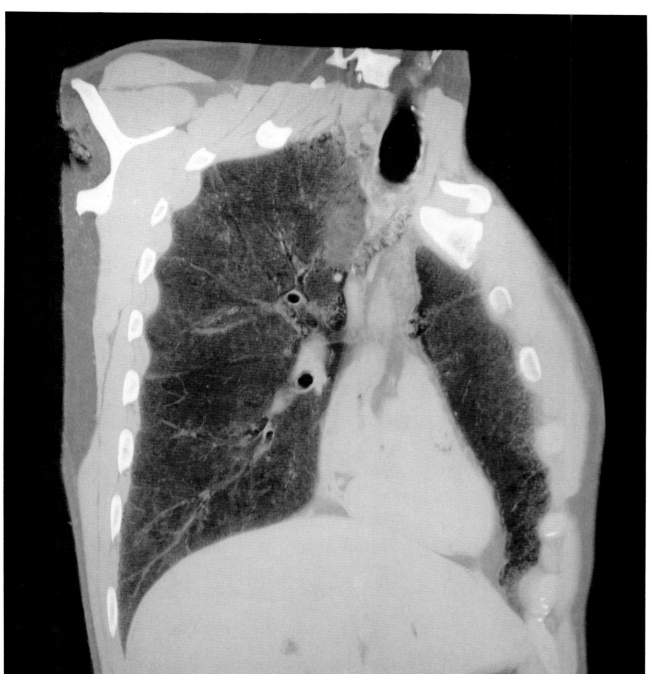

**Specimen Radiograph**

**Magnetic Resonance**

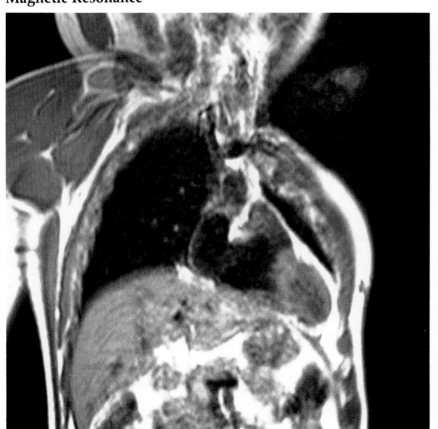

**Trispiral Tomogram**

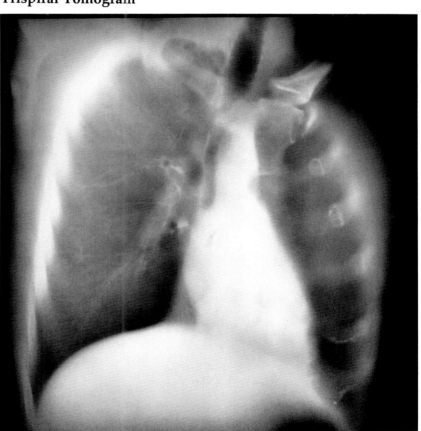

# Left Posterior Oblique

## Section 9

### Anatomic Key

#### Arteries
A 1 pulmonary trunk
A 2 main pulmonary, right
A 2* main pulmonary, right (interlobar part)
A 3 right upper lobe (RUL)
A 15 RLL, posterior basal branch
A 29 aortic arch
A 30 aorta, ascending
A 37 carotid, common, left

#### Bronchi
B 2 right upper lobe (RUL)
B 3 RUL, apical segment
B 5 RUL, anterior segment
B 6 intermediate, right
B 10 right lower lobe (RLL)
B 12 RLL, medial basal segment
B 15 RLL, posterior basal segment
B 19 LUL, anterior segment
B 21 LUL, superior lingular segment

#### Esophagus
E 1 esophagus

#### Fat
f 2 epicardial
f 3 extrapleural
f 4 mediastinal
f 7 subcutaneous

#### Fissures
F 2 right major (oblique)
F 4 aberrant (LUL)

#### Glands
G 4 thyroid

#### Heart
H 2 atrium, right
H 4 auricle, right
H 11 coronary artery, right
H 13 pericardium, parietal serous
H 14 pericardium, visceral serous
H 16 septum, interventricular
H 22 ventricle, right

#### Lobes
L 2 RUL, apical segment
L 3 RUL, posterior segment
L 4 RUL, anterior segment
L 9 RLL, superior segment
L 10 RLL, medial basal segment
L 13 RLL, posterior basal segment
L 15 LUL, apicoposterior segment
L 16 LUL, anterior segment
L 18 LUL, superior lingular segment
L 19 LUL, inferior lingular segment

#### Muscles
M 1 abdominal, external oblique
M 9 diaphragm
M 10 diaphragm, central tendon
M 15 infraspinatus
M 16 intercostal
M 21 latissimus dorsi
M 28 longus colli
M 31 pectoralis, major
M 38 scalene, anterior
M 39 scalene, middle
M 40 scalene, posterior
M 44 serratus, anterior
M 52 sternocleidomastoid
M 53 sternohyoid
M 54 sternothyroid
M 57 subscapularis
M 58 supraspinatus
M 63 trapezius

#### Nodes
N 2 bronchopulmonary (hilar)
N 6 mediastinal, anterior

#### Organs
O 3 liver

#### Pleural Structures
P 2 costal (parietal)
P 5 mediastinal (parietal)
P 6 visceral

#### Skeletal Structures
S 1 clavicle, left
S 4 rib, body, anterior (2nd left)
S 5 rib, body, lateral (9th right)
S 6 rib, body, posterior (1st right)
S 7 rib, costal cartilage (7th left)
S 10 scapula
S 15 scapula, spine
S 18 sternum, manubrium
S 20 intervertebral disc (C6–C7)

#### Trachea
T 1 trachea

#### Veins
V 4 pulmonary, inferior, RLL
V 8 brachiocephalic (innominate), left
V 16 intercostal, anterior, left
V 19 intercostal, posterior, right
V 38 vena cava, inferior
V 39 vena cava, superior

#### Nonanatomic
‡ postmortem trauma
PC pulmonary cyst

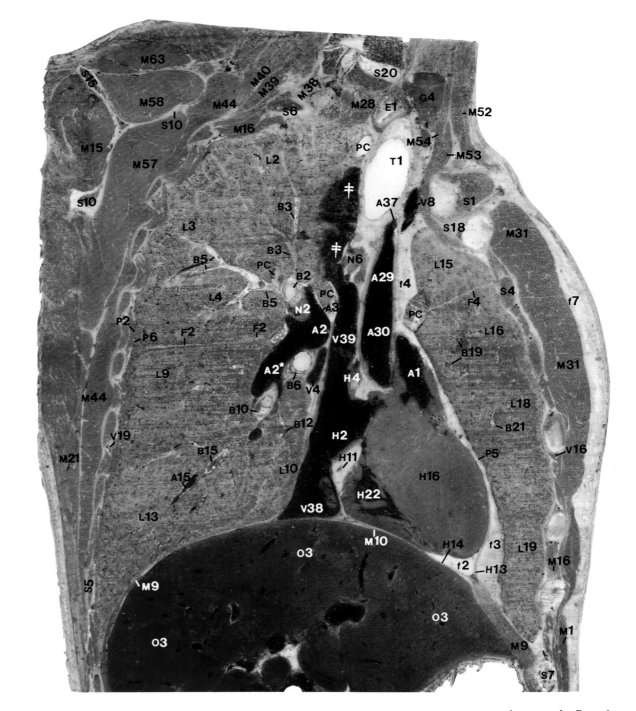

**Anatomic Specimen**

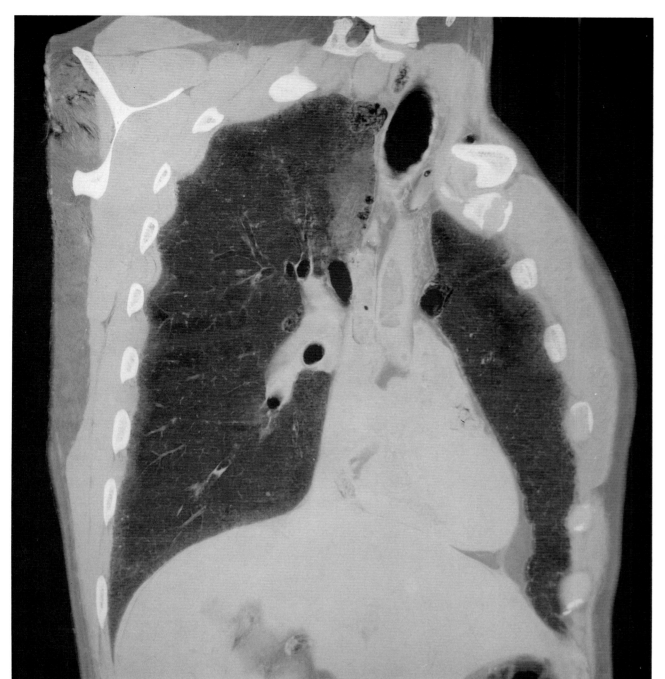

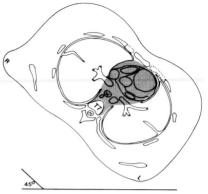

Specimen Radiograph

**Magnetic Resonance**

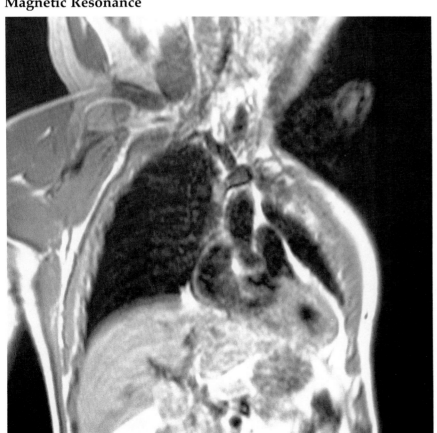

**Trispiral Tomogram**

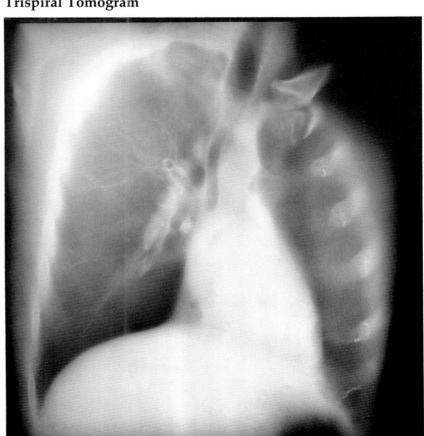

# Left Posterior Oblique

## Section 10

### Anatomic Key

**Arteries**
A  1  pulmonary trunk
A  2  main pulmonary, right
A  2* main pulmonary, right
      (interlobar part)
A  3  right upper lobe (RUL)
A 29  aortic arch
A 30  aorta, ascending

**Bronchi**
B  2  right upper lobe (RUL)
B  3  RUL, apical segment
B  5  RUL, anterior segment
B  6  intermediate, right
B 10  right lower lobe (RLL)
B 11  RLL, superior segment
B 12  RLL, medial basal segment
B 13  RLL, anterior basal segment
B 19  LUL, anterior segment

**Esophagus**
E  1  esophagus

**Fat**
f  2  epicardial

**Fissures**
F  2  right major (oblique)
F  4  aberrant (LUL)

**Glands**
G  4  thyroid

**Heart**
H  2  atrium, right
H 11  coronary artery, right
H 13  pericardium, parietal serous
H 14  pericardium, visceral serous
H 16  septum, interventricular
H 21  ventricle, left
H 22  ventricle, right

**Lobes**
L  2  RUL, apical segment
L  3  RUL, posterior segment
L  4  RUL, anterior segment
L  9  RLL, superior segment
L 10  RLL, medial basal segment
L 11  RLL, anterior basal segment
L 15  LUL, apicoposterior segment
L 16  LUL, anterior segment
L 18  LUL, superior lingular segment
L 19  LUL, inferior lingular segment

**Muscles**
M  1  abdominal, external oblique
M  9  diaphragm
M 10  diaphragm, central tendon
M 15  infraspinatus
M 16  intercostal
M 21  latissimus dorsi
M 28  longus colli
M 31  pectoralis, major
M 40  scalene, posterior
M 44  serratus, anterior
M 52  sternocleidomastoid
M 53  sternohyoid
M 54  sternothyroid
M 57  subscapularis
M 58  supraspinatus

**Nodes**
N  6  mediastinal, anterior

**Organs**
O  3  liver

**Pleural Structures**
P  5  mediastinal (parietal)

**Skeletal Structures**
S  1  clavicle, left
S  3  rib (1st right)
S  5  rib, body, lateral (9th right)
S  6  rib, body, posterior (4th right)
S  7  rib, costal cartilage (1st left)
S 10  scapula
S 15  scapula, spine
S 20  intervertebral disc (C6–C7)
S 21  vertebra, body (C7)
S 23  vertebra, inferior articular
      process (C6)
S 24  vertebra, intervertebral foramen
      (C6–C7)
S 28  vertebra, superior articular
      process (C7)

**Trachea**
T  1  trachea

**Veins**
V  4  pulmonary, inferior, RLL
V  8  brachiocephalic (innominate),
      left
V 22  jugular, anterior, left
V 38  vena cava, inferior
V 39  vena cava, superior

**Nonanatomic**
    ‡  postmortem trauma

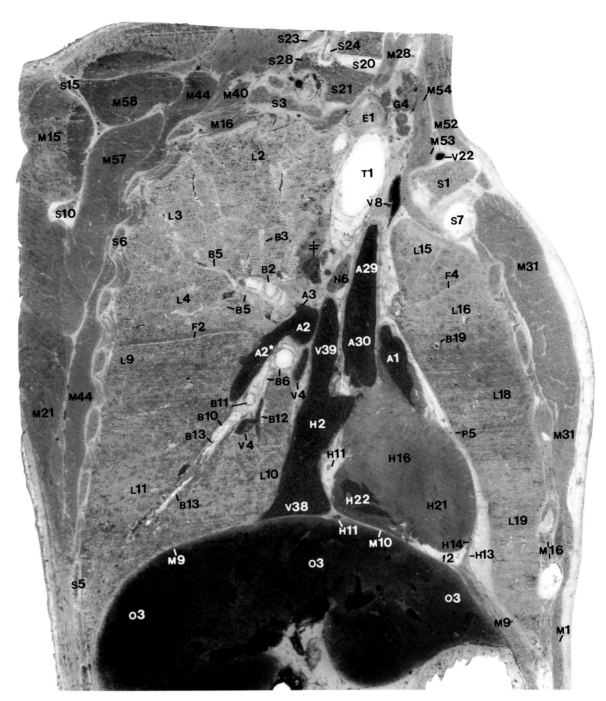

**Anatomic Specimen**

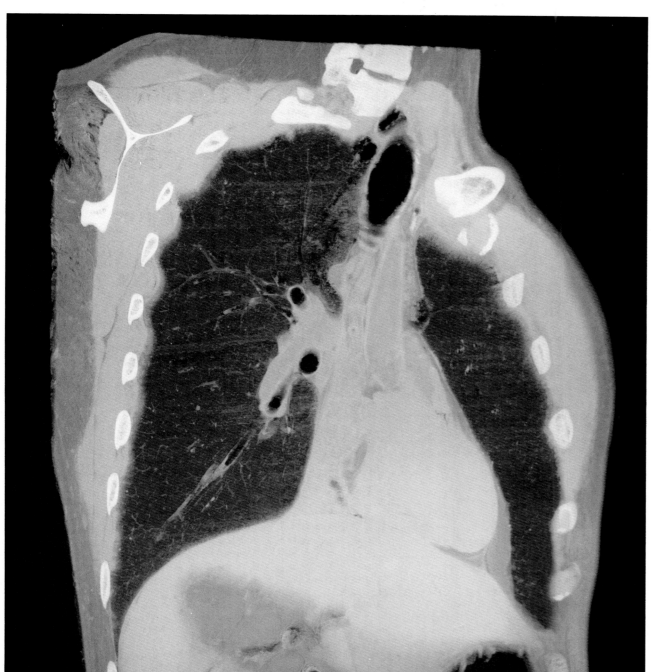

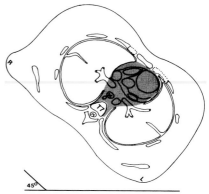

**Specimen Radiograph**

**Magnetic Resonance**

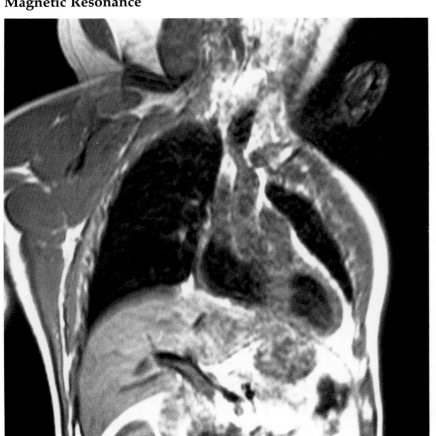

**Trispiral Tomogram**

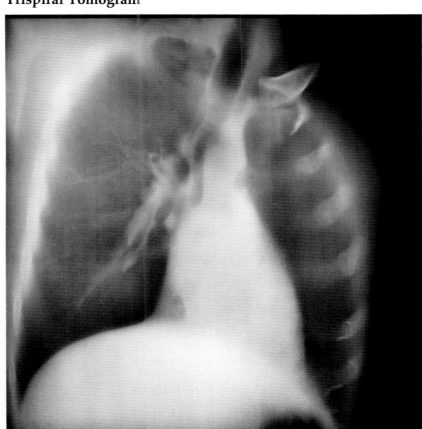

# Left Posterior Oblique

## Section 11

## Anatomic Key

### Arteries
A  1  pulmonary trunk
A  2  main pulmonary, right
A  3  right upper lobe (RUL)
A 10  right lower lobe (RLL)
A 29  aortic arch
A 30  aorta, ascending
A 37  carotid, common, left
A 54  suprascapular, right

### Bronchi
B  1  main, right
B  2  right upper lobe (RUL)
B  3  RUL, apical segment
B  4  RUL, posterior segment
B  6  intermediate, right
B 10  right lower lobe (RLL)
B 11  RLL, superior segment
B 12  RLL, medial basal segment
B 13  RLL, anterior basal segment
B 14  RLL, lateral basal segment
B 19  LUL, anterior segment

### Esophagus
E  1  esophagus

### Fat
f  2  epicardial
f  3  extrapleural

### Fissures
F  1  left major (oblique)
F  2  right major (oblique)
F  4  aberrant (LUL)

### Heart
H  1  atrium, left
H  2  atrium, right
H 11  coronary artery, right
H 16  septum, interventricular
H 21  ventricle, left
H 22  ventricle, right

### Lobes
L  2  RUL, apical segment
L  3  RUL, posterior segment
L  4  RUL, anterior segment
L  9  RLL, superior segment
L 10  RLL, medial basal segment
L 12  RLL, lateral basal segment
L 15  LUL, apicoposterior segment
L 16  LUL, anterior segment
L 18  LUL, superior lingular segment
L 19  LUL, inferior lingular segment
L 22  LLL, anteromedial basal segment

### Muscles
M  1  abdominal, external oblique
M  9  diaphragm
M 10  diaphragm, central tendon
M 15  infraspinatus
M 16  intercostal
M 21  latissimus dorsi
M 23  levator scapulae
M 28  longus colli
M 31  pectoralis, major
M 32  pectoralis, minor
M 33  platysma
M 44  serratus, anterior
M 51  splenius, cervicis
M 52  sternocleidomastoid
M 53  sternohyoid
M 54  sternothyroid
M 57  subscapularis
M 58  supraspinatus
M 63  trapezius

### Neural Structures
n  2  intercostal nerve

### Nodes
N  2  bronchopulmonary (hilar)

### Organs
O  3  liver
O  6  stomach

### Pleural Structures
P  2  costal (parietal)
P  5  mediastinal (parietal)
P  6  visceral

### Skeletal Structures
S  1  clavicle, left
S  4  rib, body, anterior (5th left)
S  6  rib, body, posterior (8th right)
S  7  rib, costal cartilage (1st left)
S  8  rib, head and neck (1st right)
S 10  scapula
S 15  scapula, spine
S 20  intervertebral disc (C6–C7)
S 21  vertebra, body (C7)
S 27  vertebra, spinous process (T1)

### Trachea
T  1  trachea

### Veins
V  4  pulmonary, inferior, RLL
V  7  azygos
V  8  brachiocephalic (innominate), left
V 22  jugular, anterior, left
V 26  jugular, internal, left
V 33  suprascapular, right
V 38  vena cava, inferior

### Nonanatomic
PC  pulmonary cyst

**Anatomic Specimen**

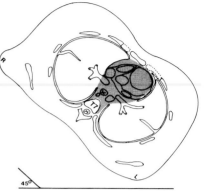

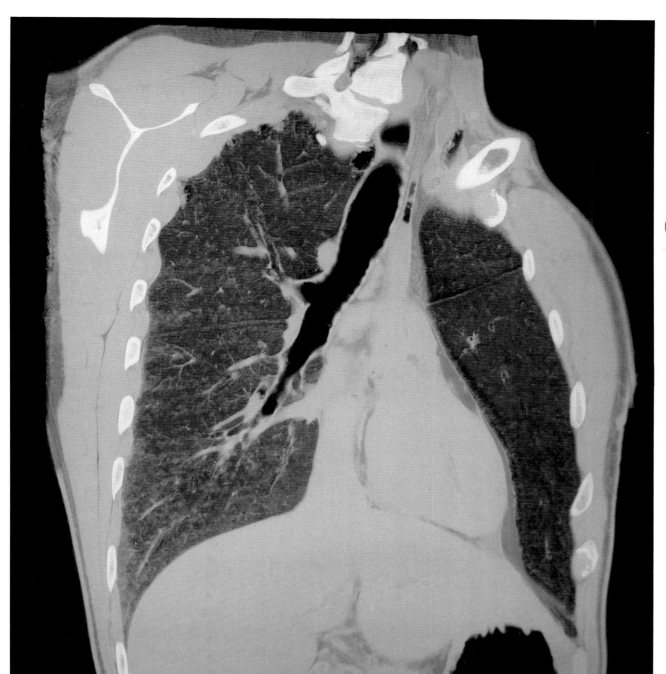

**Specimen Radiograph**

**Magnetic Resonance**

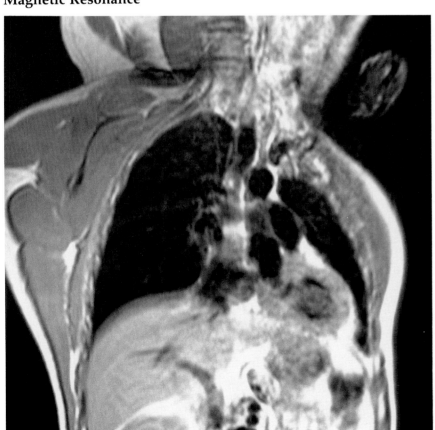

**Trispiral Tomogram**

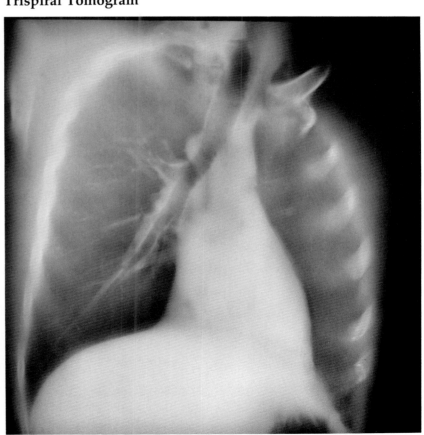

# Left Posterior Oblique

## Section 12

## Anatomic Key

### Arteries
A **1** pulmonary trunk
A **2** main pulmonary, right
A **15** RLL, posterior basal branch
A **29** aortic arch
A **30** aorta, ascending
A **37** carotid, common, left
A **49** subclavian, left
A **54** suprascapular, right
A **66** vertebral, right

### Bronchi
B **1** main, right
B **2** right upper lobe (RUL)
B **3** RUL, apical segment
B **4** RUL, posterior segment
B **6** intermediate, right
B **11** RLL, superior segment
B **12** RLL, medial basal segment
B **14** RLL, lateral basal segment
B **15** RLL, posterior basal segment
B **19** LUL, anterior segment
B **22** LUL, inferior lingular segment

### Esophagus
E **1** esophagus

### Fat
f **3** extrapleural
f **4** mediastinal

### Fissures
F **1** left major (oblique)
F **2** right major (oblique)
F **4** aberrant (LUL)

### Heart
H **1** atrium, left
H **2** atrium, right
H **6** cardiac vein, middle
H **9** coronary artery, posterior
     descending
H **11** coronary artery, right
H **12** coronary sinus
H **13** pericardium, parietal serous
H **14** pericardium, visceral serous
H **20** valve, aortic
H **21** ventricle, left

### Lobes
L **2** RUL, apical segment
L **3** RUL, posterior segment
L **9** RLL, superior segment
L **10** RLL, medial basal segment
L **12** RLL, lateral basal segment
L **15** LUL, apicoposterior segment
L **16** LUL, anterior segment
L **18** LUL, superior lingular segment
L **19** LUL, inferior lingular segment
L **22** LLL, anteromedial basal
     segment

### Muscles
M **9** diaphragm
M **10** diaphragm, central tendon
M **15** infraspinatus
M **16** intercostal
M **21** latissimus dorsi
M **23** levator scapulae
M **28** longus colli
M **31** pectoralis, major
M **32** pectoralis, minor
M **38** scalene, anterior
M **41** semispinalis, capitis
M **44** serratus, anterior
M **51** splenius, cervicis
M **52** sternocleidomastoid
M **53** sternohyoid
M **55** subclavius
M **57** subscapularis
M **58** supraspinatus
M **63** trapezius

### Neural Structures
n **5** spinal cord

### Nodes
N **2** bronchopulmonary (hilar)
N **6** mediastinal, anterior

### Organs
O **3** liver

### Pleural Structures
P **5** mediastinal (parietal)

### Skeletal Structures
S **1** clavicle, left
S **3** rib (9th right)
S **5** rib, body, lateral (2nd right)
S **7** rib, costal cartilage (1st left)
S **8** rib, head and neck (2nd right)
S **10** scapula
S **15** scapula, spine
S **21** vertebra, body (T1)
S **26** vertebra, pedicle (T1)
S **29** vertebra, transverse process
     (T1)

### Trachea
T **1** trachea

### Veins
V **4** pulmonary, inferior, RLL
V **7** azygos
V **8** brachiocephalic (innominate),
     left
V **30** subclavian, left
V **33** suprascapular, right
V **38** vena cava, inferior

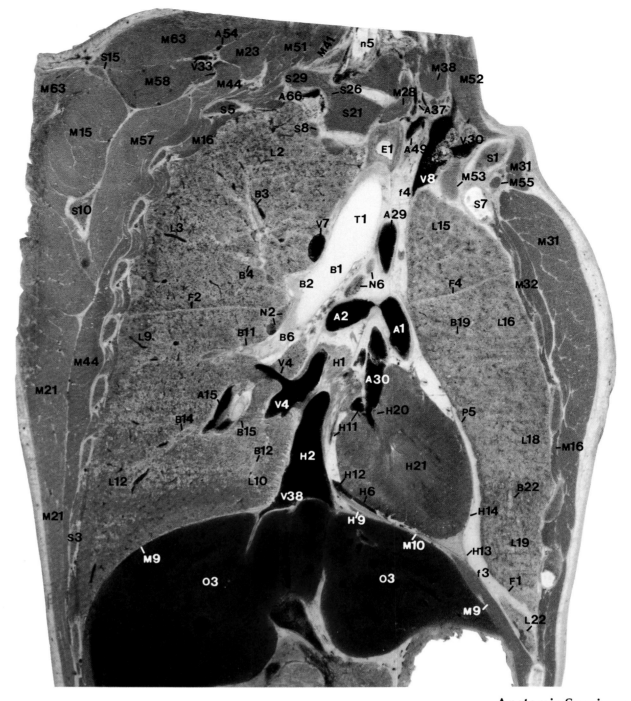

**Anatomic Specimen**

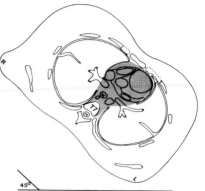

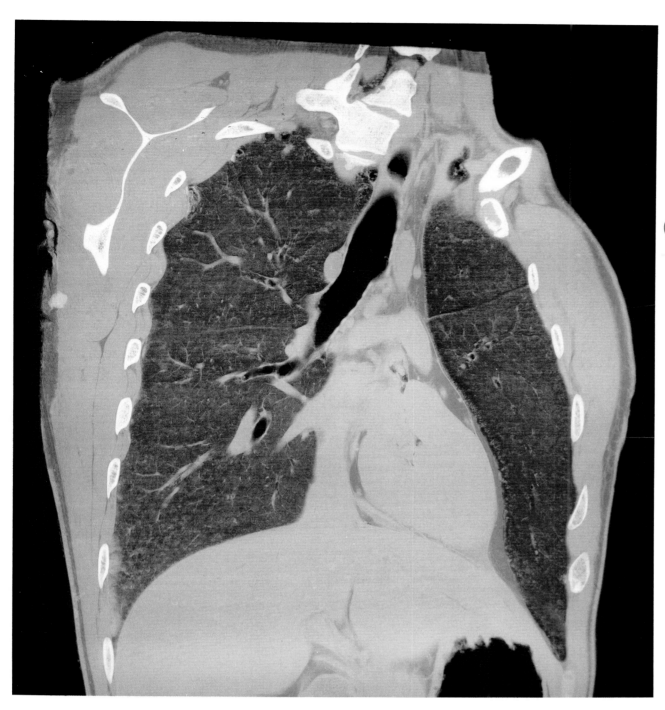

Specimen Radiograph

**Magnetic Resonance**

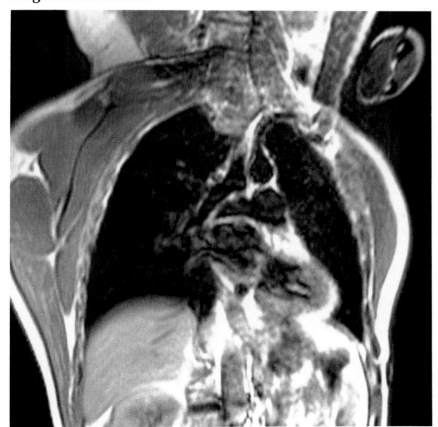

**Trispiral Tomogram**

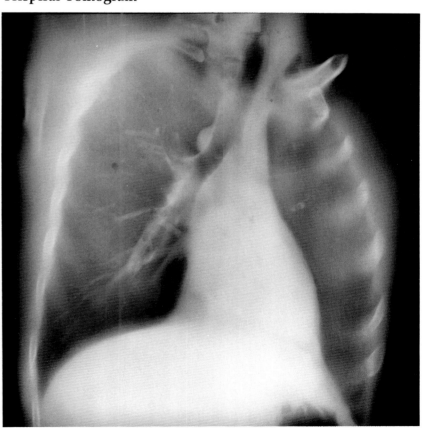

# Left Posterior Oblique

## Section 13

## Anatomic Key

### Arteries
A **1** pulmonary trunk
A **2** main pulmonary, right
A **15** RLL, posterior basal branch
A **16** main pulmonary, left
A **29** aortic arch
A **49** subclavian, left
A **54** suprascapular, right
A **66** vertebral, right

### Bronchi
B **1** main, right
B **2** right upper lobe (RUL)
B **3** RUL, apical segment
B **4** RUL, posterior segment
B **6** intermediate, right
B **11** RLL, superior segment
B **15** RLL, posterior basal segment
B **16** main, left
B **18** LUL, apicoposterior segment
B **19** LUL, anterior segment
B **21** LUL, superior lingular segment
B **22** LUL, inferior lingular segment

### Esophagus
E **1** esophagus

### Fat
f **2** epicardial
f **3** extrapleural
f **4** mediastinal
f **6** perirenal

### Fissures
F **1** left major (oblique)
F **2** right major (oblique)
F **4** aberrant (LUL)

### Glands
G **2** pancreas

### Heart
H **1** atrium, left
H **2** atrium, right
H **8** coronary artery, circumflex
H **9** coronary artery, anterior descending
H **10** coronary artery, left
H **15** septum, interatrial
H **19** valve, mitral
H **21** ventricle, left

### Lobes
L **2** RUL, apical segment
L **3** RUL, posterior segment
L **4** RUL, anterior segment
L **9** RLL, superior segment
L **10** RLL, medial basal segment
L **13** RLL, posterior basal segment
L **15** LUL, apicoposterior segment
L **16** LUL, anterior segment
L **18** LUL, superior lingular segment
L **19** LUL, inferior lingular segment
L **22** LLL, anteromedial basal segment

### Muscles
M **1** abdominal, external oblique
M **9** diaphragm
M **10** diaphragm, central tendon
M **15** infraspinatus
M **16** intercostal
M **21** latissimus dorsi
M **23** levator scapulae
M **30** omohyoid
M **31** pectoralis, major
M **32** pectoralis, minor
M **38** scalene, anterior
M **41** semispinalis, capitis
M **44** serratus, anterior
M **51** splenius, cervicis
M **52** sternocleidomastoid
M **55** subclavius
M **57** subscapularis
M **58** supraspinatus
M **63** trapezius

### Neural Structures
n **1** brachial plexus
n **2** intercostal nerve
n **5** spinal cord

### Nodes
N **2** bronchopulmonary (hilar)
N **6** mediastinal, anterior

### Organs
O **3** liver
O **6** stomach

### Skeletal Structures
s **1** clavicle, left
s **3** rib (6th left)
s **4** rib, body, anterior (1st left)
s **5** rib, body, lateral (10th right)
s **6** rib, body, posterior (2nd right)
s **10** scapula
s **15** scapula, spine
s **20** intervertebral disc (T1–T2)
s **24** vertebra, intervertebral foramen (T1–T2)

### Trachea
T **1** trachea
T **2** carina (bifurcation of trachea)

### Veins
V **4** pulmonary, inferior, RLL
V **7** azygos
V **30** subclavian, left
V **33** suprascapular, right
V **36** transverse cervical, left
V **38** vena cava, inferior
V **41** venous plexus, vertebral
V **42** vertebral, left

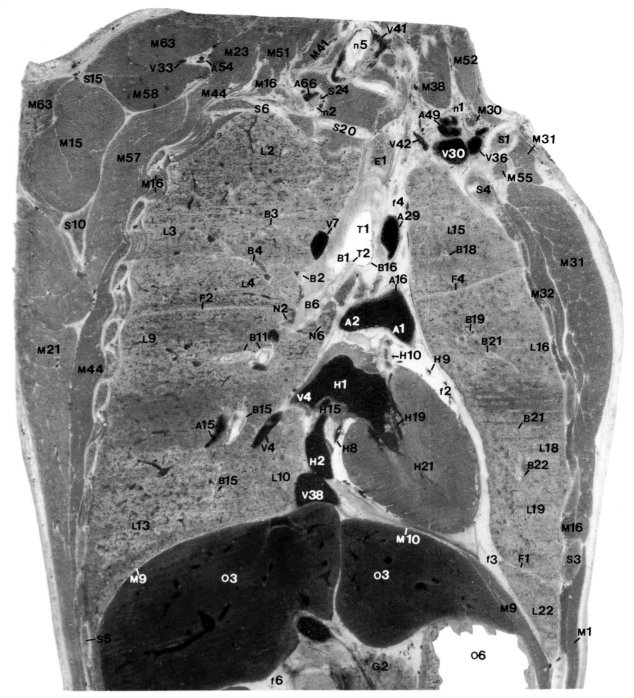

**Anatomic Specimen**

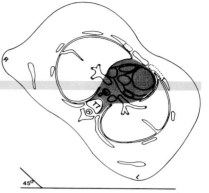

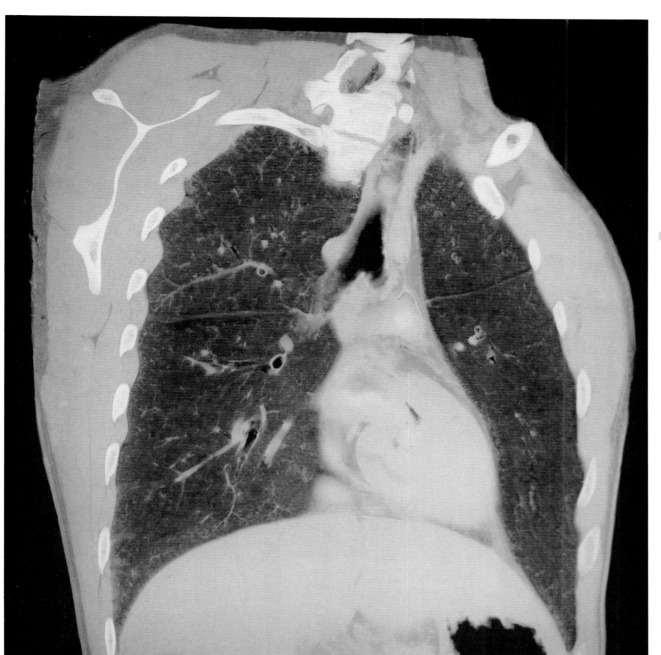

**Specimen Radiograph**

**Magnetic Resonance**

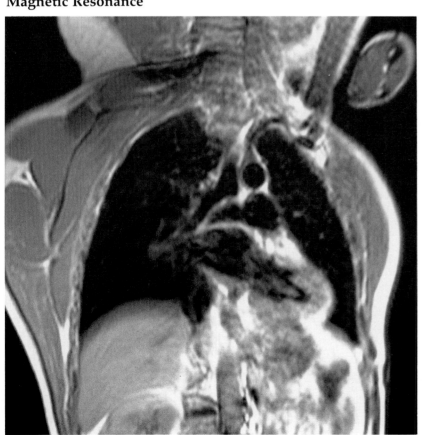

**Trispiral Tomogram**

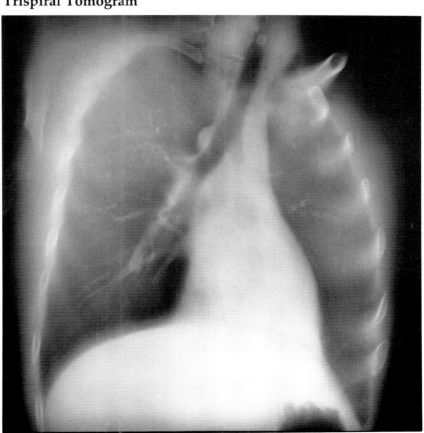

# Left Posterior Oblique

## Section 14

### Anatomic Key

**Arteries**
A 16 main pulmonary, left
A 21 LUL, superior lingular branch
A 29 aortic arch

**Bronchi**
B 3 RUL, apical segment
B 4 RUL, posterior segment
B 11 RLL, superior segment
B 12 RLL, medial basal segment
B 15 RLL, posterior basal segment
B 16 main, left
B 19 LUL, anterior segment
B 21 LUL, superior lingular segment
B 22 LUL, inferior lingular segment

**Esophagus**
E 1 esophagus

**Fat**
f 6 perirenal

**Fissures**
F 1 left major (oblique)
F 2 right major (oblique)
F 4 aberrant (LUL)

**Glands**
G 1 adrenal, right
G 2 pancreas

**Heart**
H 1 atrium, left
H 5 cardiac vein, great
H 8 coronary artery, circumflex
H 9 coronary artery, anterior descending
H 10 coronary artery, left
H 13 pericardium, parietal serous
H 14 pericardium, visceral serous
H 19 valve, mitral
H 21 ventricle, left

**Lobes**
L 2 RUL, apical segment
L 3 RUL, posterior segment
L 9 RLL, superior segment
L 10 RLL, medial basal segment
L 13 RLL, posterior basal segment
L 15 LUL, apicoposterior segment
L 16 LUL, anterior segment
L 18 LUL, superior lingular segment
L 19 LUL, inferior lingular segment
L 22 LLL, anteromedial basal segment

**Muscles**
M 1 abdominal, external oblique
M 9 diaphragm
M 12 diaphragm, crus, right
M 15 infraspinatus
M 16 intercostal
M 21 latissimus dorsi
M 23 levator scapulae
M 31 pectoralis, major
M 32 pectoralis, minor
M 38 scalene, anterior
M 41 semispinalis, capitis
M 44 serratus, anterior
M 51 splenius, cervicis
M 55 subclavius
M 57 subscapularis
M 58 supraspinatus
M 63 trapezius

**Neural Structures**
n 5 spinal cord

**Nodes**
N 6 mediastinal, anterior

**Organs**
O 2 kidney, right
O 3 liver
O 6 stomach

**Skeletal Structures**
S 1 clavicle, left
S 3 rib (1st left)
S 4 rib, body, anterior (1st left)
S 5 rib, body, lateral (10th right)
S 7 rib, costal cartilage (7th left)
S 8 rib, head and neck (2nd right)
S 10 scapula
S 15 scapula, spine
S 21 vertebra, body (T2)
S 23 vertebra, inferior articular process (T1)
S 26 vertebra, pedicle (T2)
S 28 vertebra, superior articular process (T2)

**Trachea**
T 1 trachea
T 2 carina (bifurcation of trachea)

**Veins**
V 7 azygos
V 15 intercostal, highest (superior), right
V 19 intercostal, posterior, right
V 30 subclavian, left
V 34 thoracoacromial, left
V 36 transverse cervical, left
V 38 vena cava, inferior
V 41 venous plexus, vertebral

**Nonanatomic**
PC pulmonary cyst

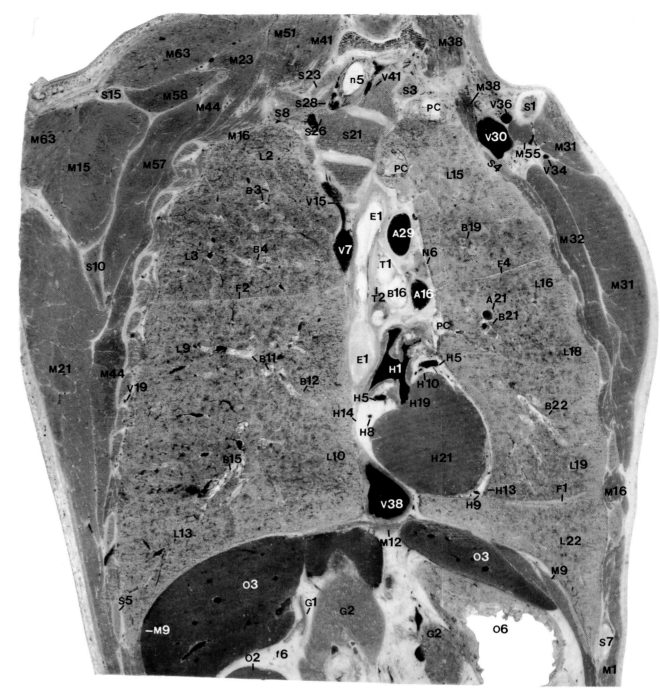

**Anatomic Specimen**

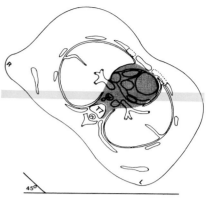

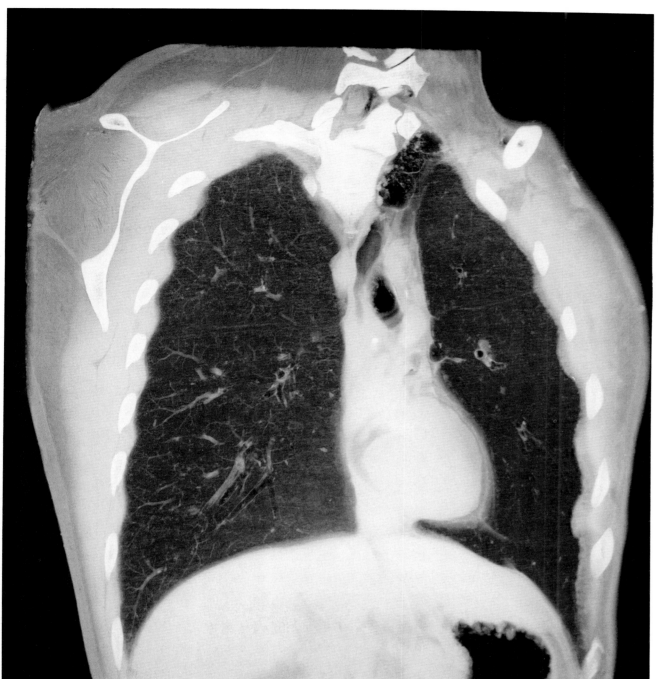

**Specimen Radiograph**

**Magnetic Resonance**

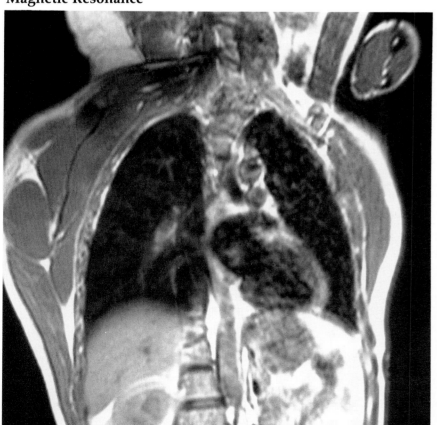

**Trispiral Tomogram**

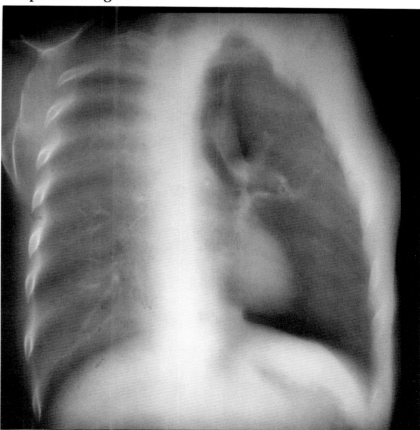

# Left Posterior Oblique

## Section 15

### Anatomic Key

#### Arteries
A 16 main pulmonary, left
A 19 LUL, anterior branch
A 22 LUL, inferior lingular branch
A 29 aortic arch
A 49 subclavian, left
A 59 thyrocervical trunk, left

#### Bronchi
B 3 RUL, apical segment
B 4 RUL, posterior segment
B 11 RLL, superior segment
B 12 RLL, medial basal segment
B 15 RLL, posterior basal segment
B 16 main, left
B 19 LUL, anterior segment
B 21 LUL, superior lingular segment
B 22 LUL, inferior lingular segment

#### Esophagus
E 1 esophagus
E 2 esophagogastric junction

#### Fissures
F 1 left major (oblique)
F 2 right major (oblique)
F 4 aberrant (LUL)

#### Glands
G 1 adrenal, right
G 2 pancreas

#### Heart
H 1 atrium, left
H 5 cardiac vein, great
H 8 coronary artery, circumflex
H 21 ventricle, left

#### Lobes
L 2 RUL, apical segment
L 3 RUL, posterior segment
L 9 RLL, superior segment
L 10 RLL, medial basal segment
L 13 RLL, posterior basal segment
L 15 LUL, apicoposterior segment
L 16 LUL, anterior segment
L 18 LUL, superior lingular segment
L 19 LUL, inferior lingular segment
L 22 LLL, anteromedial basal segment

#### Muscles
M 1 abdominal, external oblique
M 9 diaphragm
M 12 diaphragm, crus, right
M 13 erector spinae
M 15 infraspinatus
M 16 intercostal
M 21 latissimus dorsi
M 23 levator scapulae
M 30 omohyoid
M 31 pectoralis, major
M 32 pectoralis, minor
M 39 scalene, middle
M 40 scalene, posterior
M 41 semispinalis, capitis
M 44 serratus, anterior
M 51 splenius, cervicis
M 55 subclavius
M 57 subscapularis
M 58 supraspinatus
M 63 trapezius

#### Neural Structures
n 5 spinal cord

#### Nodes
N 2 bronchopulmonary (hilar)
N 11 tracheobronchial, inferior (subcarinal)

#### Organs
O 3 liver
O 6 stomach

#### Pleural Structures
P 2 costal (parietal)
P 5 mediastinal (parietal)
P 6 visceral

#### Skeletal Structures
S 1 clavicle, left
S 3 rib (3rd right)
S 4 rib, body, anterior (1st left)
S 5 rib, body, lateral (10th right)
S 7 rib, costal cartilage (7th left)
S 9 rib, tubercle (1st left)
S 10 scapula
S 15 scapula, spine
S 20 intervertebral disc (T12–L1)
S 21 vertebra, body (T3)
S 25 vertebra, lamina (T1)
S 29 vertebra, transverse process (T1)

#### Veins
V 2 pulmonary, superior, LUL
V 3 pulmonary, superior, LUL, lingula
V 4 pulmonary, inferior, RLL
V 7 azygos
V 14 intercostal, highest (superior), left
V 15 intercostal, highest (superior), right
V 18 intercostal, posterior, left
V 22 jugular, anterior, left
V 24 jugular, external, left
V 30 subclavian, left
V 34 thoracoacromial, left
V 36 transverse cervical, left
V 38 vena cava, inferior
V 41 venous plexus, vertebral

#### Nonanatomic
PC pulmonary cyst

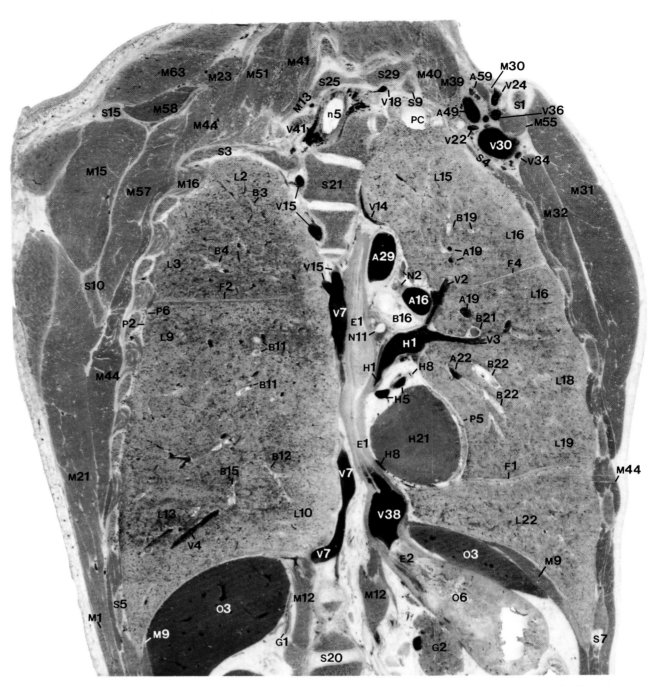

**Anatomic Specimen**

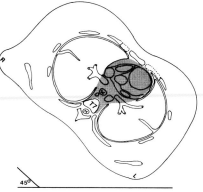

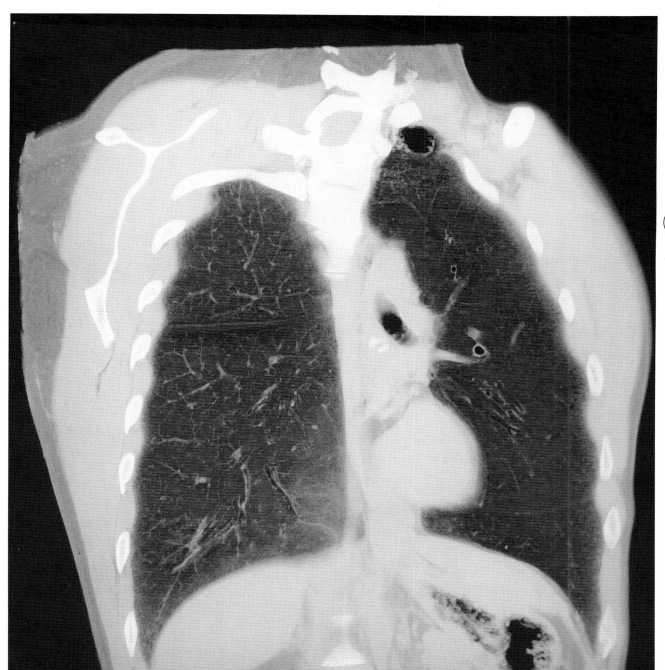

**Specimen Radiograph**

**Magnetic Resonance**

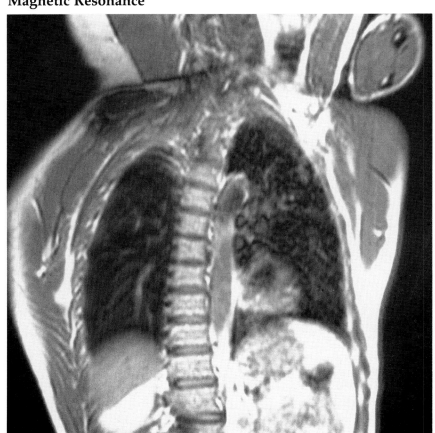

**Trispiral Tomogram**

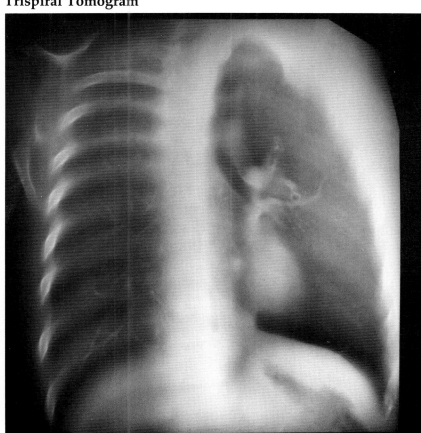

# Left Posterior Oblique

## Section 16

## Anatomic Key

### Arteries
A **16** main pulmonary, left
A **17** left upper lobe (LUL)
A **20** LUL, lingula
A **29** aortic arch
A **31** aorta, descending
A **49** subclavian, left
A **63** transverse cervical, left

### Bronchi
B **11** RLL, superior segment
B **12** RLL, medial basal segment
B **15** RLL, posterior basal segment
B **16** main, left
B **18** LUL, apicoposterior segment
B **19** LUL, anterior segment
B **20** LUL, lingula
B **21** LUL, superior lingular segment
B **22** LUL, inferior lingular segment
B **25** LLL, anteromedial basal
   segment

### Esophagus
E **2** esophagogastric junction

### Fat
f **2** epicardial
f **6** perirenal

### Fissures
F **1** left major (oblique)
F **2** right major (oblique)
F **4** aberrant (LUL)

### Glands
G **1** adrenal, right
G **2** pancreas

### Heart
H **1** atrium, left
H **5** cardiac vein, great
H **8** coronary artery, circumflex
H **21** ventricle, left

### Lobes
L **2** RUL, apical segment
L **3** RUL, posterior segment
L **9** RLL, superior segment
L **10** RLL, medial basal segment
L **13** RLL, posterior basal segment
L **15** LUL, apicoposterior segment
L **16** LUL, anterior segment
L **18** LUL, superior lingular segment
L **19** LUL, inferior lingular segment
L **22** LLL, anteromedial basal
   segment
L **23** LLL, lateral basal segment

### Muscles
M **1** abdominal, external oblique
M **9** diaphragm
M **11** diaphragm, crus, left
M **12** diaphragm, crus, right
M **13** erector spinae
M **15** infraspinatus
M **16** intercostal
M **21** latissimus dorsi
M **23** levator scapulae
M **30** omohyoid
M **31** pectoralis, major
M **32** pectoralis, minor
M **39** scalene, middle
M **44** serratus, anterior
M **50** splenius, capitis
M **51** splenius, cervicis
M **55** subclavius
M **57** subscapularis
M **58** supraspinatus
M **63** trapezius

### Neural Structures
n **1** brachial plexus
n **5** spinal cord

### Nodes
N **2** bronchopulmonary (hilar)

### Organs
O **3** liver
O **6** stomach

### Skeletal Structures
S **1** clavicle, left
S **3** rib (3rd right)
S **4** rib, body, anterior (1st left)
S **5** rib, body, lateral (5th right)
S **8** rib, head and neck (2nd left)
S **9** rib, tubercle (1st left)
S **10** scapula
S **15** scapula, spine
S **20** intervertebral disc (T3–T4)
S **21** vertebra, body (T12)
S **29** vertebra, transverse process
   (T1)

### Veins
V **5** pulmonary, inferior, LLL
V **7** azygos
V **10** cephalic, left
V **12** hemiazygos
V **14** intercostal, highest (superior),
   left
V **15** intercostal, highest (superior),
   right
V **19** intercostal, posterior, right
V **24** jugular, external, left
V **30** subclavian, left
V **34** thoracoacromial, left
V **36** transverse cervical, left
V **38** vena cava, inferior
V **41** venous plexus, vertebral

### Nonanatomic
PC pulmonary cyst

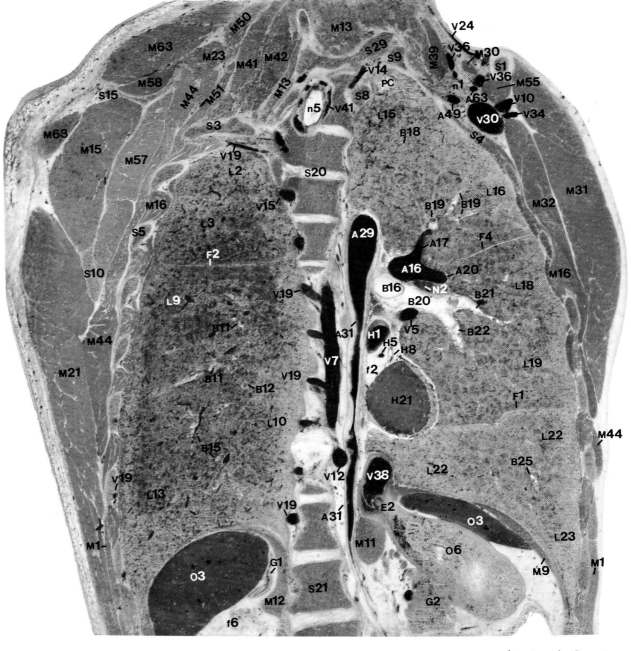

**Anatomic Specimen**

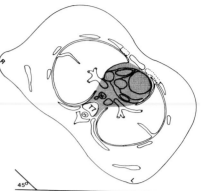

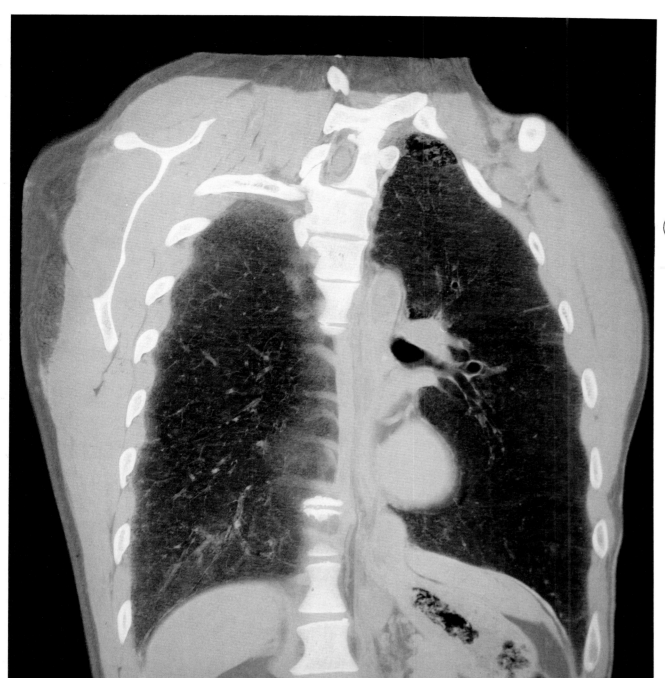

Specimen Radiograph

**Magnetic Resonance**

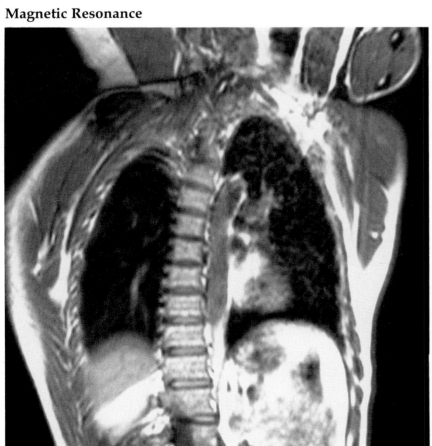

**Trispiral Tomogram**

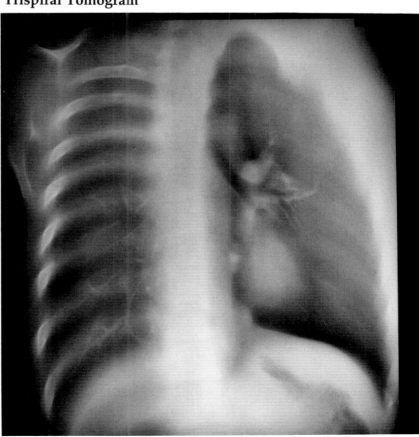

# Left Posterior Oblique

## Section 17

### Anatomic Key

**Arteries**
A 16 main pulmonary, left
A 17 left upper lobe (LUL)
A 20 LUL, lingula
A 29 aortic arch
A 31 aorta, descending
A 32 axillary, left
A 45 intercostal, posterior, left
A 53 suprascapular, left
A 57 thoracoacromial, left

**Bronchi**
B 11 RLL, superior segment
B 15 RLL, posterior basal segment
B 16 main, left
B 18 LUL, apicoposterior segment
B 19 LUL, anterior segment
B 20 LUL, lingula
B 21 LUL, superior lingular segment
B 22 LUL, inferior lingular segment
B 23 left lower lobe (LLL)
B 26 LLL, lateral basal segment

**Fat**
f 6 perirenal

**Fissures**
F 1 left major (oblique)
F 2 right major (oblique)
F 4 aberrant (LUL)

**Glands**
G 1 adrenal, left
G 2 pancreas

**Heart**
H 2 atrium, right

**Lobes**
L 3 RUL, posterior segment
L 9 RLL, superior segment
L 10 RLL, medial basal segment
L 13 RLL, posterior basal segment
L 15 LUL, apicoposterior segment
L 16 LUL, anterior segment
L 18 LUL, superior lingular segment
L 19 LUL, inferior lingular segment
L 22 LLL, anteromedial basal segment
L 23 LLL, lateral basal segment

**Muscles**
M 1 abdominal, external oblique
M 9 diaphragm
M 11 diaphragm, crus, left
M 12 diaphragm, crus, right
M 13 erector spinae
M 15 infraspinatus
M 16 intercostal
M 21 latissimus dorsi
M 23 levator scapulae
M 30 omohyoid
M 31 pectoralis, major
M 32 pectoralis, minor
M 39 scalene, middle
M 41 semispinalis, capitis
M 42 semispinalis, cervicis
M 44 serratus, anterior
M 50 splenius, capitis
M 51 splenius, cervicis
M 55 subclavius
M 57 subscapularis
M 58 supraspinatus
M 63 trapezius

**Neural Structures**
n 1 brachial plexus
n 5 spinal cord

**Nodes**
N 2 bronchopulmonary (hilar)

**Organs**
O 3 liver
O 6 stomach

**Pleural Structures**
P 2 costal (parietal)
P 6 visceral

**Skeletal Structures**
S 1 clavicle, left
S 3 rib (1st left)
S 4 rib, body, anterior (7th left)
S 5 rib, body, lateral (5th right)
S 8 rib, head and neck (4th right)
S 10 scapula
S 15 scapula, spine
S 20 intervertebral disc (T3–T4)
S 21 vertebra, body (T12)
S 22 vertebra, costal facet (T4)
S 29 vertebra, transverse process (T3)

**Veins**
V 5 pulmonary, inferior, LLL
V 6 axillary, left
V 10 cephalic, left
V 12 hemiazygos
V 18 intercostal, posterior, left
V 33 suprascapular, left
V 34 thoracoacromial, left
V 36 transverse cervical, left
V 41 venous plexus, vertebral

**Nonanatomic**
PC pulmonary cyst

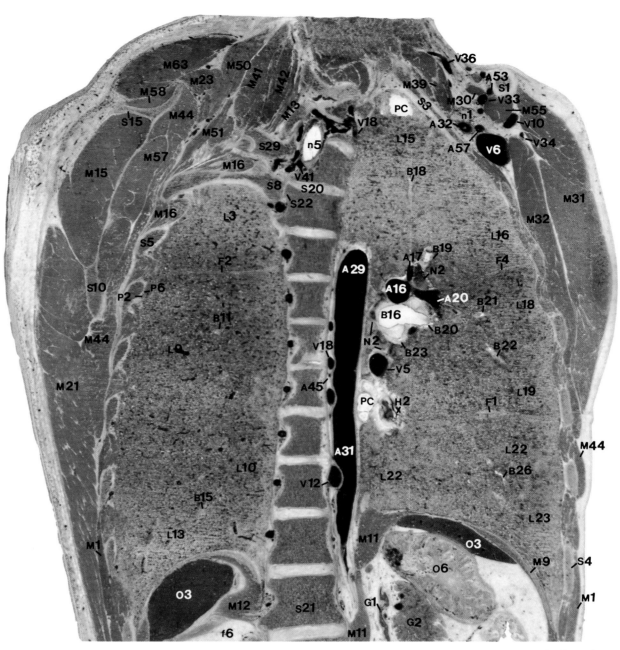

**Anatomic Specimen**

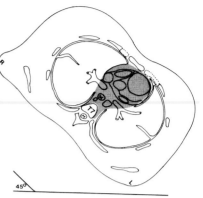

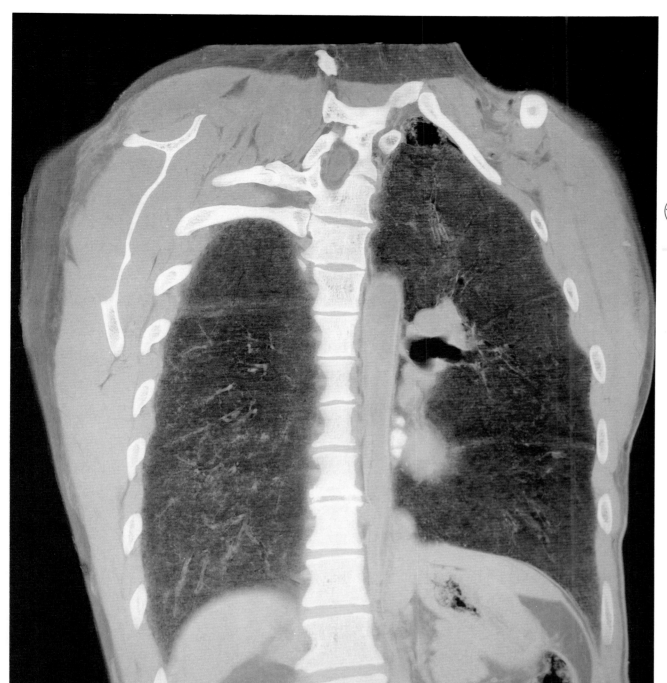

**Specimen Radiograph**

**Magnetic Resonance**

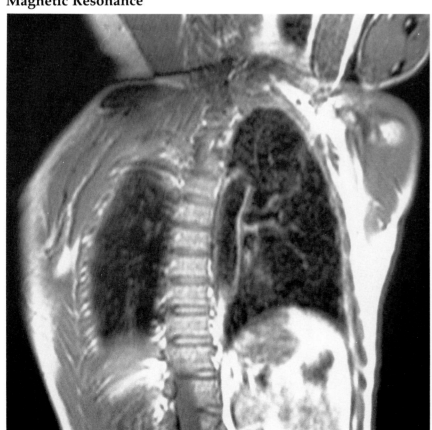

**Trispiral Tomogram**

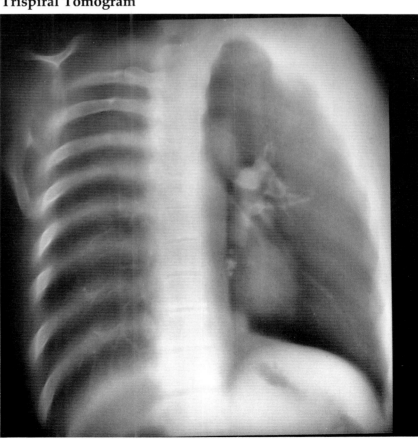

# Left Posterior Oblique

## Section 18

### Anatomic Key

#### Arteries
A 16 main pulmonary, left
A 24 LLL, superior branch
A 31 aorta, descending
A 32 axillary, left
A 43 intercostal, anterior, left
A 45 intercostal, posterior, left
A 46 intercostal, posterior, right
A 53 suprascapular, left
A 57 thoracoacromial, left

#### Bronchi
B 17 left upper lobe (LUL)
B 18 LUL, apicoposterior segment
B 19 LUL, anterior segment
B 23 left lower lobe (LLL)
B 24 LLL, superior segment
B 25 LLL, anteromedial basal segment
B 26 LLL, lateral basal segment

#### Fat
f 1 axillary
f 7 subcutaneous

#### Fissures
F 1 left major (oblique)
F 2 right major (oblique)
F 4 aberrant (LUL)

#### Glands
G 1 adrenal, left
G 2 pancreas

#### Lobes
L 3 RUL, posterior segment
L 9 RLL, superior segment
L 10 RLL, medial basal segment
L 13 RLL, posterior basal segment
L 15 LUL, apicoposterior segment
L 16 LUL, anterior segment
L 17 LUL, lingula
L 22 LLL, anteromedial basal segment
L 24 LLL, posterior basal segment

#### Muscles
M 1 abdominal, external oblique
M 9 diaphragm
M 11 diaphragm, crus, left
M 12 diaphragm, crus, right
M 13 erector spinae
M 15 infraspinatus
M 16 intercostal
M 21 latissimus dorsi
M 23 levator scapulae
M 30 omohyoid
M 31 pectoralis, major
M 32 pectoralis, minor
M 39 scalene, middle
M 40 scalene, posterior
M 41 semispinalis, capitis
M 42 semispinalis, cervicis
M 44 serratus, anterior
M 50 splenius, capitis
M 55 subclavius
M 57 subscapularis
M 58 supraspinatus
M 63 trapezius

#### Neural Structures
n 1 brachial plexus
n 5 spinal cord

#### Nodes
N 2 bronchopulmonary (hilar)

#### Organs
O 3 liver
O 6 stomach

#### Skeletal Structures
S 1 clavicle, left
S 3 rib (1st left)
S 5 rib, body, lateral (7th left)
S 6 rib, body, posterior (5th right)
S 8 rib, head and neck (5th right)
S 10 scapula
S 15 scapula, spine
S 20 intervertebral disc (T11–T12)
S 21 vertebra, body (T5)
S 22 vertebra, costal facet (T4–T5)

#### Veins
V 5 pulmonary, inferior, LLL
V 6 axillary, left
V 7 azygos
V 10 cephalic, left
V 12 hemiazygos
V 18 intercostal, posterior, left
V 19 intercostal, posterior, right
V 33 suprascapular, left
V 41 venous plexus, vertebral

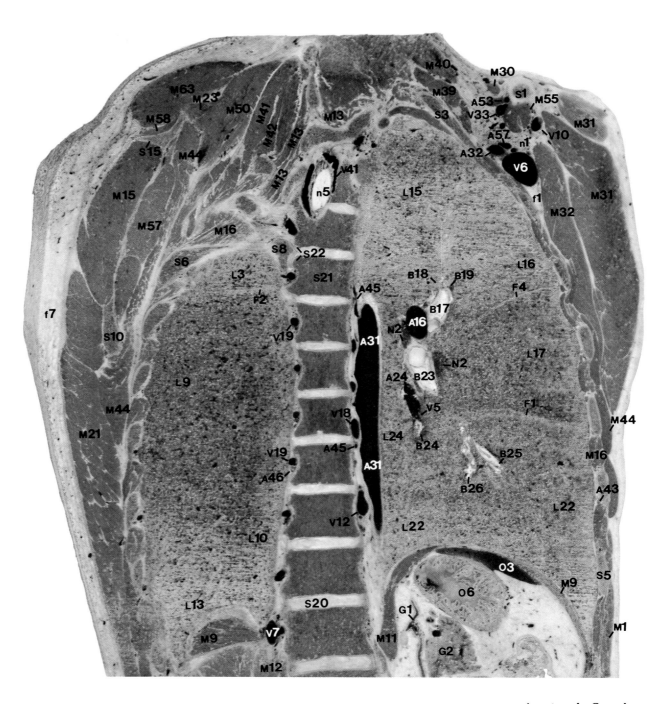

**Anatomic Specimen**

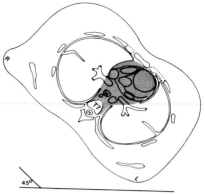

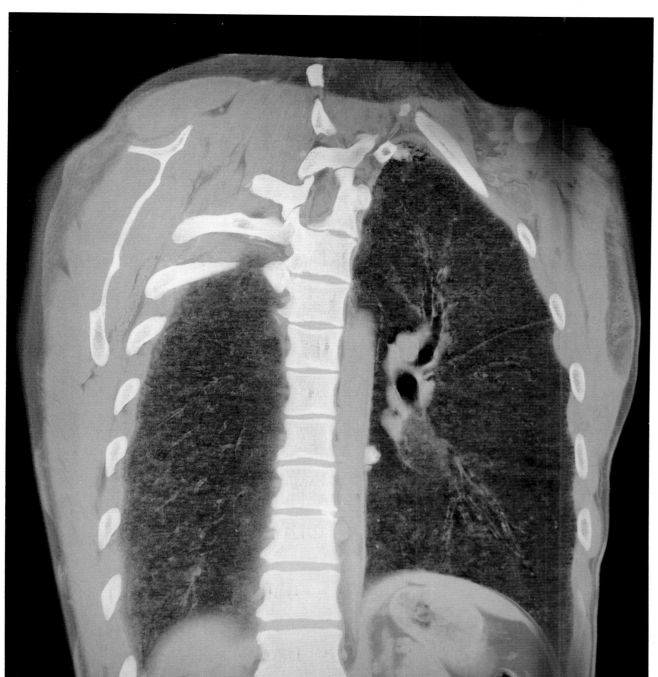

**Specimen Radiograph**

**Magnetic Resonance**

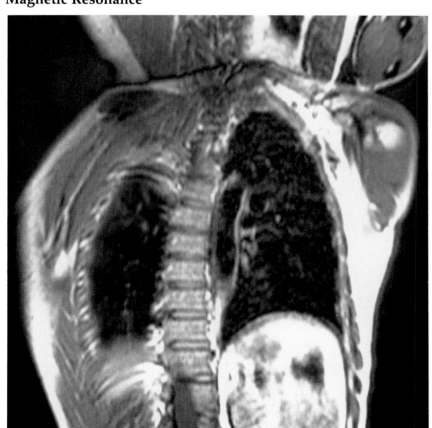

**Trispiral Tomogram**

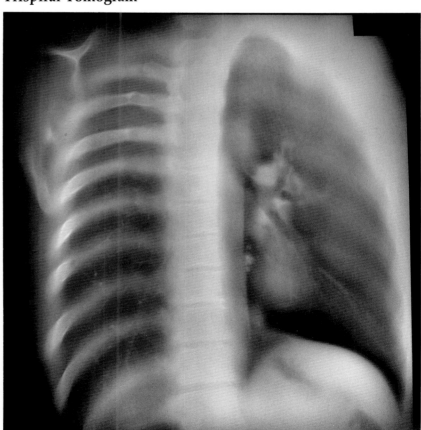

# Left Posterior Oblique

## Section 19

## Anatomic Key

### Arteries
A **16** main pulmonary, left
A **31** aorta, descending
A **32** axillary, left
A **45** intercostal, posterior, left
A **46** intercostal, posterior, right
A **53** suprascapular, left
A **57** thoracoacromial, left

### Bronchi
B **18** LUL, apicoposterior segment
B **19** LUL, anterior segment
B **23** left lower lobe (LLL)
B **24** LLL, superior segment
B **25** LLL, anteromedial basal segment
B **26** LLL, lateral basal segment

### Fat
f **1** axillary
f **3** extrapleural

### Fissures
F **1** left major (oblique)
F **2** right major (oblique)
F **4** aberrant (LUL)

### Glands
G **1** adrenal, left
G **2** pancreas

### Lobes
L **3** RUL, posterior segment
L **9** RLL, superior segment
L **10** RLL, medial basal segment
L **13** RLL, posterior basal segment
L **15** LUL, apicoposterior segment
L **16** LUL, anterior segment
L **17** LUL, lingula
L **22** LLL, anteromedial basal segment
L **23** LLL, lateral basal segment
L **24** LLL, posterior basal segment

### Muscles
M **1** abdominal, external oblique
M **9** diaphragm
M **11** diaphragm, crus, left
M **12** diaphragm, crus, right
M **13** erector spinae
M **15** infraspinatus
M **16** intercostal
M **30** omohyoid
M **31** pectoralis, major
M **32** pectoralis, minor
M **33** platysma
M **39** scalene, middle
M **40** scalene, posterior
M **41** semispinalis, capitis
M **42** semispinalis, cervicis
M **44** serratus, anterior
M **50** splenius, capitis
M **55** subclavius
M **57** subscapularis
M **61** transversospinalis
M **63** trapezius

### Neural Structures
n **1** brachial plexus
n **5** spinal cord

### Nodes
N **2** bronchopulmonary (hilar)

### Organs
O **6** stomach

### Pleural Structures
P **2** costal (parietal)
P **6** visceral

### Skeletal Structures
S **1** clavicle, left
S **3** rib (7th left)
S **4** rib, body, anterior (3rd left)
S **5** rib, body, lateral (11th right)
S **6** rib, body, posterior (1st left)
S **8** rib, head and neck (5th right)
S **9** rib, tubercle (4th right)
S **10** scapula
S **15** scapula, spine
S **20** intervertebral disc (T12–L1)
S **21** vertebra, body (T5)
S **29** vertebra, transverse process (T4)

### Veins
V **5** pulmonary, inferior, LLL
V **6** axillary, left
V **10** cephalic, left
V **12** hemiazygos
V **18** intercostal, posterior, left
V **19** intercostal, posterior, right
V **33** suprascapular, left
V **41** venous plexus, vertebral

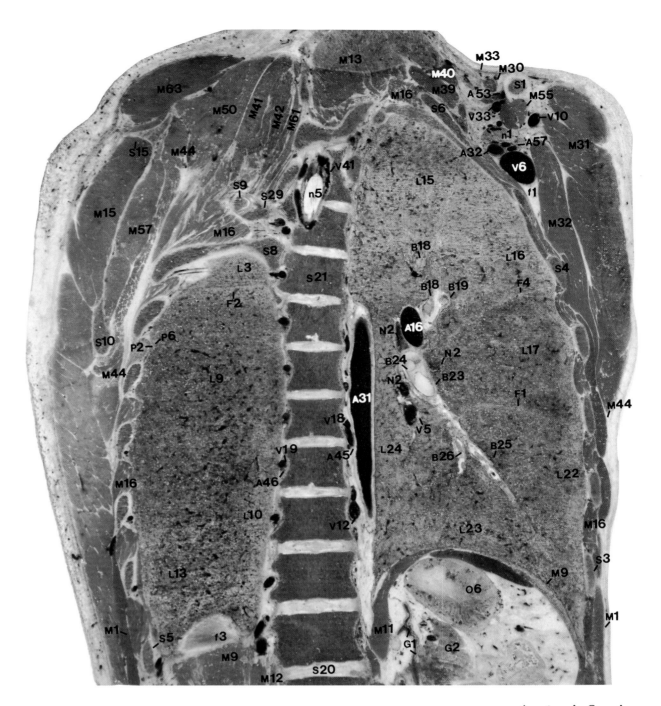

**Anatomic Specimen**

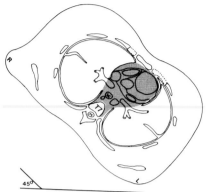

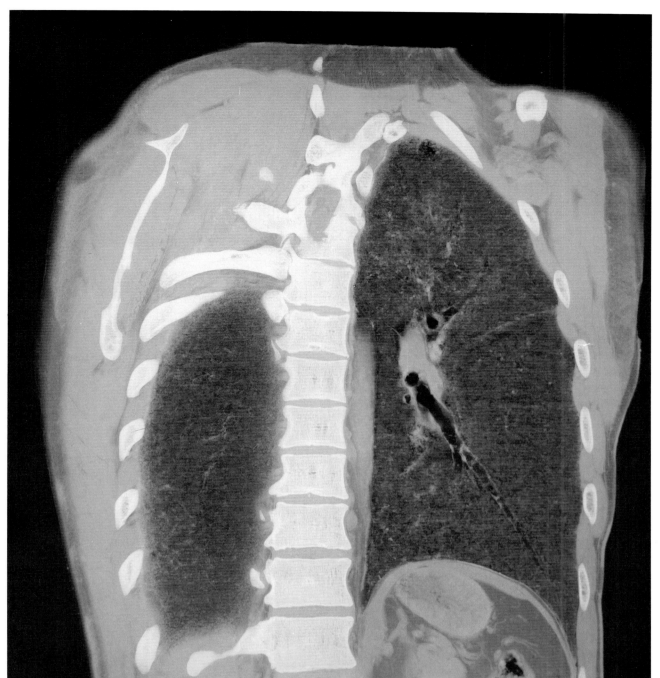

**Specimen Radiograph**

**Magnetic Resonance**

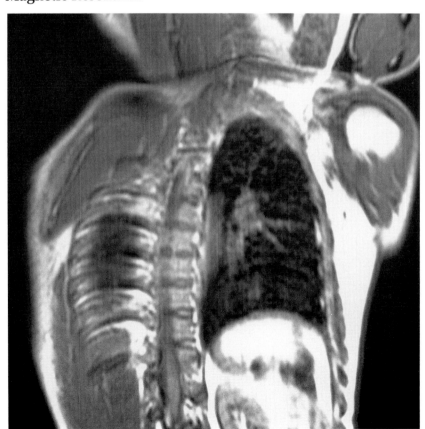

**Trispiral Tomogram**

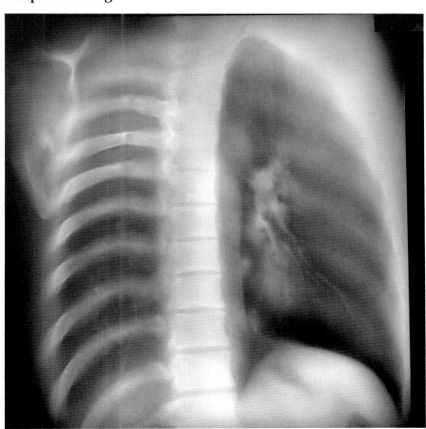

# Left Posterior Oblique

## Section 20

### Anatomic Key

**Arteries**
A 23 left lower lobe (LLL)
A 24 LLL, superior branch
A 25 LLL, anteromedial basal branch
A 32 axillary, left
A 45 intercostal, posterior, left
A 46 intercostal, posterior, right
A 57 thoracoacromial, left

**Bronchi**
B 18 LUL, apicoposterior segment
B 19 LUL, anterior segment
B 23 left lower lobe (LLL)
B 24 LLL, superior segment
B 25 LLL, anteromedial basal
segment
B 26 LLL, lateral basal segment
B 27 LLL, posterior basal segment

**Fat**
f 1 axillary
f 6 perirenal

**Fissures**
F 1 left major (oblique)
F 4 aberrant (LUL)

**Glands**
G 1 adrenal, left
G 2 pancreas

**Lobes**
L 9 RLL, superior segment
L 13 RLL, posterior basal segment
L 15 LUL, apicoposterior segment
L 17 LUL, lingula
L 21 LLL, superior segment
L 22 LLL, anteromedial basal
segment
L 23 LLL, lateral basal segment
L 24 LLL, posterior basal segment

**Muscles**
M 1 abdominal, external oblique
M 9 diaphragm
M 11 diaphragm, crus, left
M 15 infraspinatus
M 16 intercostal
M 21 latissimus dorsi
M 31 pectoralis, major
M 32 pectoralis, minor
M 35 rhomboid, major
M 44 serratus, anterior
M 55 subclavius
M 57 subscapularis
M 61 transversospinalis
M 63 trapezius

**Neural Structures**
n 1 brachial plexus
n 5 spinal cord

**Nodes**
N 2 bronchopulmonary (hilar)

**Organs**
O 2 kidney, left
O 5 spleen
O 6 stomach

**Pleural Structures**
P 2 costal (parietal)
P 6 visceral

**Skeletal Structures**
S 1 clavicle, left
S 3 rib (6th right)
S 4 rib, body, anterior (8th left)
S 5 rib, body, lateral (2nd left)
S 8 rib, head and neck (7th right)
S 10 scapula
S 20 intervertebral disc (T11–T12)
S 21 vertebra, body (T7)
S 22 vertebra, costal facet (T6–T7)

**Veins**
V 5 pulmonary, inferior, LLL
V 6 axillary, left
V 7 azygos
V 10 cephalic, left
V 12 hemiazygos
V 18 intercostal, posterior, left
V 19 intercostal, posterior, right
V 33 suprascapular, left
V 41 venous plexus, vertebral

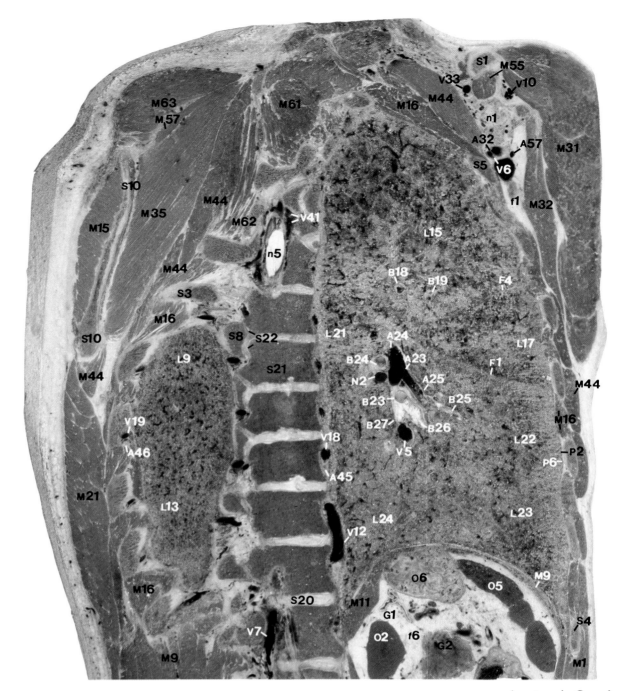

**Anatomic Specimen**

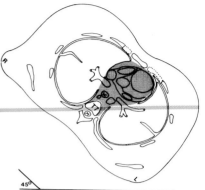

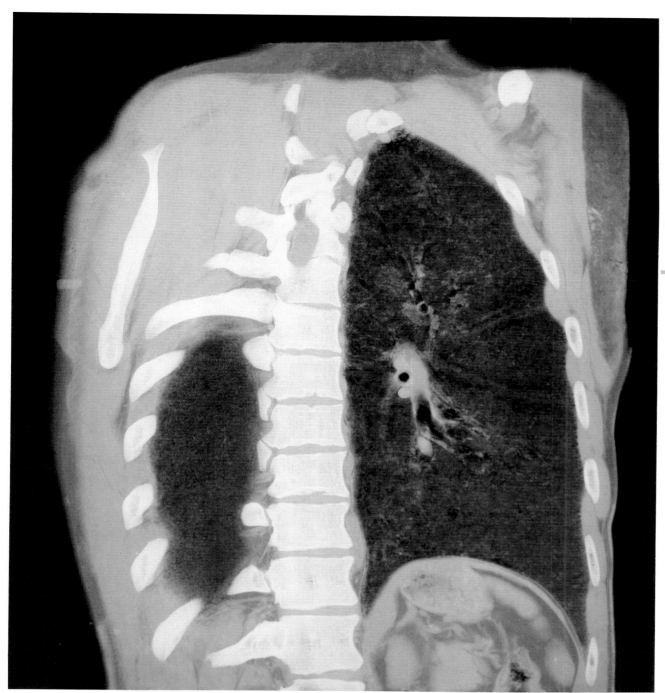

**Specimen Radiograph**

**Magnetic Resonance**

**Trispiral Tomogram**

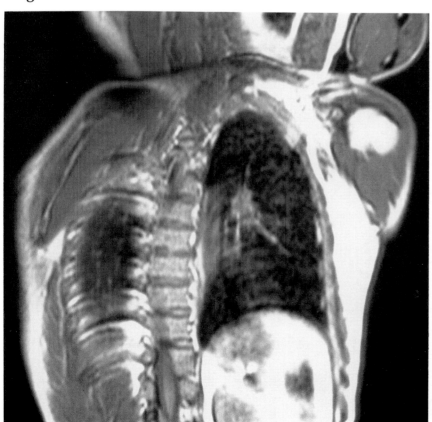

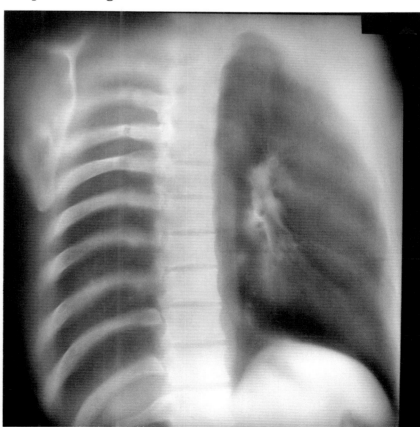

# Left Posterior Oblique

## Section 21

### Anatomic Key

**Arteries**
A 32 axillary, left

**Bronchi**
B 18 LUL, apicoposterior segment
B 24 LLL, superior segment
B 26 LLL, lateral basal segment
B 27 LLL, posterior basal segment

**Fat**
f 1 axillary
f 6 perirenal
f 7 subcutaneous

**Fissures**
F 1 left major (oblique)
F 4 aberrant (LUL)

**Glands**
G 2 pancreas

**Lobes**
L 15 LUL, apicoposterior segment
L 17 LUL, lingula
L 21 LLL, superior segment
L 23 LLL, lateral basal segment
L 24 LLL, posterior basal segment

**Muscles**
M 1 abdominal, external oblique
M 9 diaphragm
M 11 diaphragm, crus, left
M 13 erector spinae
M 16 intercostal
M 21 latissimus dorsi
M 27 longissimus, thoracis
M 31 pectoralis, major
M 32 pectoralis, minor
M 35 rhomboid, major
M 44 serratus, anterior
M 46 serratus, posterior superior
M 61 transversospinalis
M 63 trapezius

**Neural Structures**
n 1 brachial plexus
n 5 spinal cord

**Organs**
O 2 kidney, left
O 5 spleen
O 6 stomach

**Skeletal Structures**
S 1 clavicle, left
S 3 rib (9th left)
S 5 rib, body, lateral (2nd left)
S 6 rib, body, posterior (10th right)
S 8 rib, head and neck (8th right)
S 10 scapula
S 12 scapula, coracoid process
S 20 intervertebral disc (T7–T8)
S 21 vertebra, body (T12)

**Veins**
V 6 axillary, left
V 12 hemiazygos
V 18 intercostal, posterior, left
V 19 intercostal, posterior, right
V 34 thoracoacromial, left
V 41 venous plexus, vertebral

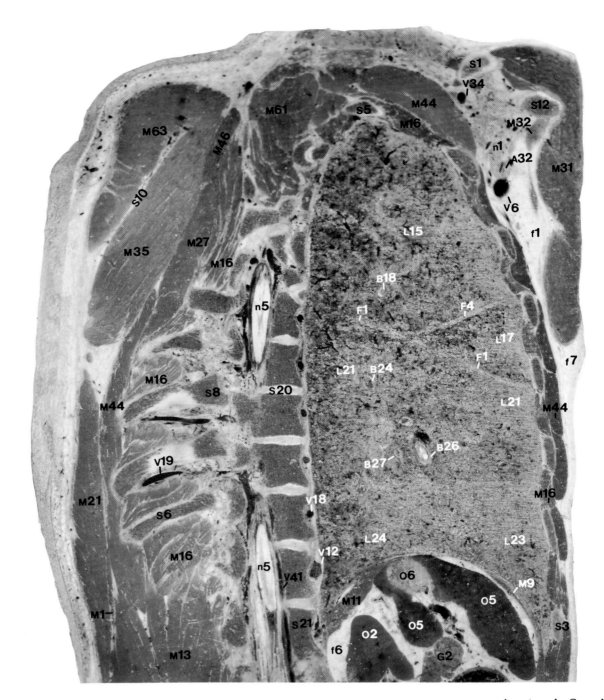

**Anatomic Specimen**

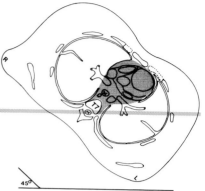

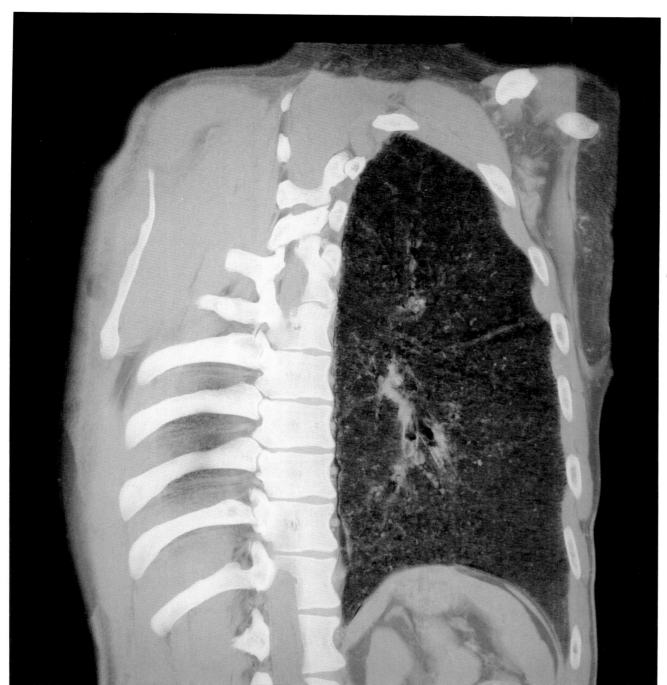

**Specimen Radiograph**

**Magnetic Resonance**

**Trispiral Tomogram**

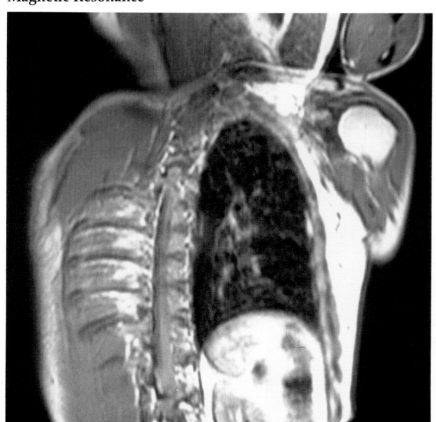

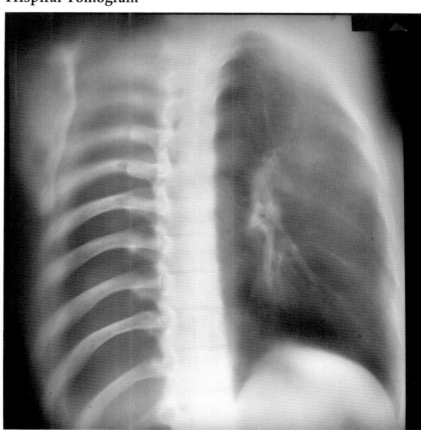

# Left Posterior Oblique

## Section 22

### Anatomic Key

**Arteries**
A 32 axillary, left

**Bronchi**
B 18 LUL, apicoposterior segment
B 24 LLL, superior segment
B 26 LLL, lateral basal segment
B 27 LLL, posterior basal segment

**Fat**
f 1 axillary

**Fissures**
F 1 left major (oblique)
F 4 aberrant (LUL)

**Glands**
G 2 pancreas

**Lobes**
L 15 LUL, apicoposterior segment
L 17 LUL, lingula
L 21 LLL, superior segment
L 23 LLL, lateral basal segment
L 24 LLL, posterior basal segment

**Muscles**
M 1 abdominal, external oblique
M 7 coracobrachialis
M 9 diaphragm
M 11 diaphragm, crus, left
M 13 erector spinae
M 14 iliocostalis, thoracis
M 16 intercostal
M 21 latissimus dorsi
M 27 longissimus, thoracis
M 31 pectoralis, major
M 35 rhomboid, major
M 44 serratus, anterior
M 46 serratus, posterior superior
M 61 transversospinalis
M 63 trapezius

**Neural Structures**
n 1 brachial plexus
n 5 spinal cord

**Organs**
O 2 kidney, left
O 5 spleen

**Skeletal Structures**
s 1 clavicle, left
s 3 rib (3rd right)
s 5 rib, body, lateral (9th left)
s 8 rib, head and neck (8th right)
s 12 scapula, coracoid process
s 21 vertebra, body (T12)

**Veins**
V 6 axillary, left
V 12 hemiazygos
V 18 intercostal, posterior, left
V 19 intercostal, posterior, right
V 34 thoracoacromial, left
V 41 venous plexus, vertebral

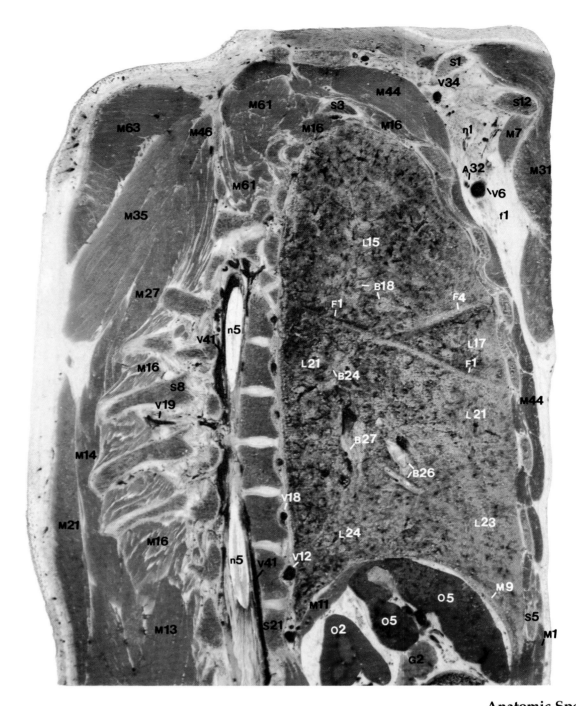

**Anatomic Specimen**

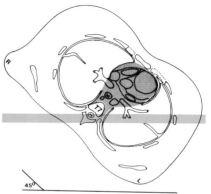

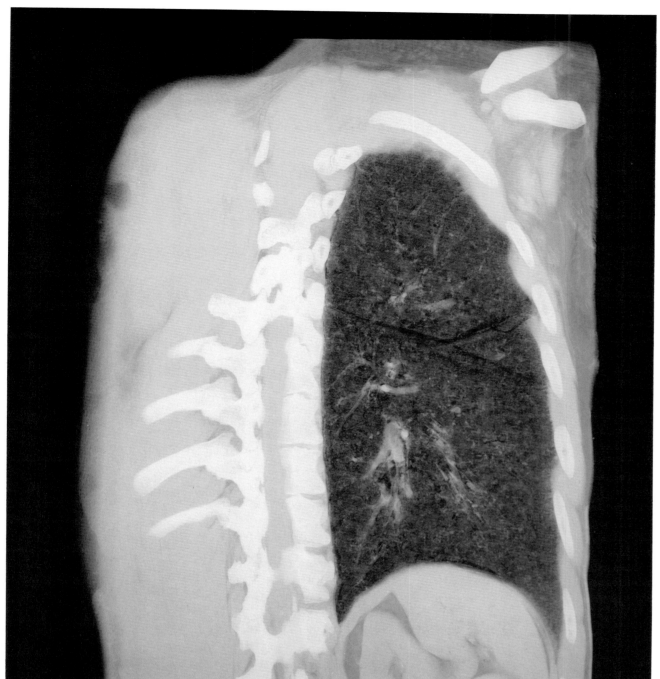

**Specimen Radiograph**

**Magnetic Resonance**

**Trispiral Tomogram**

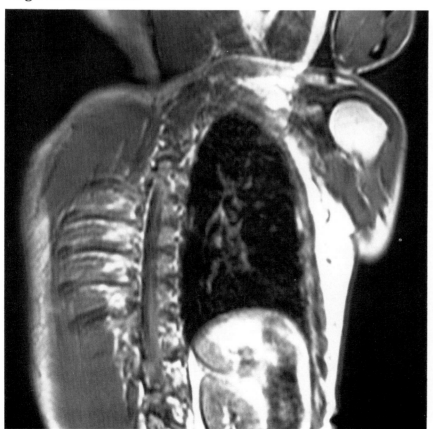

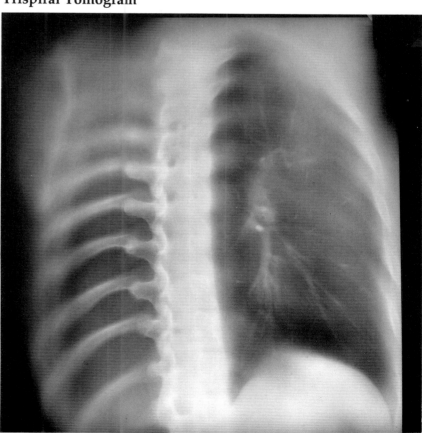

# Left Posterior Oblique

## Section 23

### Anatomic Key

**Bronchi**

B 18 LUL, apicoposterior segment
B 24 LLL, superior segment
B 26 LLL, lateral basal segment
B 27 LLL, posterior basal segment

**Fat**

f 6 perirenal
f 7 subcutaneous

**Fissures**

F 1 left major (oblique)
F 4 aberrant (LUL)

**Lobes**

L 15 LUL, apicoposterior segment
L 17 LUL, lingula
L 21 LLL, superior segment
L 23 LLL, lateral basal segment
L 24 LLL, posterior basal segment

**Muscles**

M 1 abdominal, external oblique
M 7 coracobrachialis
M 9 diaphragm
M 11 diaphragm, crus, left
M 13 erector spinae
M 16 intercostal
M 21 latissimus dorsi
M 41 semispinalis, capitis
M 42 semispinalis, cervicis
M 44 serratus, anterior
M 58 supraspinatus
M 61 transversospinalis
M 63 trapezius

**Organs**

O 2 kidney, left
O 5 spleen

**Skeletal Structures**

S 1 clavicle, left
S 2 humerus, left
S 3 rib (3rd left)
S 5 rib, body, lateral (9th left)
S 6 rib, body, posterior (3rd left)
S 11 scapula, acromion
S 14 scapula, neck
S 15 scapula, spine

**Veins**

V 18 intercostal, posterior, left

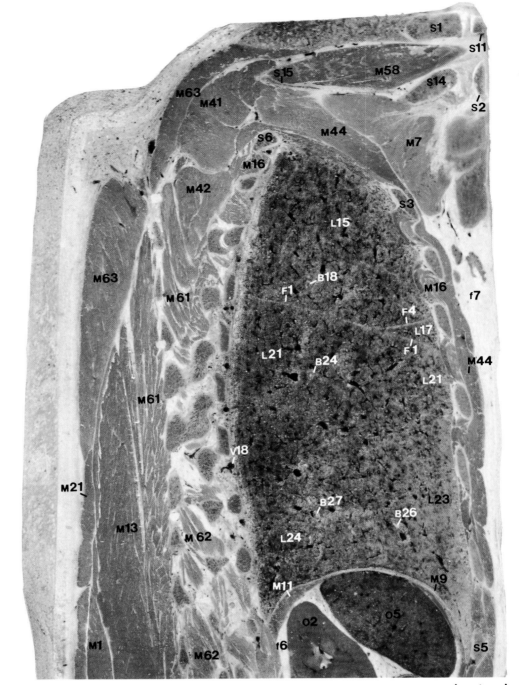

**Anatomic Specimen**

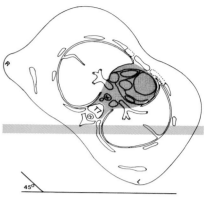

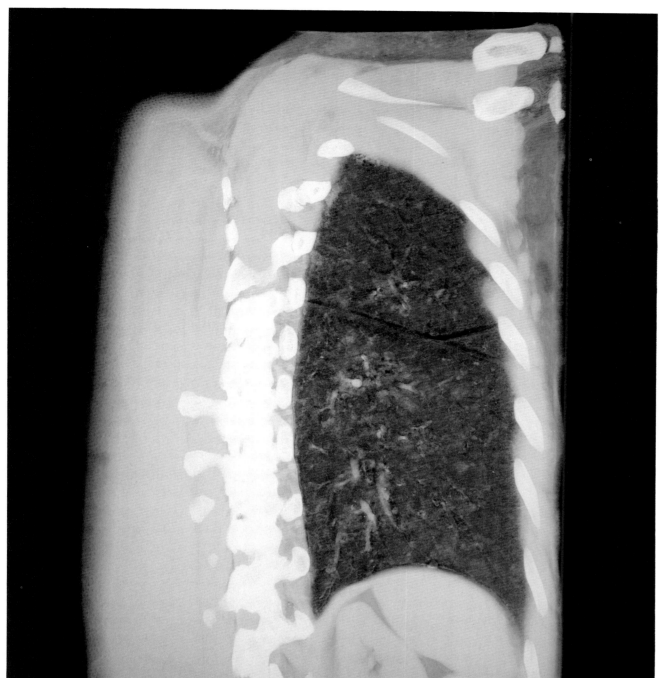

**Specimen Radiograph**

**Magnetic Resonance**

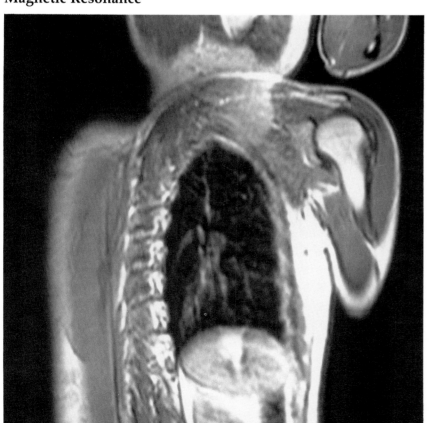

**Trispiral Tomogram**

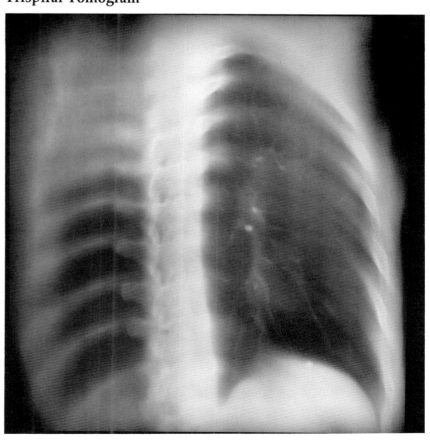

# Left Posterior Oblique

## Section 24

### Anatomic Key

**Fat**
f  6  perirenal

**Fissures**
F  1  left major (oblique)

**Lobes**
L  15  LUL, apicoposterior segment
L  21  LLL, superior segment
L  24  LLL, posterior basal segment

**Muscles**
M  9  diaphragm
M 13  erector spinae
M 16  intercostal
M 21  latissimus dorsi
M 35  rhomboid, major
M 44  serratus, anterior
M 57  subscapularis
M 58  supraspinatus
M 61  transversospinalis
M 63  trapezius

**Organs**
O  2  kidney, left
O  5  spleen

**Pleural Structures**
P  2  costal
P  6  visceral

**Skeletal Structures**
S  1  clavicle, left
S  2  humerus, left
S  3  rib (3rd left)
S  5  rib, body, lateral (9th left)
S 13  scapula, glenoid cavity
S 15  scapula, spine

**Nonanatomic**
NB  nylon bolt

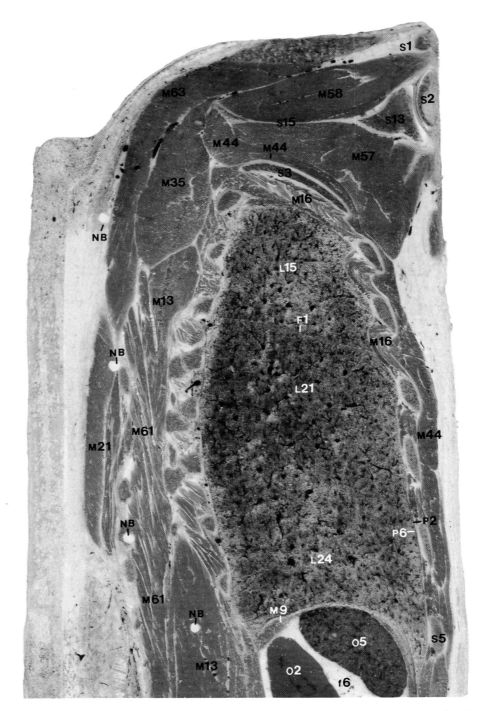

**Anatomic Specimen**

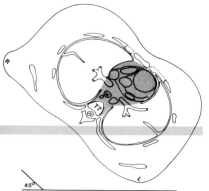

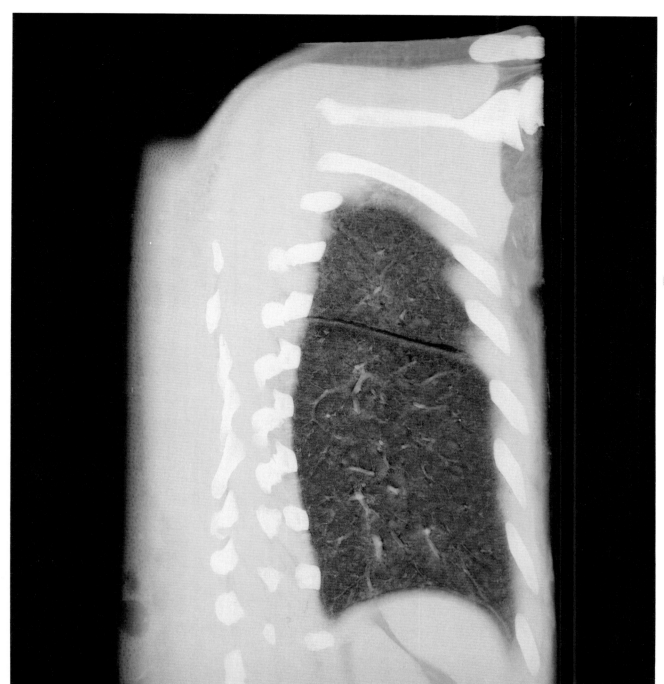

**Specimen Radiograph**

**Magnetic Resonance**

**Trispiral Tomogram**

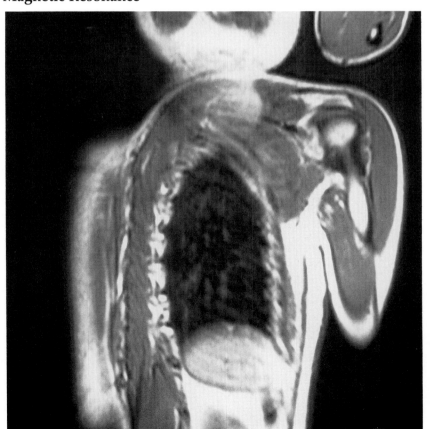

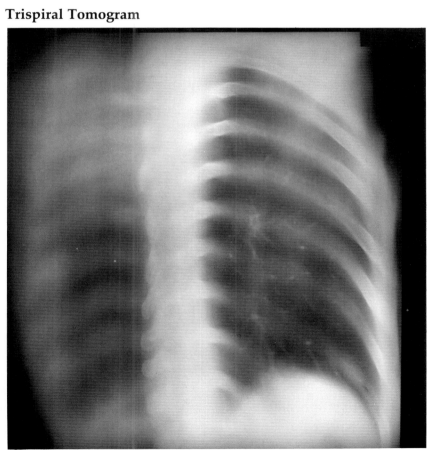

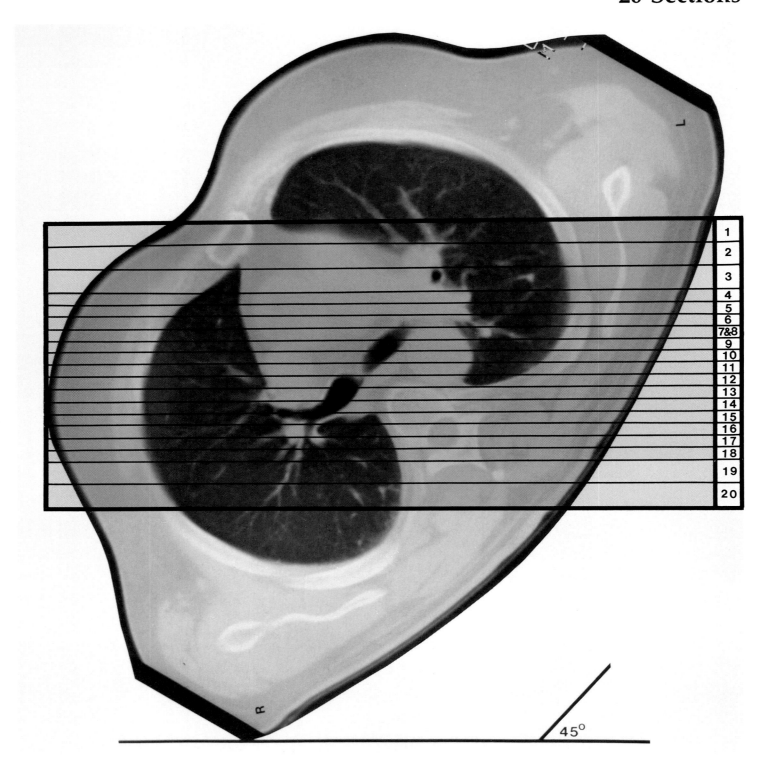

1
2
3
4
5
6
7&8
9
10
11
12
13
14
15
16
17
18
19
20

45°

# Chapter 6
# Right Posterior Oblique Plane

## The Anatomic Specimen

The cadaver for the right posterior oblique sections was that of a 20-year-old white male who died as the result of a close-range shotgun wound to the head. There was extensive soft tissue maceration of the skull base and neck with transsection of major vessels (jugular and carotid).

The lungs remained fully expanded following attempted life support. The immediate postmortem chest radiograph was normal, and the body was placed in a neutral supine position in the freezer. The thorax was sectioned 50 months later.

During the making of the trispiral tomograms, it became apparent that postmortem air filled the major mediastinal vessels and the cardiac chambers. After careful planning from the tomograms and CT scans, a plane was selected that would permit either direct entry or transcardiac catheterization of the air-filled structures similar to that used for the left posterior oblique specimen in Chapter 5. With the frozen specimen securely bolted to a 45° plastic frame, a single cut was made through the thorax. Under fluoroscopic control, the vessels and cardiac chambers were filled with packed red cells to prevent subsequent sludging, and the blood was allowed to freeze.

Normal radiographic density of the heart and great vessels was restored, and the presection trispiral tomograms were considered representative of a living subject. Prior to the bivalve procedure, the thorax was anchored to the plastic frame so that it could be returned to tomography and to the saw for ultimate sectioning in exactly the same orientation.

## Section Plan

The plan of section for the right posterior oblique series (facing page) is shown in colored strips over an axial CT hard copy radiograph of the chest representing the precise charting of the positions of the gross sections. Each section is color-coded according to its thickness: **blue**, 10 mm, and **pink**, 5 mm. The level and thickness of section is indicated throughout the chapter by the color bar overlying the miniature reference key at the top right of each double-page layout.

The most anterior sections (Sections 1–3) are 10 mm thick. The first section includes portions of the LUL, lingula, LLL, and anterolateral margin of the left ventricle. Sections 4–18 are 5 mm thick and extend through the critical medial portions of both lungs, the hila, and mediastinum. The most posterior

sections (Sections 19 and 20) are 10 mm thick. These sections show a wide margin of skin on the right side of each section where the skin surface is oriented obliquely. The final section (Section 20) includes only the lateral, peripheral portion of the right lung.

Two critical sections (Sections 7 and 8) were lost at the site of the bivalve procedure; they are shown on the section plan as a clear zone. There are no matching specimen radiographs of these sections, since surface planes were damaged during the filling of the blood vessels. So that essential anatomic structures could be identified, the opposing surfaces, which had been photographed immediately after the specimen was bivalved, are reproduced as colored prints, Sections 7 and 8. Representative clinical images from the MR and trispiral tomogram series were made at such close increments that they could be matched to the anatomic specimen surfaces shown.

## Orientation of Illustrations

Structures are viewed as seen from front to back, with the thorax positioned obliquely at a 45° rotation from the coronal plane and with the right side tilted posteriorly. Throughout the atlas this projection is referred to as a *right posterior oblique*, but it is the same as, and can also be called, a *left anterior oblique* projection. Structures on the right side of the chest appear on the reader's left, and structures on the left side of the chest appear on the reader's right. The anterior structures appear on the reader's left in all illustrations. The left lung appears in Section 1 and a small segment of the right lung first appears in Section 2 and persists throughout the remaining sections.

## Anatomic–Radiographic Correlation

Several structures are seen to excellent advantage in this projection.

### Bronchial Structures
The bronchial structures in the right posterior oblique orientation are not as completely visualized as they are in the three other nonaxial projections. Several bronchi, however, are seen to excellent advantage. The complete pattern of bronchial imaging in the right posterior oblique projection in both lungs is summarized graphically in Table 6.1.

**Right Lung**: The origin of the RML bronchus (B7) shows to excellent advantage in this projection, and can be seen in the specimen radiographs and trispiral tomograms in Sections 14 and 15. The medial segmental bronchus to the RML (B9) is partly seen in Section 14 but is beautifully demonstrated in Section 15 in both the specimen radiograph and the trispiral tomogram. The profile of the anterior bronchus to the RUL (B5) is also well seen in Section 15. The RLL bronchus (B10), the medial basal bronchus to the RLL (B12), and the apical bronchus to the RUL (B3) are clearly seen in Section 16. A short segment of the RLL bronchus (B10) and the anterior basal segment to the RLL (B13) are seen in Section 17.

The right mainstem (B1), RUL (B2), and intermediate (B6) bronchi are cut in cross section and hence are not as well seen in the right posterior oblique projection.

**Left Lung**: The profiles of several bronchi are seen to excellent advantage in Section 6: the LUL bronchus (B17) with its anterior (B19) and apicoposterior (B18) divisions, as well as the LLL bronchus (B23) dividing into the anteromedial basal (B25) and lateral basal (B26) divisions, are clearly visualized in both the specimen radiograph and the trispiral tomogram. The LLL bronchus (B23) is seen extending inferiorly in the trispiral tomograms of Sections 7 and 8. There were no matching specimen radiographs at these levels. The distal trachea (T1) and the distal left mainstem bronchus (B16) extending into the LLL (B23) are unusually well seen in the specimen radiographs and trispiral tomograms of Sections 9 and 10. The superior bronchus to the LLL (B24) is better seen in Section 9 in this projection than in any of the other projections. The carina (T2) is clearly seen in the trispiral tomogram in Sections 10 and 11.

### Fissures

The major (F2) and horizontal (F3) fissures in the right lung and the major fissure (F1) in the left lung are not as well seen in the specimen radiographs and trispiral tomograms as they are in the gross anatomic specimens.

### Neural Structures

The brachial plexus (n1) is well seen on the left side in Anatomic Sections 1–6 and on the right side in Anatomic Sections 15–20. The thoracic portion of the spinal cord (n5) is nicely seen extending from T5 to T12 in Anatomic Section 16.

### Vascular Structures

Many cardiac structures are best seen in oblique planes. For example, in this right posterior oblique projection the ventricular (Sections 2–6) and atrial (Sections 9–13) septa are optimally oriented to separate the four heart chambers. This projection also demonstrates the relationship of the brachiocephalic vessels to the aortic arch to good advantage. The left inferior pulmonary vein (V5) is well seen in profile in Section 9, and the left pulmonary (A16) and the left lower lobe (A23) arteries in Sections 7 and 8. The superior vena cava (V39) descending in the right anterior mediastinum can be well seen in Section 10 in both the gross anatomic section and the specimen radiograph. The confluence of the superior (V2) and inferior (V5) left pulmonary veins is also well seen in this section.

Special attention is drawn to the profile of the inferior pulmonary vein (V4) in Sections 14 and 15. The outline of this vessel in oblique and lateral projections represents the confluence of the inferior (V4) and superior (V0) right pulmonary veins entering the left auricle. This shadow has on occasion been mistaken for a "tumor" at or behind the right cardiac margin in plain films and tomograms of the chest. Lack of anatomic correlation has resulted in surgical exploration. The smooth margins of the mass, resembling a "powder flask," and its proximity to the RML bronchus are diagnostic.

In Section 18 the external jugular vein (V25) is clearly seen coursing anterior to the belly of the omohyoid muscle (M30), which is the normal course of this vessel, in contrast to the variant course of the external jugular veins seen in sagittal Sections 9 and 31 (Chapter 4).

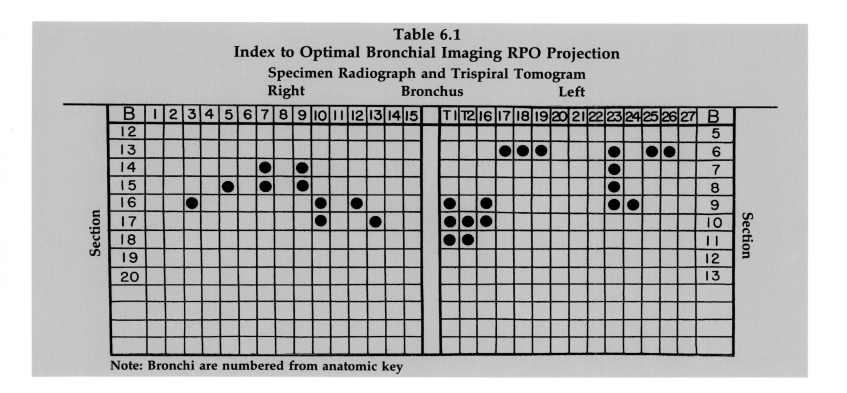

**Table 6.1**
**Index to Optimal Bronchial Imaging RPO Projection**
Specimen Radiograph and Trispiral Tomogram
Right          Bronchus          Left

Note: Bronchi are numbered from anatomic key

## Other Structures
The esophagogastric junction (E2) is seen in detail in Anatomic Sections 7–10.

## Magnetic Resonance Images
The vascular structures are outstanding in the MR images of the size-related volunteer and correlate well with the anatomic specimens. The right internal thoracic artery (A48) and vein (V21) are seen in MR Sections 4 and 5. The ascending aorta (A30) and pulmonary trunk (A1) are seen in Section 6, while the images of the ascending aorta (A30), brachiocephalic trunk (A34), and left common carotid artery (A37) are seen in beautiful profile in Section 7. The pulmonary trunk (A1) and the left pulmonary artery (A16) are seen in Section 8, as well as the ascending aorta (A30), aortic arch (A29), and left subclavian artery (A49). Although to date MR techniques do not image the bronchi well, the trachea (T1) and left mainstem bronchus (B16) are well outlined in Sections 9 and 10. The superior vena cava (V39) is sharply seen in MR Section 10, and the descending aorta (A31) in Section 11.

The cardiac chambers are well seen in Sections 1–12.

# Right Posterior Oblique

## Section 1

### Anatomic Key

**Arteries**
A 32 axillary, left
A 51 subscapular, left
A 53 suprascapular, left

**Bronchi**
B 21 LUL, superior lingular segment
B 25 LLL, anteromedial basal
      segment

**Fat**
f 1 axillary
f 2 epicardial
f 3 extrapleural

**Fissures**
F 1 left major (oblique)

**Heart**
H 13 pericardium, parietal serous
H 14 pericardium, visceral serous
H 16 septum, interventricular
H 21 ventricle, left
H 22 ventricle, right

**Lobes**
L 16 LUL, anterior segment
L 18 LUL, superior lingular segment
L 19 LUL, inferior lingular segment
L 22 LLL, anteromedial basal
      segment

**Muscles**
M 1 abdominal, external oblique
M 3 abdominis, rectus
M 9 diaphragm
M 10 diaphragm, central tendon
M 15 infraspinatus
M 16 intercostal
M 21 latissimus dorsi
M 31 pectoralis, major
M 33 platysma
M 44 serratus, anterior
M 52 sternocleidomastoid
M 55 subclavius
M 57 subscapularis
M 58 supraspinatus
M 59 teres major
M 60 teres minor
M 63 trapezius

**Neural Structures**
n 1 brachial plexus

**Nodes**
N 1 axillary

**Organs**
O 1 colon
O 3 liver
O 4 small bowel
O 5 spleen
O 6 stomach

**Pleural Structures**
P 2 costal (parietal)
P 5 mediastinal (parietal)
P 6 visceral

**Skeletal Structures**
S 1 clavicle, left
S 3 rib (2nd left)
S 5 rib, body, lateral (9th left)
S 7 rib, costal cartilage (2nd left)
S 10 scapula
S 15 scapula, spine
S 17 sternum, body
S 19 sternum, xiphoid

**Veins**
V 6 axillary, left
V 10 cephalic, left
V 16 intercostal, anterior, left
V 24 jugular, external, left
V 32 subscapular, left
V 33 suprascapular, left
V 34 thoracoacromial, left

**Nonanatomic**
NB nylon bolt

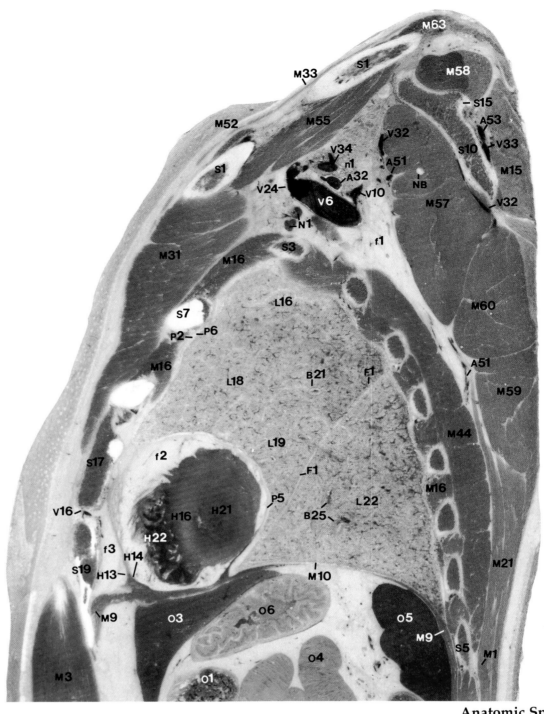

**Anatomic Specimen**

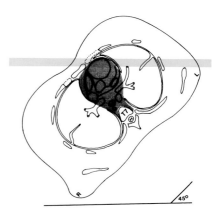

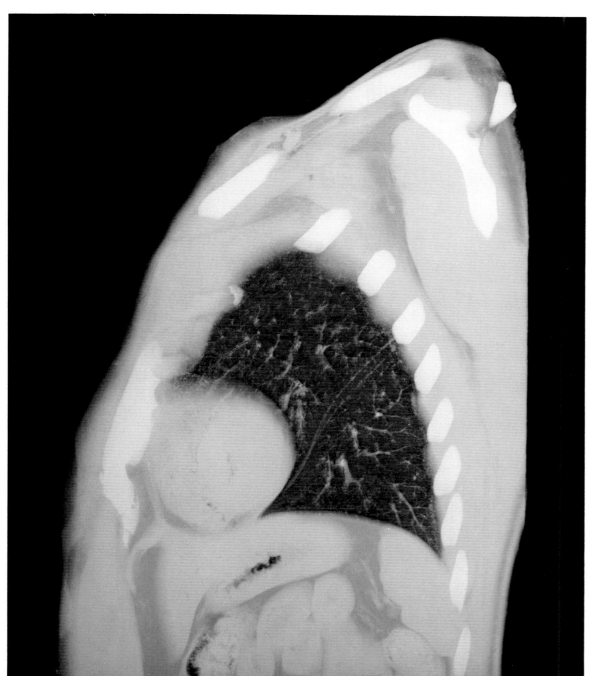

**Specimen Radiograph**

**Magnetic Resonance**

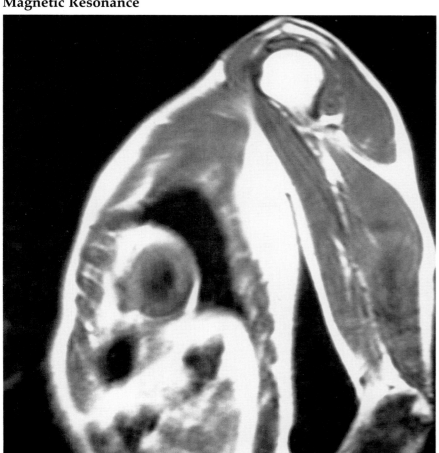

**Trispiral Tomogram**

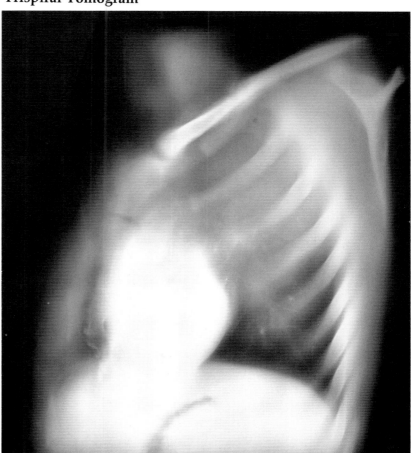

# Right Posterior Oblique

## Section 2

### Anatomic Key

**Arteries**
A 49 subclavian, left

**Bronchi**
B 21 LUL, superior lingular segment
B 22 LUL, inferior lingular segment
B 24 LLL, superior segment
B 25 LLL, anteromedial basal
     segment

**Fat**
f 1 axillary
f 2 epicardial
f 3 extrapleural
f 4 mediastinal

**Fissures**
F 1 left major (oblique)

**Heart**
H 13 pericardium, parietal serous
H 14 pericardium, visceral serous
H 16 septum, interventricular
H 21 ventricle, left
H 22 ventricle, right

**Lobes**
L 7 RML, medial segment
L 16 LUL, anterior segment
L 18 LUL, superior lingular segment
L 21 LLL, superior segment
L 22 LLL, anteromedial basal
     segment

**Muscles**
M 1 abdominal, external oblique
M 3 abdominis, rectus
M 9 diaphragm
M 10 diaphragm, central tendon
M 15 infraspinatus
M 16 intercostal
M 21 latissimus dorsi
M 31 pectoralis, major
M 33 platysma
M 44 serratus, anterior
M 52 sternocleidomastoid
M 55 subclavius
M 57 subscapularis
M 58 supraspinatus
M 59 teres major
M 63 trapezius

**Neural Structures**
n 1 brachial plexus

**Organs**
O 1 colon
O 3 liver
O 4 small bowel
O 5 spleen
O 6 stomach

**Pleural Structures**
P 2 costal (parietal)
P 5 mediastinal (parietal)
P 6 visceral

**Skeletal Structures**
S 1 clavicle, left
S 5 rib, body, lateral (2nd left)
S 6 rib, body, posterior (8th left)
S 7 rib, costal cartilage (1st left)
S 10 scapula
S 15 scapula, spine
S 17 sternum, body

**Veins**
V 24 jugular, external, left
V 30 subclavian, left
V 33 suprascapular, left
V 36 transverse cervical, left

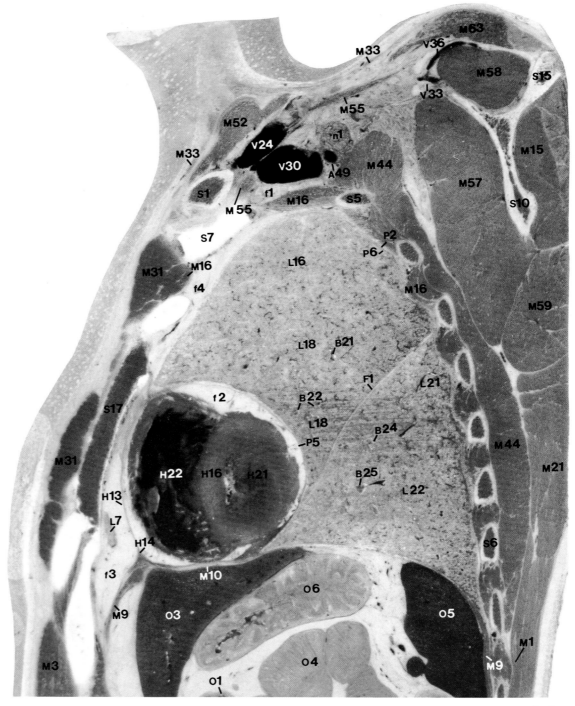

**Anatomic Specimen**

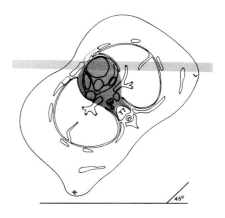

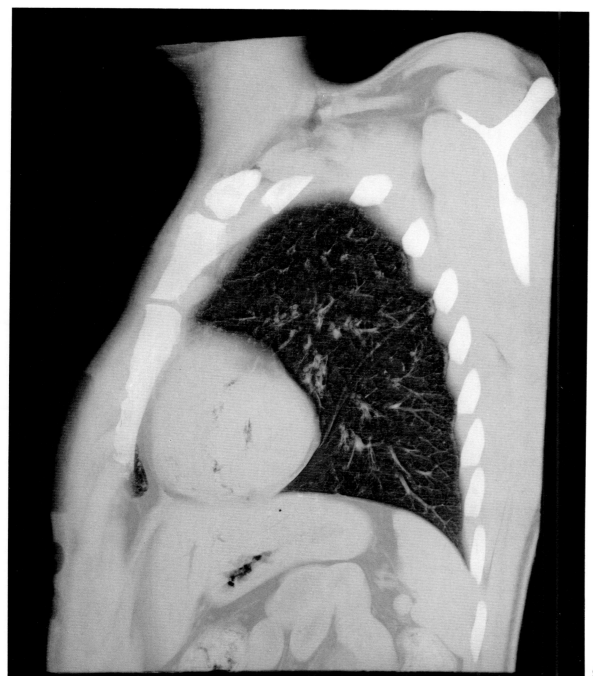

**Specimen Radiograph**

**Magnetic Resonance**

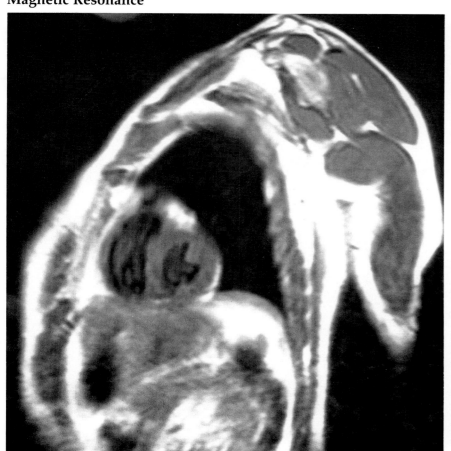

**Trispiral Tomogram**

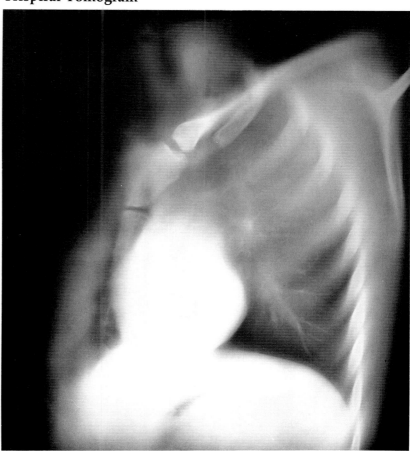

# Right Posterior Oblique

## Section 3

## Anatomic Key

### Arteries
A **47** internal thoracic (mammary), left
A **49** subclavian, left

### Bronchi
B **19** LUL, anterior segment
B **21** LUL, superior lingular segment
B **22** LUL, inferior lingular segment
B **24** LLL, superior segment
B **25** LLL, anteromedial basal segment
B **26** LLL, lateral basal segment

### Fat
f **2** epicardial
f **3** extrapleural

### Fissures
F **1** left major (oblique)

### Heart
H **8** coronary artery, circumflex
H **13** pericardium, parietal serous
H **14** pericardium, visceral serous
H **16** septum, interventricular
H **21** ventricle, left
H **22** ventricle, right

### Lobes
L **7** RML, medial segment
L **16** LUL, anterior segment
L **18** LUL, superior lingular segment
L **19** LUL, inferior lingular segment
L **21** LLL, superior segment
L **22** LLL, anteromedial basal segment
L **23** LLL, lateral basal segment

### Muscles
M **1** abdominal, external oblique
M **9** diaphragm
M **10** diaphragm, central tendon
M **15** infraspinatus
M **16** intercostal
M **21** latissimus dorsi
M **30** omohyoid
M **31** pectoralis, major
M **33** platysma
M **39** scalene, middle
M **44** serratus, anterior
M **52** sternocleidomastoid
M **53** sternohyoid
M **54** sternothyroid
M **57** subscapularis
M **58** supraspinatus
M **59** teres major
M **62** transversus thoracis
M **63** trapezius

### Neural Structures
n **1** brachial plexus

### Organs
O **3** liver
O **4** small bowel
O **5** spleen
O **6** stomach

### Pleural Structures
P **2** costal (parietal)
P **3** costodiaphragmatic recess
P **5** mediastinal (parietal)
P **6** visceral

### Skeletal Structures
S **1** clavicle, left
S **3** rib (2nd left)
S **4** rib, body, anterior (1st left)
S **5** rib, body, lateral (9th left)
S **6** rib, body, posterior (2nd left)
S **7** rib, costal cartilage (4th right)
S **10** scapula
S **15** scapula, spine
S **17** sternum, body
S **18** sternum, manubrium

### Veins
v **2** pulmonary, superior, LUL
v **3** pulmonary, superior, LUL, lingula
v **5** pulmonary, inferior, LLL
v **17** intercostal, anterior, right
v **18** intercostal, posterior, left
v **24** jugular, external, left
v **30** subclavian, left
v **33** suprascapular, left

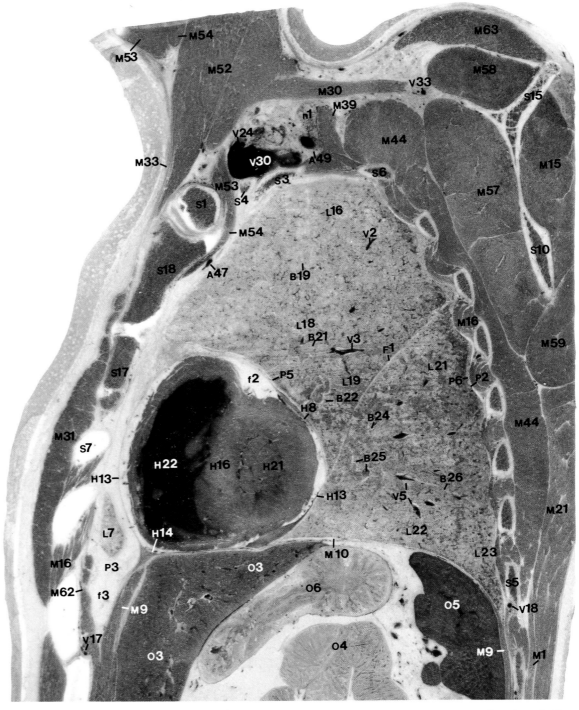

**Anatomic Specimen**

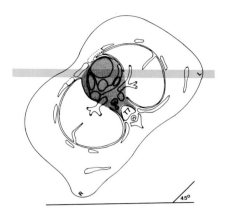

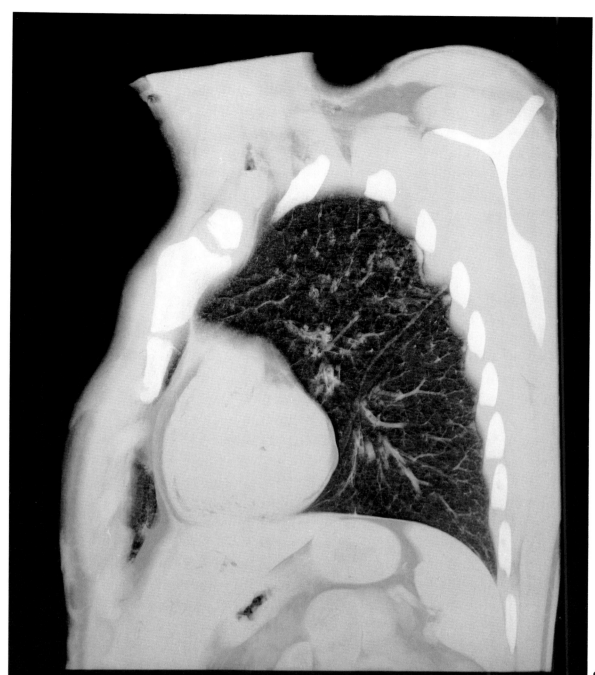

Specimen Radiograph

**Magnetic Resonance**

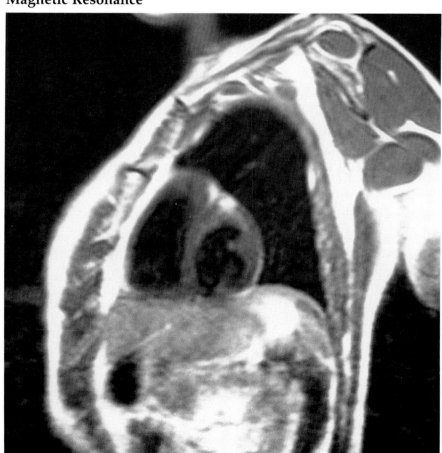

**Trispiral Tomogram**

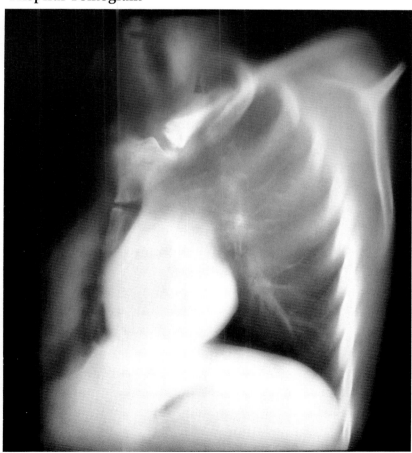

# Right Posterior Oblique

## Section 4

## Anatomic Key

### Arteries
A 1 pulmonary trunk
A 49 subclavian, left

### Bronchi
B 18 LUL, apicoposterior segment
B 19 LUL, anterior segment
B 21 LUL, superior lingular segment
B 22 LUL, inferior lingular segment
B 24 LLL, superior segment
B 25 LLL, anteromedial basal segment
B 26 LLL, lateral basal segment

### Fat
f 2 epicardial
f 3 extrapleural
f 4 mediastinal

### Fissures
F 1 left major (oblique)

### Glands
G 2 pancreas
G 4 thyroid

### Heart
H 8 coronary artery, circumflex
H 9 coronary artery, anterior descending
H 13 pericardium, parietal serous
H 14 pericardium, visceral serous
H 16 septum, interventricular
H 18 valve, pulmonary
H 21 ventricle, left
H 22 ventricle, right

### Lobes
L 4 RUL, anterior segment
L 7 RML, medial segment
L 15 LUL, apicoposterior segment
L 16 LUL, anterior segment
L 18 LUL, superior lingular segment
L 19 LUL, inferior lingular segment
L 21 LLL, superior segment
L 22 LLL, anteromedial basal segment
L 23 LLL, lateral basal segment

### Muscles
M 1 abdominal, external oblique
M 9 diaphragm
M 10 diaphragm, central tendon
M 15 infraspinatus
M 16 intercostal
M 21 latissimus dorsi
M 31 pectoralis, major
M 38 scalene, anterior
M 39 scalene, middle
M 40 scalene, posterior
M 44 serratus, anterior
M 53 sternohyoid
M 54 sternothyroid
M 57 subscapularis
M 58 supraspinatus
M 59 teres major
M 62 transversus thoracis
M 63 trapezius

### Neural Structures
n 1 brachial plexus

### Organs
o 3 liver
o 4 small bowel
o 5 spleen
o 6 stomach

### Pleural Structures
P 2 costal (parietal)
P 5 mediastinal (parietal)
P 6 visceral

### Skeletal Structures
s 1 clavicle, left
s 3 rib (1st left)
s 5 rib, body, lateral (10th left)
s 7 rib, costal cartilage (6th right)
s 10 scapula
s 15 scapula, spine
s 18 sternum, manubrium

### Trachea and Larynx
T 3 larynx

### Veins
v 2 pulmonary, superior, LUL
v 3 pulmonary, superior, LUL, lingula
v 5 pulmonary, inferior, LLL
v 17 intercostal, anterior, right
v 18 intercostal, posterior, left
v 22 jugular, anterior, left
v 26 jugular, internal, left
v 30 subclavian, left
v 33 suprascapular, left
v 35 thyroid, inferior, left
v 36 transverse cervical, left

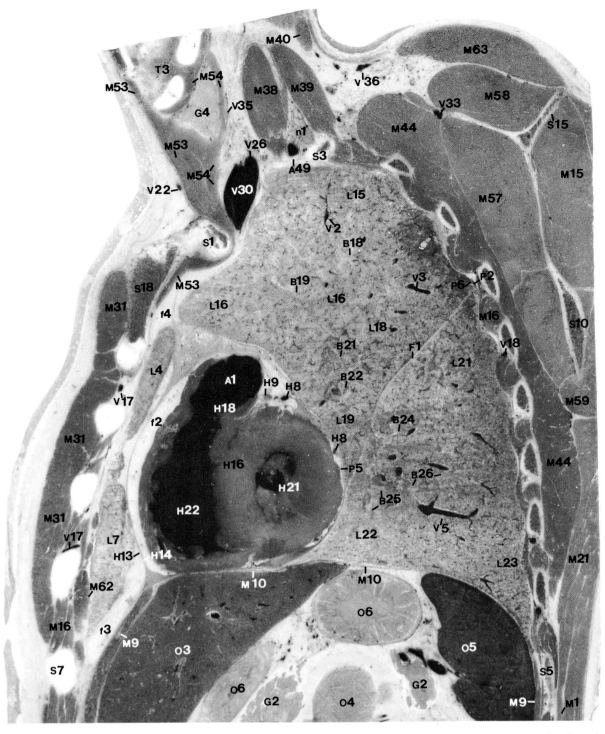

**Anatomic Specimen**

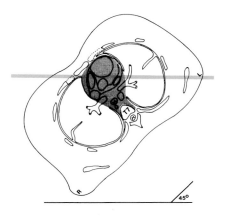

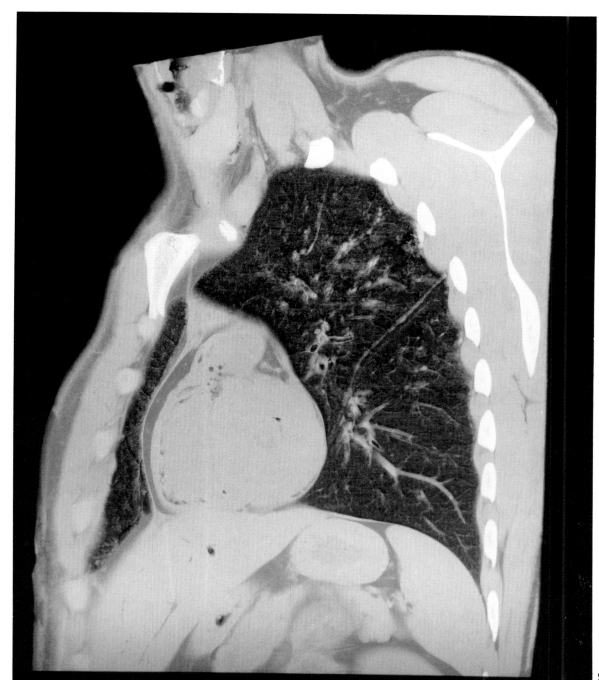

Specimen Radiograph

Magnetic Resonance

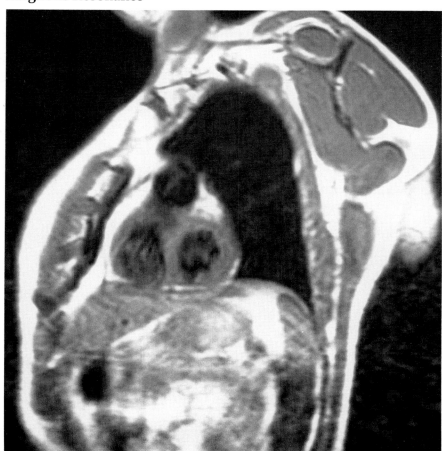

Trispiral Tomogram

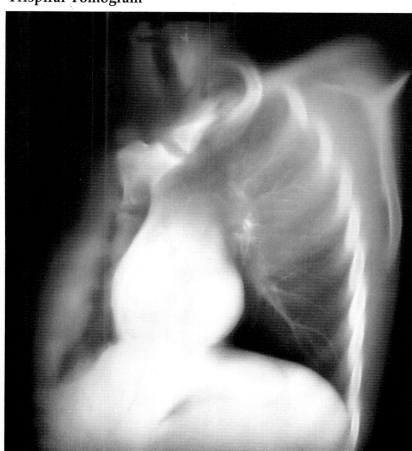

# Right Posterior Oblique

## Section 5

## Anatomic Key

### Arteries
A 1 pulmonary trunk
A 39 costocervical trunk, left
A 49 subclavian, left
A 63 transverse cervical, left

### Bronchi
B 18 LUL, apicoposterior segment
B 19 LUL, anterior segment
B 21 LUL, superior lingular segment
B 22 LUL, inferior lingular segment
B 24 LLL, superior segment
B 25 LLL, anteromedial basal segment
B 26 LLL, lateral basal segment

### Fat
f 2 epicardial
f 3 extrapleural
f 4 mediastinal

### Fissures
F 1 left major (oblique)
F 3 right minor (horizontal)

### Glands
G 2 pancreas
G 4 thyroid

### Heart
H 3 auricle, left
H 8 coronary artery, circumflex
H 9 coronary artery, posterior descending
H 10 coronary artery, left
H 13 pericardium, parietal serous
H 14 pericardium, visceral serous
H 16 septum, interventricular
H 21 ventricle, left
H 22 ventricle, right

### Lobes
L 4 RUL, anterior segment
L 7 RML, medial segment
L 15 LUL, apicoposterior segment
L 16 LUL, anterior segment
L 18 LUL, superior lingular segment
L 19 LUL, inferior lingular segment
L 21 LLL, superior segment
L 22 LLL, anteromedial basal segment
L 23 LLL, lateral basal segment

### Muscles
M 3 abdominis, rectus
M 9 diaphragm
M 10 diaphragm, central tendon
M 15 infraspinatus
M 16 intercostal
M 21 latissimus dorsi
M 31 pectoralis, major
M 38 scalene, anterior
M 39 scalene, middle
M 40 scalene, posterior
M 44 serratus, anterior
M 53 sternohyoid
M 54 sternothyroid
M 57 subscapularis
M 58 supraspinatus
M 59 teres major
M 62 transversus thoracis
M 63 trapezius

### Neural Structures
n 1 brachial plexus

### Organs
O 3 liver
O 5 spleen
O 6 stomach

### Pleural Structures
P 2 costal (parietal)
P 3 costodiaphragmatic recess
P 5 mediastinal (parietal)
P 6 visceral

### Skeletal Structures
S 3 rib (6th right)
S 5 rib, body, lateral (10th left)
S 6 rib, body, posterior (1st left)
S 7 rib, costal cartilage (2nd right)
S 10 scapula
S 15 scapula, spine
S 18 sternum, manubrium

### Trachea and Larynx
T 3 larynx

### Veins
V 2 pulmonary, superior, LUL
V 3 pulmonary, superior, LUL, lingula
V 5 pulmonary, inferior, LLL
V 8 brachiocephalic (innominate), left
V 17 intercostal, anterior, right
V 18 intercostal, posterior, left
V 21 internal thoracic (mammary), right
V 26 jugular, internal, left
V 35 thyroid, inferior, left
V 36 transverse cervical, left

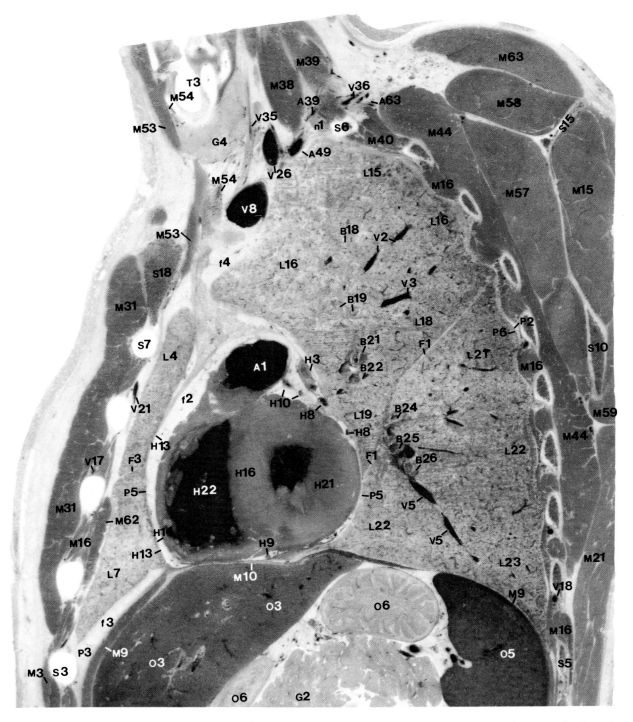

**Anatomic Specimen**

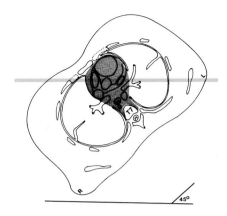

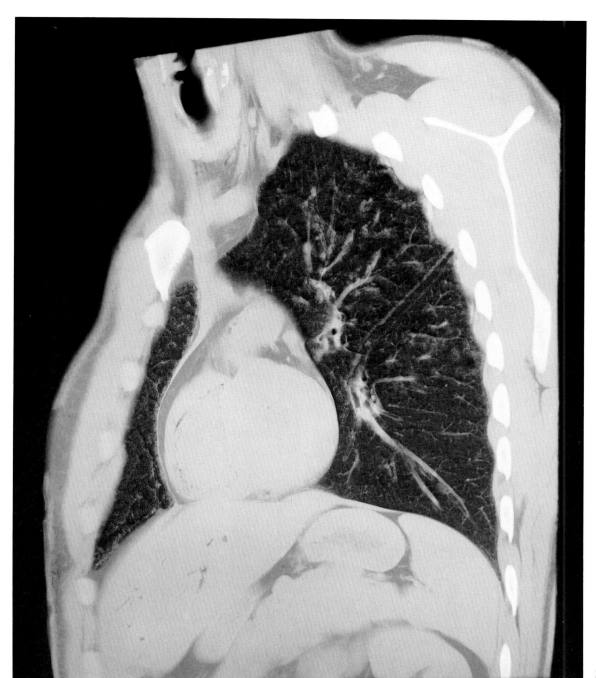

Specimen Radiograph

**Magnetic Resonance**

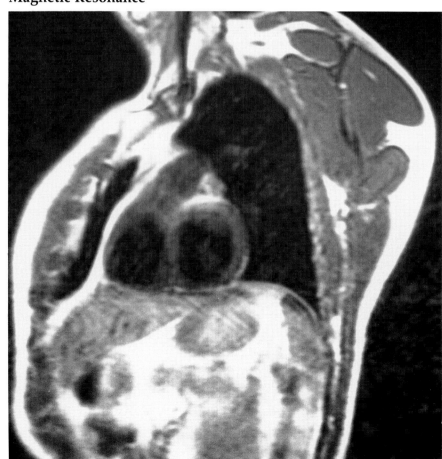

**Trispiral Tomogram**

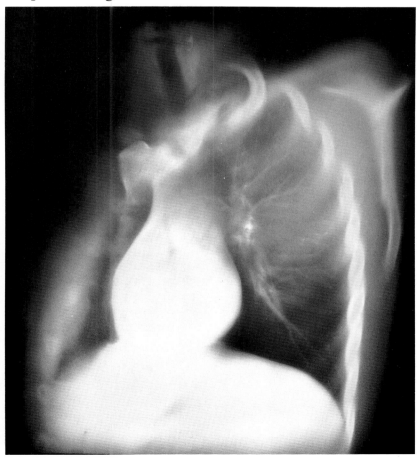

# Right Posterior Oblique

## Section 6

### Anatomic Key

**Arteries**
A 1 pulmonary, trunk
A 30 aorta, ascending
A 37 carotid, common, left
A 49 subclavian, left

**Bronchi**
B 17 left upper lobe (LUL)
B 18 LUL, apicoposterior segment
B 19 LUL, anterior segment
B 21 LUL, superior lingular segment
B 22 LUL, inferior lingular segment
B 24 LLL, superior segment
B 25 LLL, anteromedial basal
    segment
B 26 LLL, lateral basal segment

**Fat**
f 2 epicardial
f 3 extrapleural
f 4 mediastinal
f 7 subcutaneous

**Fissures**
F 1 left major (oblique)
F 3 right minor (horizontal)

**Glands**
G 1 adrenal, left
G 2 pancreas
G 4 thyroid

**Heart**
H 2 atrium, right
H 3 auricle, left
H 5 cardiac vein, great
H 6 cardiac vein, middle
H 8 coronary artery, circumflex
H 9 coronary artery, posterior
    descending
H 13 pericardium, parietal serous
H 14 pericardium, visceral serous
H 16 septum, interventricular
H 20 valve, aortic
H 21 ventricle, left
H 22 ventricle, right

**Lobes**
L 4 RUL, anterior segment
L 7 RML, medial segment
L 15 LUL, apicoposterior segment
L 16 LUL, anterior segment
L 18 LUL, superior lingular segment
L 19 LUL, inferior lingular segment
L 21 LLL, superior segment
L 22 LLL, anteromedial basal
    segment
L 23 LLL, lateral basal segment

**Muscles**
M 1 abdominal, external oblique
M 3 abdominis, rectus
M 9 diaphragm
M 10 diaphragm, central tendon
M 15 infraspinatus
M 16 intercostal
M 21 latissimus dorsi
M 28 longus colli
M 31 pectoralis, major
M 38 scalene, anterior
M 39 scalene, middle
M 40 scalene, posterior
M 44 serratus, anterior
M 52 sternocleidomastoid
M 53 sternohyoid
M 54 sternothyroid
M 57 subscapularis
M 58 supraspinatus
M 59 teres major
M 62 transversus thoracis
M 63 trapezius

**Neural Structures**
n 1 brachial plexus

**Nodes**
N 2 bronchopulmonary (hilar)
N 10 supraclavicular

**Organs**
O 2 kidney, left
O 3 liver
O 5 spleen
O 6 stomach

**Pleural Structures**
P 1 anterior junction line
P 3 costodiaphragmatic recess
P 5 mediastinal (parietal)

**Skeletal Structures**
S 1 clavicle, right
S 5 rib, body, lateral (4th left)
S 6 rib, body, posterior (1st left)
S 7 rib, costal cartilage (2nd right)
S 10 scapula
S 15 scapula, spine
S 18 sternum, manubrium

**Trachea and Larnyx**
T 1 trachea
T 3 larynx

**Veins**
V 2 pulmonary, superior, LUL
V 3 pulmonary, superior, LUL,
    lingula
V 5 pulmonary, inferior, LLL
V 8 brachiocephalic (innominate),
    left
V 18 intercostal, posterior, left
V 26 jugular, internal, left
V 36 transverse cervical, left

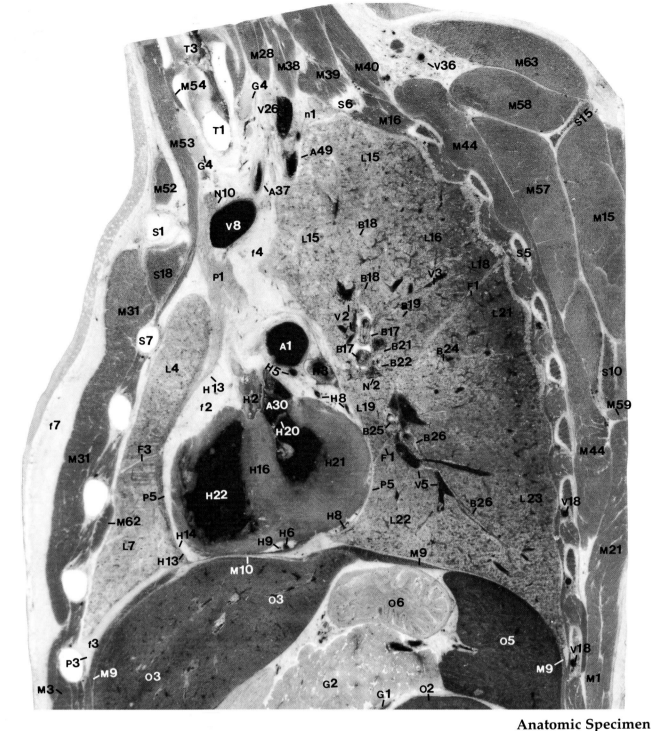

**Anatomic Specimen**

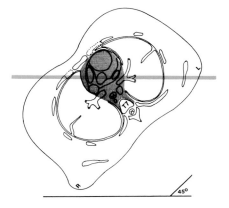

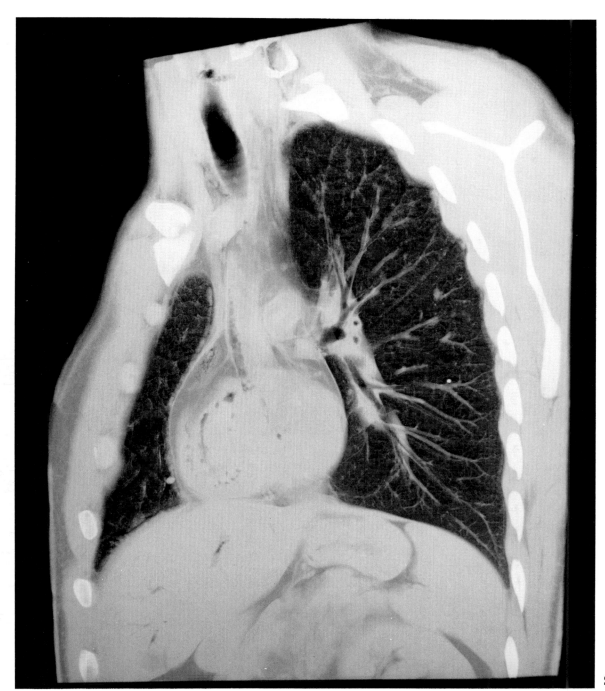

Specimen Radiograph

**Magnetic Resonance**

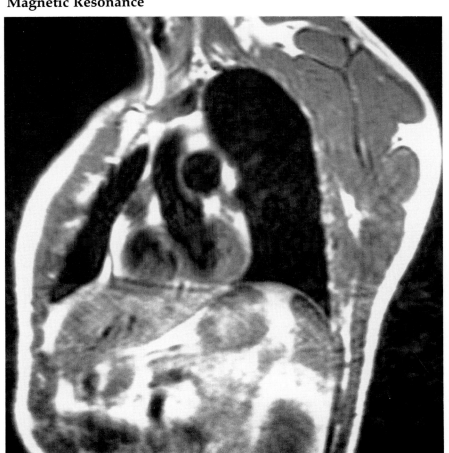

**Trispiral Tomogram**

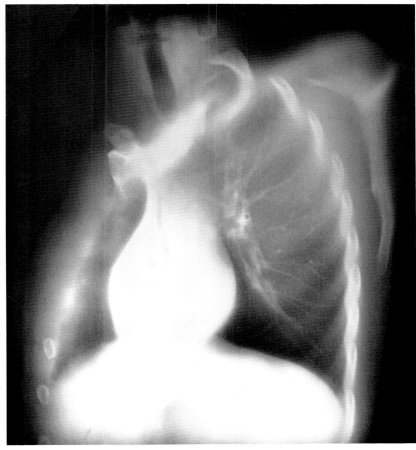

# Right Posterior Oblique

## Section 7

### Anatomic Key

**Arteries**
A 1 pulmonary trunk
A 16 main pulmonary, left
A 18 LUL, apicoposterior branch
A 19 LUL, anterior branch
A 23 left lower lobe (LLL)
A 24 LLL, superior branch
A 26 LLL, lateral basal branch
A 29 aortic arch
A 30 aorta, ascending
A 31 aorta, descending
A 34 brachiocephalic (innominate) trunk
A 37 carotid, common, left
A 45 intercostal, posterior, left
A 49 subclavian, left
A 63 transverse cervical, left

**Bronchi**
B 17 left upper lobe (LUL)
B 18 LUL, apicoposterior segment
B 23 left lower lobe (LLL)
B 25 LLL, anteromedial basal segment
B 26 LLL, lateral basal segment

**Esophagus**
E 1 esophagus
E 2 esophagogastric junction

**Fat**
f 2 epicardial

**Fissures**
F 1 left major (oblique)
F 3 right minor (horizontal)

**Glands**
G 1 adrenal, left
G 2 pancreas
G 4 thyroid

**Heart**
H 1 atrium, left
H 2 atrium, right
H 3 auricle, left
H 5 cardiac vein, great
H 6 cardiac vein, middle
H 8 coronary artery, circumflex
H 9 coronary artery, posterior descending
H 14 pericardium, visceral serous
H 16 septum, interventricular
H 17 valve, tricuspid
H 19 valve, mitral
H 21 ventricle, left
H 22 ventricle, right

**Lobes**
L 4 RUL, anterior segment
L 7 RML, medial segment
L 15 LUL, apicoposterior segment
L 16 LUL, anterior segment
L 21 LLL, superior segment
L 22 LLL, anteromedial basal segment
L 23 LLL, lateral basal segment

**Muscles**
M 3 abdominis, rectus
M 9 diaphragm
M 10 diaphragm, central tendon
M 11 diaphragm, crus, left
M 15 infraspinatus
M 16 intercostal
M 21 latissimus dorsi
M 23 levator scapulae
M 28 longus colli
M 31 pectoralis, major
M 39 scalene, middle
M 40 scalene, posterior
M 44 serratus, anterior
M 52 sternocleidomastoid
M 53 sternohyoid
M 54 sternothyroid
M 57 subscapularis
M 63 trapezius

**Organs**
O 2 kidney, left
O 3 liver
O 5 spleen
O 6 stomach

**Pleural Structures**
P 2 costal (parietal)
P 6 visceral

**Skeletal Structures**
S 1 clavicle, right
S 4 rib, body, anterior (6th right)
S 6 rib, body, posterior (10th left)
S 7 rib, costal cartilage (2nd right)
S 9 rib, tubercle (1st left)
S 10 scapula
S 15 scapula, spine
S 18 sternum, manubrium
S 21 vertebra, body (C7)
S 29 vertebra, transverse process (T1)

**Trachea and Larynx**
T 1 trachea
T 3 larynx

**Veins**
V 2 pulmonary, superior, LUL
V 5 pulmonary, inferior, LLL
V 9 brachiocephalic (innominate), right
V 18 intercostal, posterior, left
V 27 jugular, internal, right
V 33 suprascapular, left
V 36 transverse cervical, left

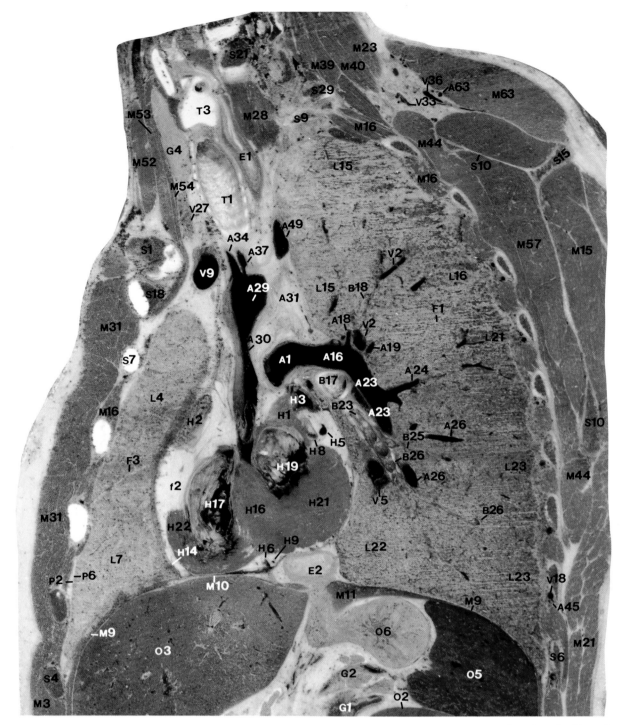

**Anatomic Specimen**

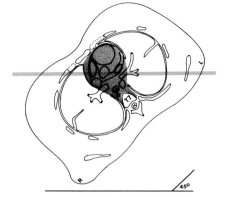

No
Specimen Radiograph
made

**Specimen Radiograph**

**Magnetic Resonance**

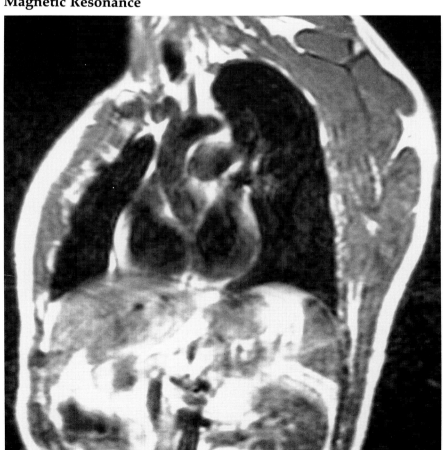

**Trispiral Tomogram**

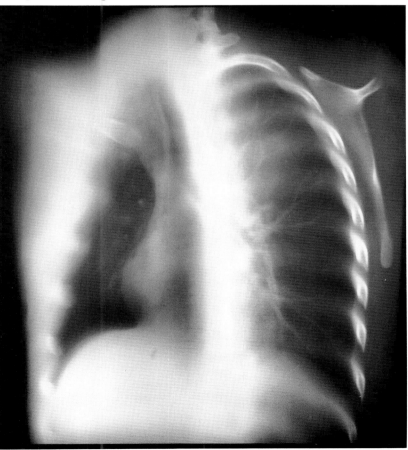

# Right Posterior Oblique

## Section 8

## Anatomic Key

### Arteries
A  2  main pulmonary, right
A  16  main pulmonary, left
A  23  left lower lobe (LLL)
A  24  LLL, superior branch
A  25  LLL, anteromedial basal branch
A  26  LLL, lateral basal branch
A  29  aortic arch
A  30  aorta, ascending
A  31  aorta, descending
A  34  brachiocephalic (innominate) trunk
A  37  carotid, common, left
A  49  subclavian, left
A  63  transverse cervical, left

### Bronchi
B  17  left upper lobe (LUL)
B  18  LUL, apicoposterior segment
B  23  left lower lobe (LLL)
B  24  LLL, superior segment
B  25  LLL, anteromedial basal segment
B  26  LLL, lateral basal segment

### Esophagus
E  1  esophagus
E  2  esophagogastric junction

### Fat
f  2  epicardial

### Fissures
F  1  left major (oblique)
F  3  right minor (horizontal)

### Glands
G  1  adrenal, left
G  4  thyroid

### Heart
H  1  atrium, left
H  2  atrium, right
H  5  cardiac vein, great
H  6  cardiac vein, middle
H  8  coronary artery, circumflex
H  9  coronary artery, posterior descending
H  13  pericardium, parietal serous
H  14  pericardium, visceral serous
H  16  septum, interventricular
H  21  ventricle, left
H  22  ventricle, right

### Lobes
L  4  RUL, anterior segment
L  7  RML, medial segment
L  15  LUL, apicoposterior segment
L  16  LUL, anterior segment
L  21  LLL, superior segment
L  22  LLL, anteromedial basal segment
L  23  LLL, lateral basal segment

### Muscles
M  1  abdominal, external oblique
M  3  abdominis, rectus
M  9  diaphragm
M  10  diaphragm, central tendon
M  11  diaphragm, crus, left
M  12  diaphragm, crus, right
M  15  infraspinatus
M  16  intercostal
M  21  latissimus dorsi
M  28  longus colli
M  31  pectoralis, major
M  33  platysma
M  40  scalene, posterior
M  43  semispinalis, thoracis
M  44  serratus, anterior
M  51  splenius, cervicis
M  52  sternocleidomastoid
M  53  sternohyoid
M  54  sternothyroid
M  57  subscapularis
M  58  supraspinatus
M  63  trapezius

### Neural Structures
n  1  brachial plexus
n  7  spinal nerve

### Nodes
N  7  mediastinal, posterior

### Organs
O  2  kidney, left
O  3  liver
O  5  spleen
O  6  stomach

### Skeletal Structures
S  1  clavicle, right
S  4  rib, body, anterior (6th right)
S  5  rib, body, lateral (10th left)
S  6  rib, body, posterior (2nd left)
S  7  rib, costal cartilage (2nd right)
S  10  scapula
S  15  scapula, spine
S  18  sternum, manubrium
S  20  intervertebral disc (C7–T1)
S  29  vertebra, transverse process (T1)

### Trachea and Larynx
T  1  trachea
T  3  larynx

### Veins
V  2  pulmonary, superior, LUL
V  5  pulmonary, inferior, LLL
V  9  brachiocephalic (innominate), right
V  17  intercostal, anterior, right
V  18  intercostal, posterior, left
V  27  jugular, internal, right
V  33  suprascapular, left
V  36  transverse cervical, left
V  39  vena cava, superior

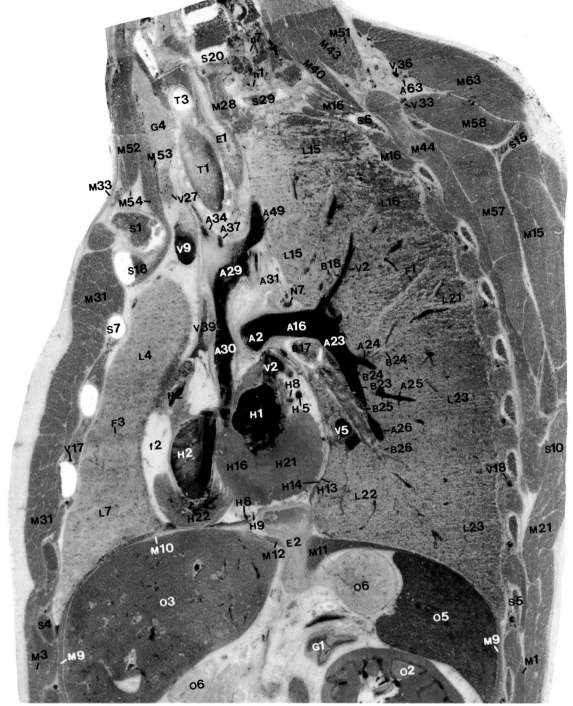

**Anatomic Specimen**

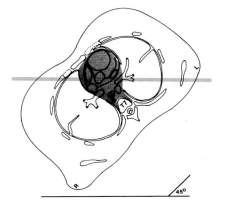

No
Specimen Radiograph
made

**Specimen Radiograph**

**Magnetic Resonance**

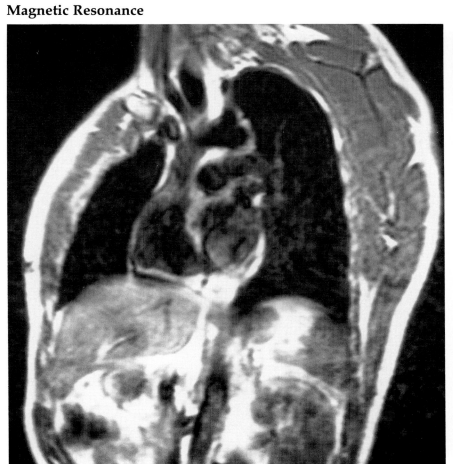

**Trispiral Tomogram**

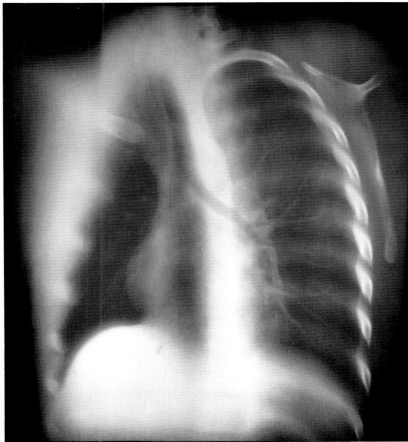

# Right Posterior Oblique

## Section 9

## Anatomic Key

### Arteries
A  2  main pulmonary, right
A 24  LLL, superior branch
A 27  LLL, posterior basal branch
V 30  aorta, ascending
A 31  aorta, descending
A 38  carotid, common, right
A 45  intercostal, posterior, left
A 50  subclavian, right

### Bronchi
B 16  main, left
B 17  left upper lobe (LUL)
B 18  LUL, apicoposterior segment
B 23  left lower lobe (LLL)
B 24  LLL, superior segment
B 27  LLL, posterior basal segment

### Esophagus
E  1  esophagus
E  2  esophagogastric junction

### Fat
f  2  epicardial
f  4  mediastinal
f  6  perirenal

### Fissures
F  1  left major (oblique)
F  3  right minor (horizontal)

### Glands
G  1  adrenal, left
G  4  thyroid

### Heart
H  1  atrium, left
H  2  atrium, right
H  6  cardiac vein, middle
H  8  coronary artery, circumflex
H  9  coronary artery, posterior descending
H 12  coronary sinus
H 15  septum, interatrial
H 21  ventricle, left
H 22  ventricle, right

### Lobes
L  4  RUL, anterior segment
L  7  RML, medial segment
L 15  LUL, apicoposterior segment
L 21  LLL, superior segment
L 22  LLL, anteromedial basal segment
L 24  LLL, posterior basal segment

### Muscles
M  1  abdominal, external oblique
M  3  abdominis, rectus
M  9  diaphragm
M 10  diaphragm, central tendon
M 11  diaphragm, crus, left
M 12  diaphragm, crus, right
M 15  infraspinatus
M 16  intercostal
M 21  latissimus dorsi
M 23  levator scapulae
M 28  longus colli
M 31  pectoralis, major
M 32  pectoralis, minor
M 33  platysma
M 35  rhomboid, major
M 40  scalene, posterior
M 41  semispinalis, capitis
M 44  serratus, anterior
M 51  splenius, cervicis
M 52  sternocleidomastoid
M 53  sternohyoid
M 54  sternothyroid
M 57  subscapularis
M 58  supraspinatus
M 63  trapezius

### Neural Structures
n  1  brachial plexus

### Organs
O  2  kidney, left
O  3  liver
O  5  spleen
O  6  stomach

### Pleural Structures
P  2  costal (parietal)
P  5  mediastinal (parietal)
P  6  visceral

### Skeletal Structures
S  1  clavicle, right
S  3  rib (2nd left)
S  4  rib, body, anterior (6th right)
S  6  rib, body, posterior (10th left)
S  7  rib, costal cartilage (1st right)
S 10  scapula
S 15  scapula, spine
S 21  vertebra, body (T1)

### Trachea
T  1  trachea

### Veins
V  2  pulmonary, superior, LUL
V  5  pulmonary, inferior, LLL
V  9  brachiocephalic (innominate), right
V 18  intercostal, posterior, left
V 27  jugular, internal, right
V 33  suprascapular, left
V 36  transverse cervical, left
V 39  vena cava, superior

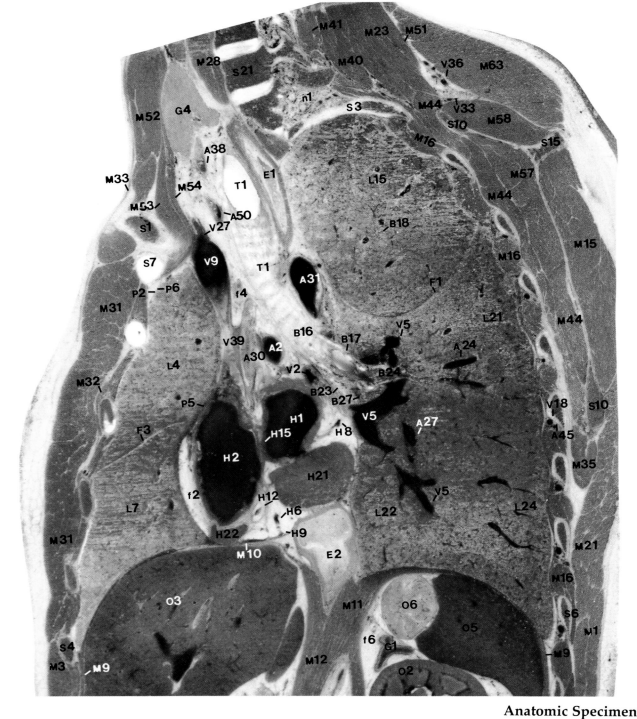

**Anatomic Specimen**

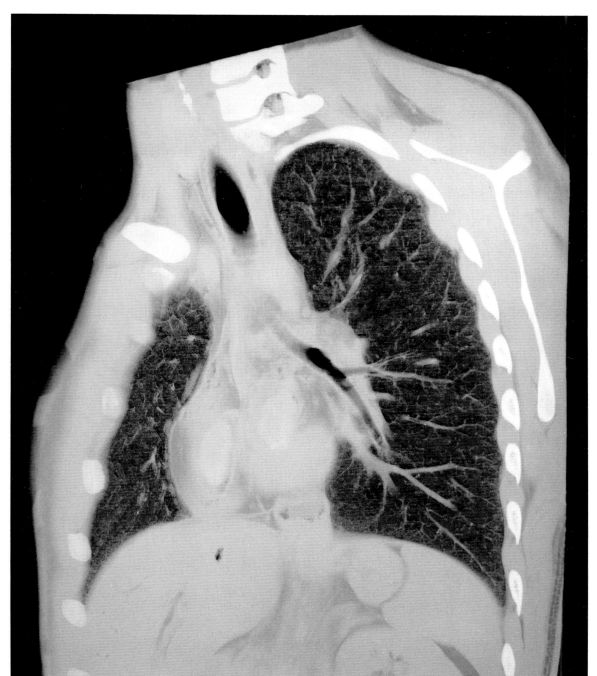

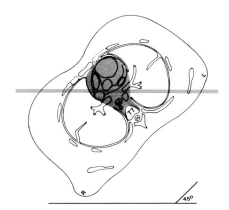

Specimen Radiograph

**Magnetic Resonance**

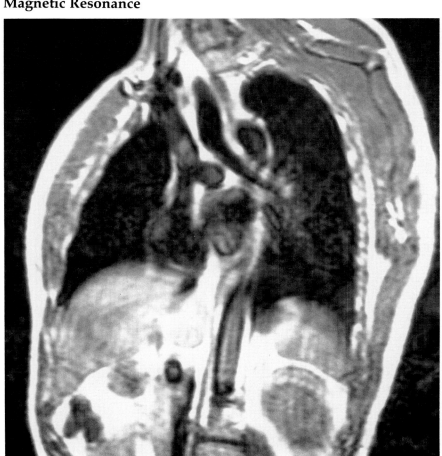

**Trispiral Tomogram**

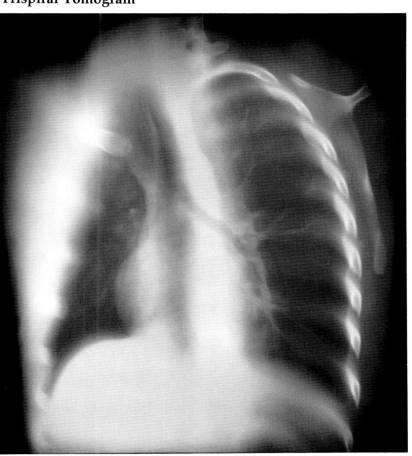

# Right Posterior Oblique

## Section 10

## Anatomic Key

### Arteries
A   2  main pulmonary, right
A  24  LLL, superior branch
A  27  LLL, posterior basal branch
A  31  aorta, descending
A  38  carotid, common, right
A  45  intercostal, posterior, left
A  50  subclavian, right

### Bronchi
B  16  main, left
B  23  left lower lobe (LLL)
B  24  LLL, superior segment
B  27  LLL, posterior basal segment

### Esophagus
E   1  esophagus
E   2  esophagogastric junction

### Fat
f   2  epicardial
f   4  mediastinal
f   6  perirenal

### Fissures
F   1  left major (oblique)
F   3  right minor (horizontal)

### Glands
G   1  adrenal, left
G   4  thyroid

### Heart
H   1  atrium, left
H   2  atrium, right
H  15  septum, interatrial
H  21  ventricle, left

### Lobes
L   4  RUL, anterior segment
L   7  RML, medial segment
L  15  LUL, apicoposterior segment
L  21  LLL, superior segment
L  22  LLL, anteromedial basal
       segment
L  24  LLL, posterior basal segment

### Muscles
M   3  abdominis, rectus
M   9  diaphragm
M  10  diaphragm, central tendon
M  11  diaphragm, crus, left
M  12  diaphragm, crus, right
M  15  infraspinatus
M  16  intercostal
M  21  latissimus dorsi
M  23  levator scapulae
M  31  pectoralis, major
M  32  pectoralis, minor
M  33  platysma
M  35  rhomboid, major
M  40  scalene, posterior
M  41  semispinalis, capitis
M  44  serratus, anterior
M  52  sternocleidomastoid
M  53  sternohyoid
M  57  subscapularis
M  58  supraspinatus
M  63  trapezius

### Neural Structures
n   7  spinal nerve

### Organs
O   2  kidney, left
O   3  liver
O   5  spleen
O   6  stomach

### Pleural Structures
P   2  costal (parietal)
P   5  mediastinal (parietal)
P   6  visceral

### Skeletal Structures
S   1  clavicle, right
S   4  rib, body, anterior (6th right)
S   6  rib, body, posterior (11th left)
S   7  rib, costal cartilage (2nd right)
S   8  rib, head and neck (3rd left)
S  10  scapula
S  15  scapula, spine
S  20  intervertebral disc (C7–T1)
S  21  vertebra, body (T2)
S  22  vertebra, costal facet (T3)
S  26  vertebra, pedicle (T2)

### Trachea
T   1  trachea
T   2  carina (bifurcation of trachea)

### Veins
V   2  pulmonary, superior, LUL
V   5  pulmonary, inferior, LLL
V   9  brachiocephalic (innominate),
       right
V  18  intercostal, posterior, left
V  27  jugular, internal, right
V  36  transverse cervical, left
V  39  vena cava, superior

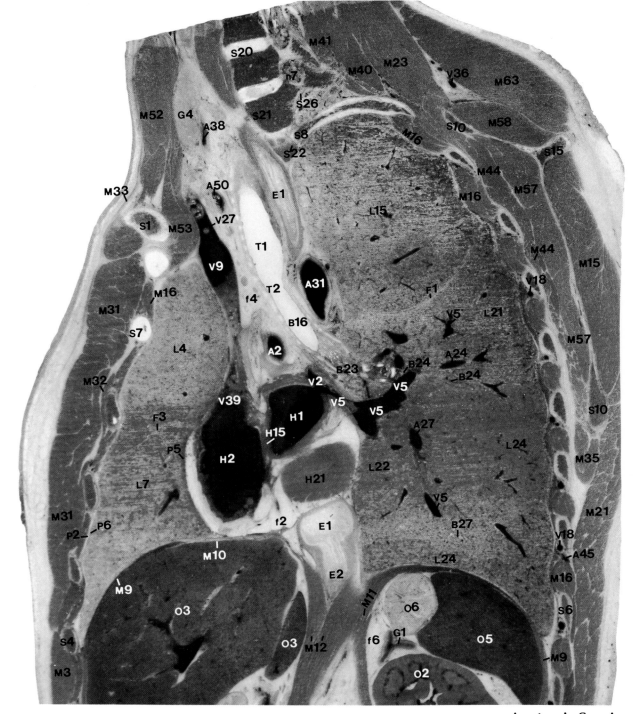

**Anatomic Specimen**

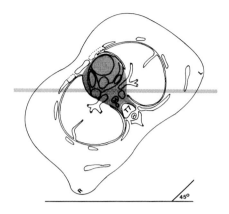

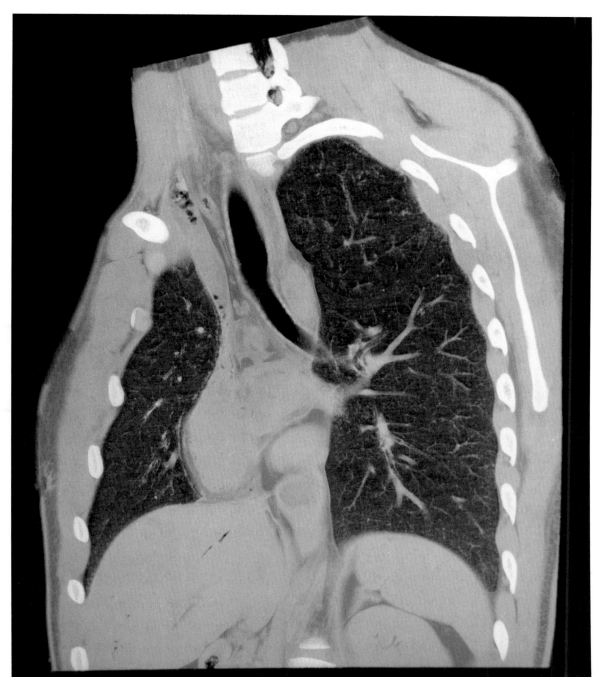

**Specimen Radiograph**

**Magnetic Resonance**

**Trispiral Tomogram**

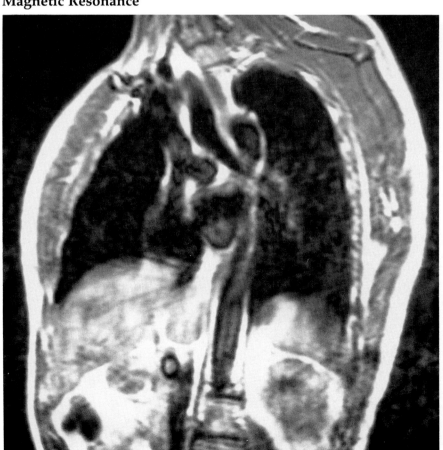

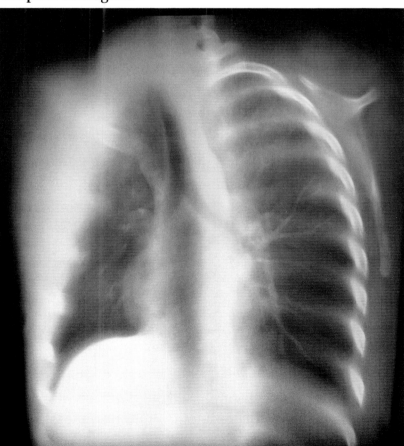

# Right Posterior Oblique

## Section 11

## Anatomic Key

### Arteries
A 2 main pulmonary, right
A 7 right middle lobe (RML)
A 24 LLL, superior branch
A 27 LLL, posterior basal branch
A 31 aorta, descending
A 45 intercostal, posterior, left
A 50 subclavian, right
A 66 vertebral, right

### Bronchi
B 3 RUL, apical segment
B 5 RUL, anterior segment
B 8 RML, lateral segment
B 9 RML, medial segment
B 24 LLL, superior segment
B 27 LLL, posterior basal segment

### Esophagus
E 1 esophagus

### Fat
f 6 perirenal

### Fissures
F 1 left major (oblique)
F 2 right major (oblique)
F 3 right minor (horizontal)

### Heart
H 1 atrium, left
H 2 atrium, right
H 15 septum, interatrial

### Lobes
L 2 RUL, apical segment
L 4 RUL, anterior segment
L 6 RML, lateral segment
L 7 RML, medial segment
L 10 RLL, medial basal segment
L 11 RLL, anterior basal segment
L 15 LUL, apicoposterior segment
L 21 LLL, superior segment
L 24 LLL, posterior basal segment

### Muscles
M 9 diaphragm
M 10 diaphragm, central tendon
M 11 diaphragm, crus, left
M 12 diaphragm, crus, right
M 13 erector spinae
M 15 infraspinatus
M 16 intercostal
M 21 latissimus dorsi
M 23 levator scapulae
M 28 longus colli
M 31 pectoralis, major
M 32 pectoralis, minor
M 33 platysma
M 35 rhomboid, major
M 38 scalene, anterior
M 40 scalene, posterior
M 41 semispinalis, capitis
M 44 serratus, anterior
M 50 splenius, capitis
M 52 sternocleidomastoid
M 57 subscapularis
M 58 supraspinatus
M 61 transversospinalis
M 63 trapezius

### Neural Structures
n 5 spinal cord
n 7 spinal nerve

### Organs
O 2 kidney, left
O 3 liver
O 5 spleen

### Pleural Structures
P 2 costal (parietal)
P 5 mediastinal (parietal)
P 6 visceral

### Skeletal Structures
S 1 clavicle, right
S 4 rib, body, anterior (2nd right)
S 5 rib, body, lateral (10th left)
S 6 rib, body, posterior (3rd left)
S 7 rib, costal cartilage (1st right)
S 8 rib, head and neck (3rd left)
S 10 scapula
S 20 intervertebral disc (T11–T12)
S 21 vertebra, body (T1)
S 22 vertebra, costal facet (T2)
S 23 vertebra, inferior articular process (T1)
S 29 vertebra, transverse process (T2)

### Trachea
T 1 trachea
T 2 carina (bifurcation of trachea)

### Veins
V 0 pulmonary, superior, RUL
V 1 pulmonary, superior, RML
V 9 brachiocephalic (innominate), right
V 12 hemiazygos
V 18 intercostal, posterior, left
V 23 jugular, anterior, right
V 27 jugular, internal, right
V 36 transverse cervical, left
V 38 vena cava, inferior
V 39 vena cava, superior
V 41 venous plexus, vertebral

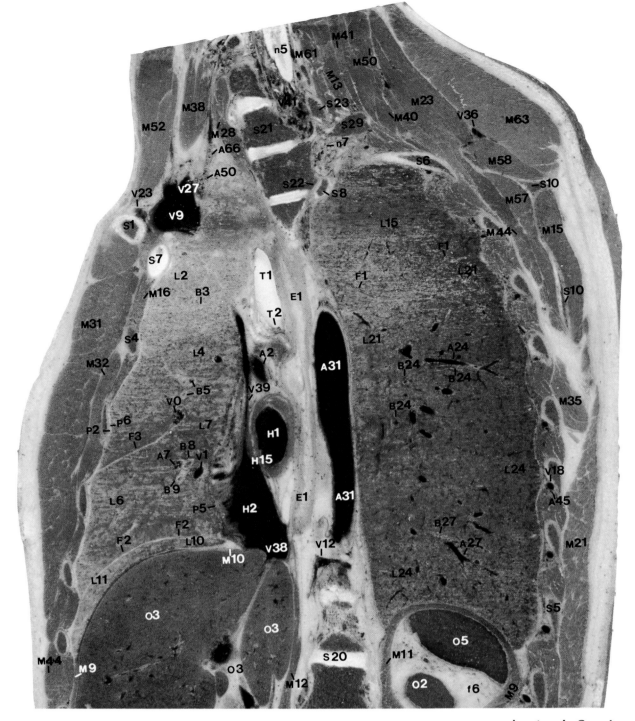

**Anatomic Specimen**

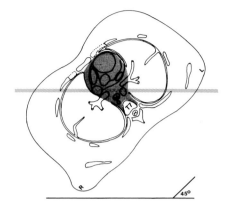

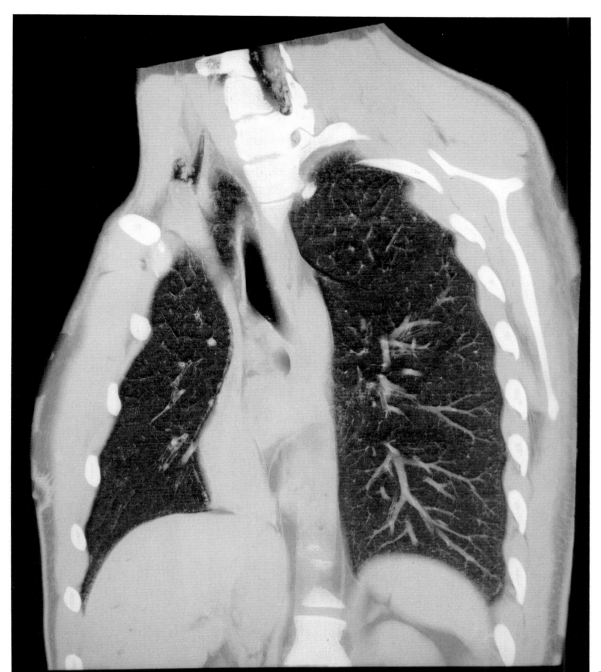

**Specimen Radiograph**

**Magnetic Resonance**

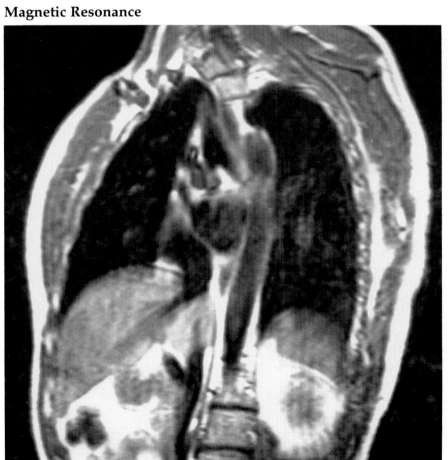

**Trispiral Tomogram**

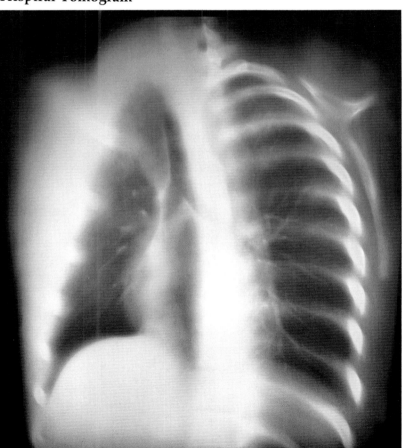

291

# Right Posterior Oblique

## Section 12

### Anatomic Key

**Arteries**
A 2 main pulmonary, right
A 7 right middle lobe (RML)
A 27 LLL, posterior basal branch
A 45 intercostal, posterior, left
A 50 subclavian, right

**Bronchi**
B 3 RUL, apical segment
B 5 RUL, anterior segment
B 8 RML, lateral segment
B 9 RML, medial segment
B 24 LLL, superior segment

**Esophagus**
E 1 esophagus

**Fat**
f 7 subcutaneous

**Fissures**
F 1 left major (oblique)
F 2 right major (oblique)
F 3 right minor (horizontal)

**Heart**
H 1 atrium, left
H 2 atrium, right
H 15 septum, interatrial

**Lobes**
L 2 RUL, apical segment
L 4 RUL, anterior segment
L 6 RML, lateral segment
L 7 RML, medial segment
L 10 RLL, medial basal segment
L 11 RLL, anterior basal segment
L 15 LUL, apicoposterior segment
L 21 LLL, superior segment
L 24 LLL, posterior basal segment

**Muscles**
M 1 abdominal, external oblique
M 9 diaphragm
M 11 diaphragm, crus, left
M 12 diaphragm, crus, right
M 13 erector spinae
M 15 infraspinatus
M 16 intercostal
M 21 latissimus dorsi
M 23 levator scapulae
M 28 longus colli
M 31 pectoralis, major
M 32 pectoralis, minor
M 33 platysma
M 35 rhomboid, major
M 36 rhomboid, minor
M 38 scalene, anterior
M 41 semispinalis, capitis
M 44 serratus, anterior
M 46 serratus, posterior superior
M 50 splenius, capitis
M 51 splenius, cervicis
M 52 sternocleidomastoid
M 55 subclavius
M 57 subscapularis
M 58 supraspinatus
M 61 transversospinalis
M 63 trapezius

**Neural Structures**
n 5 spinal cord

**Nodes**
N 7 mediastinal, posterior
N 9 paratracheal
N 11 tracheobronchial, inferior (subcarinal)

**Organs**
O 3 liver
O 6 stomach

**Skeletal Structures**
s 1 clavicle
s 3 rib (3rd left)
s 4 rib, body, anterior (1st right)
s 5 rib, body, lateral (5th right)
s 6 rib, body, posterior (9th left)
s 10 scapula
s 20 intervertebral disc (T11–T12)
s 21 vertebra, body (T2)

**Trachea**
T 1 trachea

**Veins**
V 0 pulmonary, superior, RUL
V 1 pulmonary, superior, RML
V 7 azygos
V 18 intercostal, posterior, left
V 31 subclavian, right
V 37 transverse cervical, right
V 38 vena cava, inferior

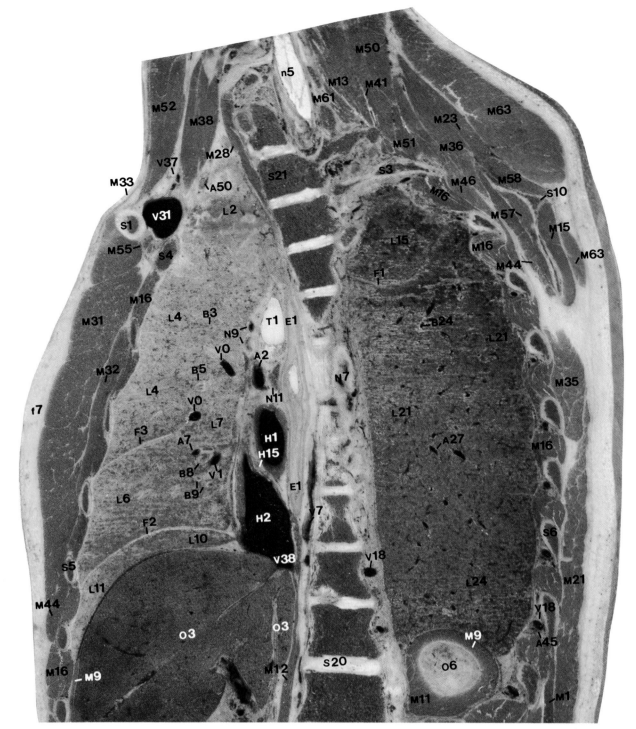

**Anatomic Specimen**

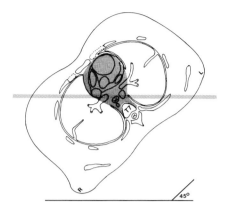

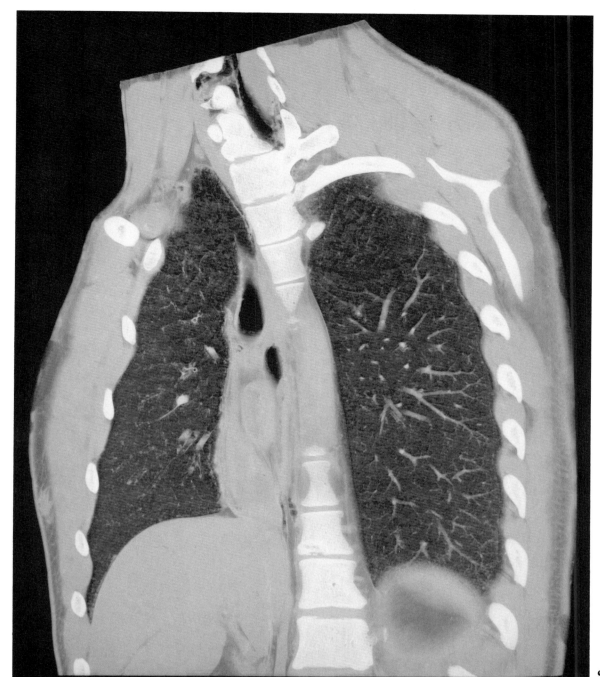

**Specimen Radiograph**

**Magnetic Resonance**

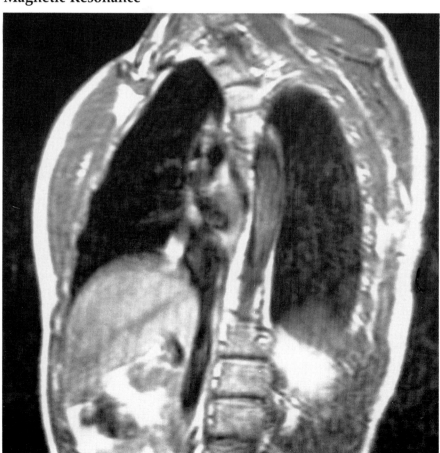

**Trispiral Tomogram**

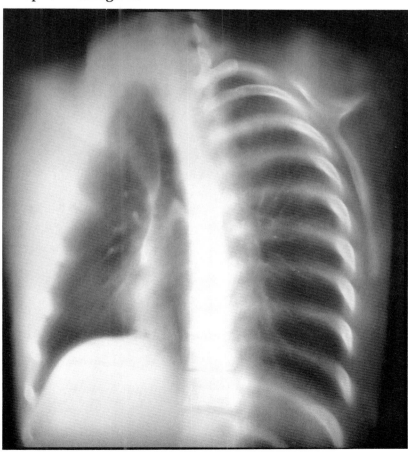

# Right Posterior Oblique

## Section 13

## Anatomic Key

### Arteries
A 2* main pulmonary, right (interlobar part)
A 7 right middle lobe (RML)
A 45 intercostal, posterior, left
A 50 subclavian, right
A 66 vertebral, right

### Bronchi
B 1 main, right
B 3 RUL, apical segment
B 5 RUL, anterior segment
B 6 intermediate, right
B 8 RML, lateral segment
B 9 RML, medial segment

### Esophagus
E 1 esophagus

### Fissures
F 1 left major (oblique)
F 2 right major (oblique)
F 3 right minor (horizontal)

### Heart
H 1 atrium, left
H 2 atrium, right
H 15 septum, interatrial

### Lobes
L 2 RUL, apical segment
L 4 RUL, anterior segment
L 6 RML, lateral segment
L 7 RML, medial segment
L 10 RLL, medial basal segment
L 11 RLL, anterior basal segment
L 15 LUL, apicoposterior segment
L 21 LLL, superior segment
L 24 LLL, posterior basal segment

### Muscles
M 1 abdominal, external oblique
M 9 diaphragm
M 12 diaphragm, crus, right
M 13 erector spinae
M 14 iliocostalis, thoracis
M 15 infraspinatus
M 16 intercostal
M 21 latissimus dorsi
M 23 levator scapulae
M 28 longus colli
M 31 pectoralis, major
M 32 pectoralis, minor
M 35 rhomboid, major
M 36 rhomboid, minor
M 38 scalene, anterior
M 41 semispinalis, capitis
M 44 serratus, anterior
M 46 serratus, posterior superior
M 50 splenius, capitis
M 51 splenius, cervicis
M 52 sternocleidomastoid
M 55 subclavius
M 58 supraspinatus
M 61 transversospinalis
M 63 trapezius

### Neural Structures
n 2 intercostal nerve
n 5 spinal cord

### Organs
O 3 liver

### Pleural Structures
P 2 costal (parietal)
P 6 visceral

### Skeletal Structures
S 1 clavicle, right
S 3 rib (4th left)
S 4 rib, body, anterior (1st right)
S 5 rib, body, lateral (6th right)
S 6 rib, body, posterior (1st right)
S 8 rib, head and neck (2nd right)
S 9 rib, tubercle (12th left)
S 10 scapula
S 20 intervertebral disc (T11–T12)
S 21 vertebra, body (T4)
S 23 vertebra, inferior articular process (T2)
S 25 vertebra, lamina (T1)
S 26 vertebra, pedicle (T3)
S 28 vertebra, superior articular process (T3)
S 29 vertebra, transverse process (T3)

### Veins
V 0 pulmonary, superior, RUL
V 1 pulmonary, superior, RML
V 12 hemiazygos
V 18 intercostal, posterior, left
V 19 intercostal, posterior, right
V 23 jugular, anterior, right
V 25 jugular, external, right
V 31 subclavian, right
V 37 transverse cervical, right
V 38 vena cava, inferior

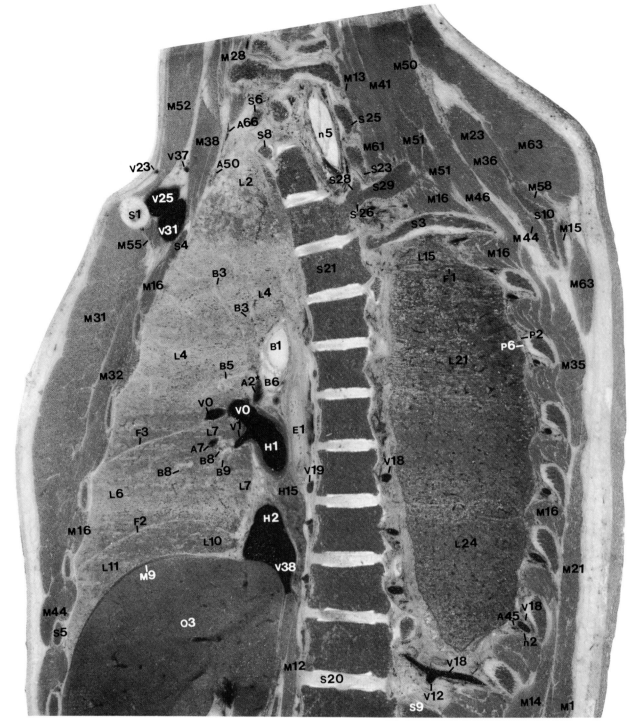

**Anatomic Specimen**

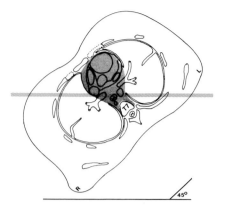

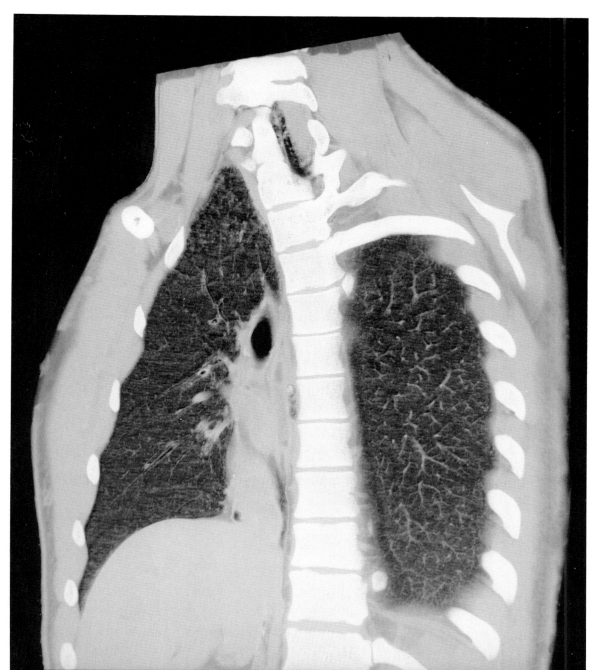

Specimen Radiograph

**Magnetic Resonance**

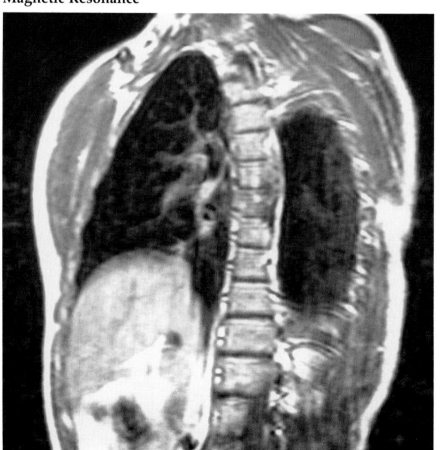

**Trispiral Tomogram**

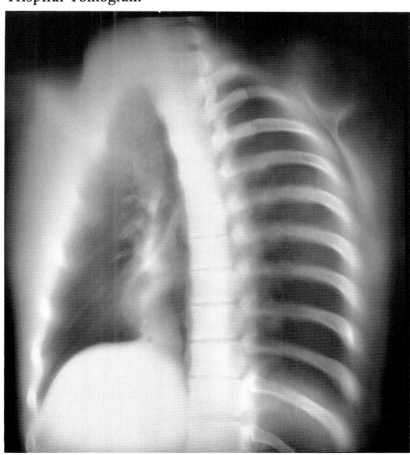

### Anatomic Key

**Arteries**
A 2* main pulmonary, right (interlobar part)
A 7 right middle lobe (RML)
A 45 intercostal, posterior, left
A 50 subclavian, right

**Bronchi**
B 1 main, right
B 3 RUL, apical segment
B 5 RUL, anterior segment
B 6 intermediate, right
B 7 right middle lobe (RML)
B 8 RML, lateral segment
B 9 RML, medial segment

**Fissures**
F 2 right major (oblique)
F 3 right minor (horizontal)

**Glands**
G 1 adrenal, right

**Lobes**
L 2 RUL, apical segment
L 4 RUL, anterior segment
L 6 RML, lateral segment
L 7 RML, medial segment
L 10 RLL, medial basal segment
L 11 RLL, anterior basal segment
L 21 LLL, superior segment
L 24 LLL, posterior basal segment

**Muscles**
M 9 diaphragm
M 12 diaphragm, crus, right
M 13 erector spinae
M 14 iliocostalis, thoracis
M 16 intercostal
M 31 pectoralis, major
M 32 pectoralis, minor
M 35 rhomboid, major
M 36 rhomboid, minor
M 38 scalene, anterior
M 39 scalene, middle
M 40 scalene, posterior
M 41 semispinalis, capitis
M 44 serratus, anterior
M 46 serratus, posterior superior
M 50 splenius, capitis
M 51 splenius, cervicis
M 52 sternocleidomastoid
M 55 subclavius
M 61 transversospinalis
M 63 trapezius

**Neural Structures**
n 5 spinal cord

**Nodes**
N 2 bronchopulmonary (hilar)

**Organs**
O 3 liver

**Skeletal Structures**
s 1 clavicle, right
s 3 rib (7th right)
s 4 rib, body, anterior (1st right)
s 5 rib, body, lateral (5th left)
s 10 scapula
s 20 intervertebral disc (T11–T12)
s 21 vertebra, body (T2)

**Veins**
v 0 pulmonary, superior, RUL
v 4 pulmonary, inferior, RLL
v 7 azygos
v 12 hemiazygos
v 15 intercostal, highest (superior), right
v 18 intercostal, posterior, left
v 19 intercostal, posterior, right
v 25 jugular, external, right
v 31 subclavian, right
v 36 transverse cervical, left
v 37 transverse cervical, right
v 38 vena cava, inferior
v 43 vertebral, right

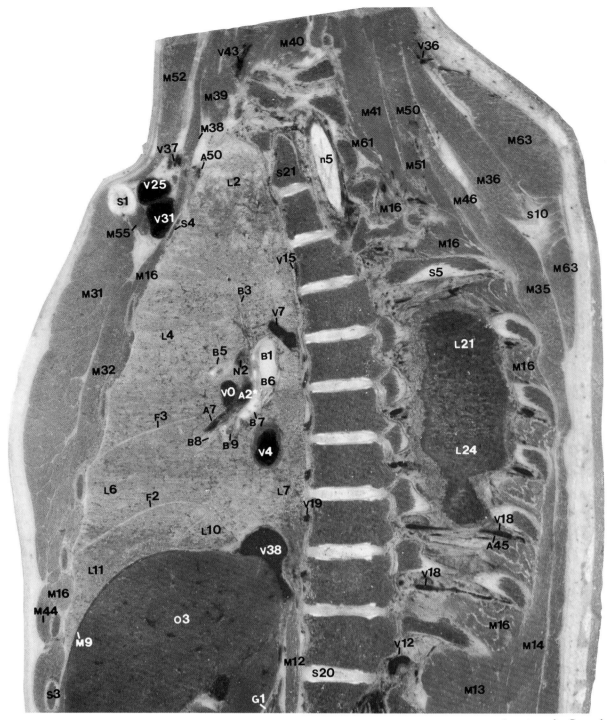

**Anatomic Specimen**

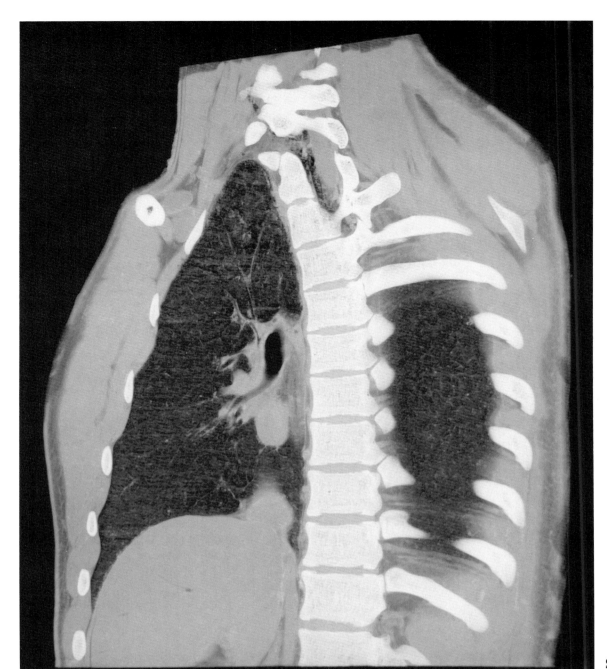

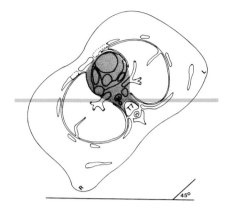

Specimen Radiograph

Magnetic Resonance

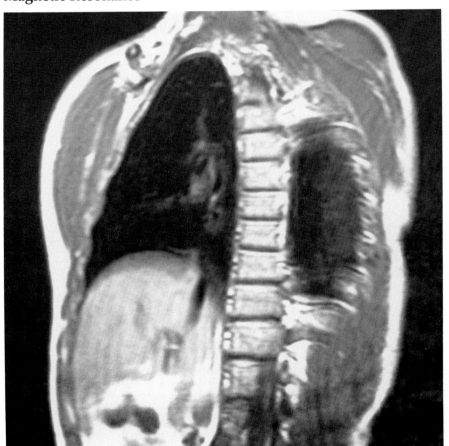

Trispiral Tomogram

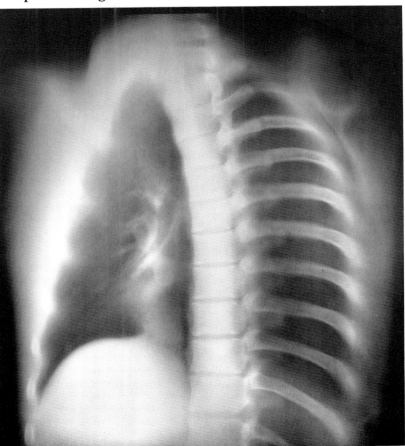

### Anatomic Key

**Arteries**
A  7  right middle lobe (RML)
A 50  subclavian, right
A 63  transverse cervical, left

**Bronchi**
B  2  right upper lobe (RUL)
B  3  RUL, apical segment
B  4  RUL, posterior segment
B  5  RUL, anterior segment
B  6  intermediate, right
B  7  right middle lobe (RML)
B  8  RML, lateral segment
B  9  RML, medial segment

**Fissures**
F  2  right major (oblique)
F  3  right minor (horizontal)

**Glands**
G  1  adrenal, right

**Lobes**
L  2  RUL, apical segment
L  4  RUL, anterior segment
L  6  RML, lateral segment
L  7  RML, medial segment
L 10  RLL, medial basal segment
L 11  RLL, anterior basal segment
L 21  LLL, superior segment
L 24  LLL, posterior basal segment

**Muscles**
M  9  diaphragm
M 12  diaphragm, crus, right
M 13  erector spinae
M 14  iliocostalis, thoracis
M 16  intercostal
M 21  latissimus dorsi
M 31  pectoralis, major
M 32  pectoralis, minor
M 33  platysma
M 35  rhomboid, major
M 36  rhomboid, minor
M 38  scalene, anterior
M 39  scalene, middle
M 40  scalene, posterior
M 41  semispinalis, capitis
M 44  serratus, anterior
M 46  serratus, posterior superior
M 50  splenius, capitis
M 51  splenius, cervicis
M 52  sternocleidomastoid
M 55  subclavius
M 63  trapezius

**Neural Structures**
n  1  brachial plexus
n  5  spinal cord

**Nodes**
N 10  supraclavicular

**Organs**
O  3  liver

**Pleural Structures**
P  2  costal (parietal)
P  6  visceral

**Skeletal Structures**
S  1  clavicle, right
S  3  rib (7th right)
S  4  rib, body, anterior (1st right)
S  5  rib, body, lateral (5th left)
S  6  rib, body, posterior (10th left)
S  8  rib, head and neck (7th left)
S 20  intervertebral disc (T9–T10)
S 21  vertebra, body (T12)
S 22  vertebra, costal facet (T6–T7)
S 27  vertebra, spinous process (T1)
S 29  vertebra, transverse process (T4)

**Veins**
V  0  pulmonary, superior, RUL
V  4  pulmonary, inferior, RLL
V  7  azygos
V 12  hemiazygos
V 14  intercostal, highest (superior), left
V 15  intercostal, highest (superior), right
V 18  intercostal, posterior, left
V 19  intercostal, posterior, right
V 25  jugular, external, right
V 31  subclavian, right
V 36  transverse cervical, left
V 37  transverse cervical, right
V 38  vena cava, inferior
V 41  venous plexus, vertebral
V 43  vertebral, right

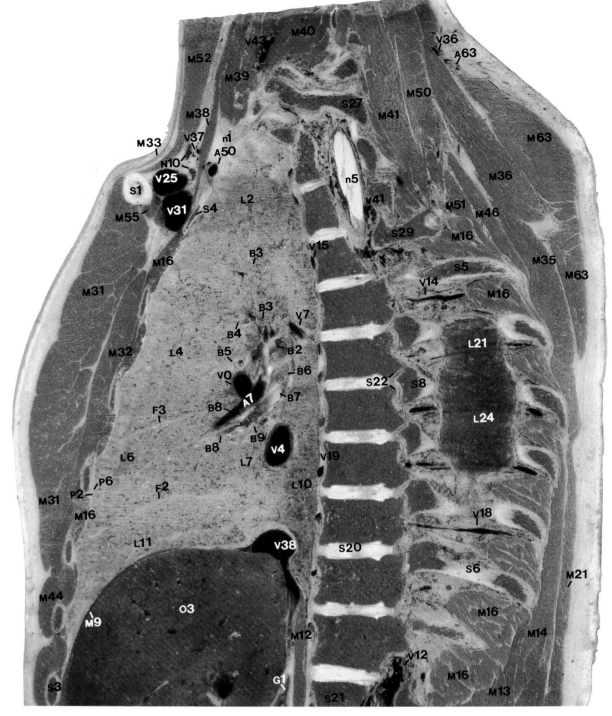

**Anatomic Specimen**

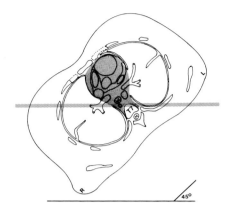

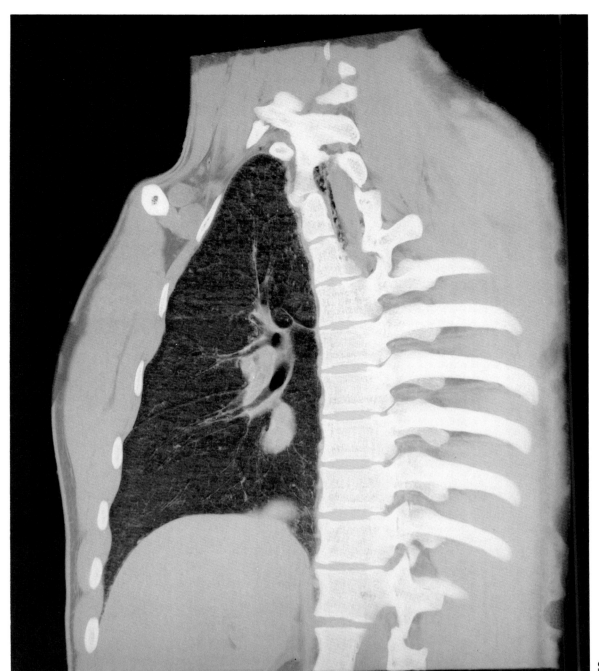

Specimen Radiograph

Magnetic Resonance

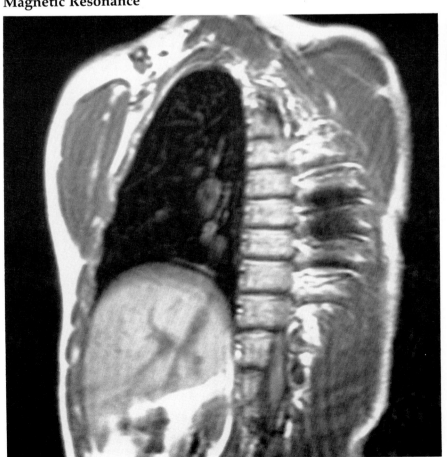

Trispiral Tomogram

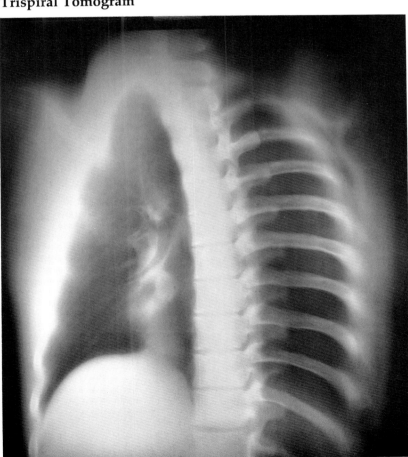

# Right Posterior Oblique

## Section 16

## Anatomic Key

### Arteries
A 4 RUL, apical branch
A 7 right middle lobe (RML)
A 50 subclavian, right

### Bronchi
B 3 RUL, apical segment
B 5 RUL, anterior segment
B 6 intermediate, right
B 10 right lower lobe (RLL)
B 12 RLL, medial basal segment
B 13 RLL, anterior basal segment

### Fissures
F 2 right major (oblique)
F 3 right minor (horizontal)

### Glands
G 1 adrenal, right

### Lobes
L 2 RUL, apical segment
L 4 RUL, anterior segment
L 6 RML, lateral segment
L 7 RML, medial segment
L 9 RLL, superior segment
L 10 RLL, medial basal segment
L 11 RLL, anterior basal segment
L 12 RLL, lateral basal segment

### Muscles
M 9 diaphragm
M 12 diaphragm, crus, right
M 14 iliocostalis, thoracis
M 16 intercostal
M 21 latissimus dorsi
M 27 longissimus, thoracis
M 31 pectoralis, major
M 32 pectoralis, minor
M 33 platysma
M 35 rhomboid, major
M 36 rhomboid, minor
M 38 scalene, anterior
M 39 scalene, middle
M 40 scalene, posterior
M 43 semispinalis, thoracis
M 44 serratus, anterior
M 50 splenius, capitis
M 52 sternocleidomastoid
M 55 subclavius
M 61 transversospinalis
M 63 trapezius

### Neural Structures
n 1 brachial plexus
n 5 spinal cord

### Organs
O 3 liver

### Pleural Structures
P 2 costal (parietal)
P 6 visceral

### Skeletal Structures
S 1 clavicle, right
S 3 rib (1st right)
S 4 rib, body, anterior (7th right)
S 8 rib, head and neck (5th right)
S 20 intervertebral disc (T7–T8)
S 21 vertebra, body (T11)
S 22 vertebra, costal facet (T4–T5)
S 27 vertebra, spinous process (T2)

### Veins
V 0 pulmonary, superior, RUL
V 4 pulmonary, inferior, RLL
V 15 intercostal, highest (superior), right
V 19 intercostal, posterior, right
V 25 jugular, external, right
V 31 subclavian, right
V 33 suprascapular, right
V 41 venous plexus, vertebral

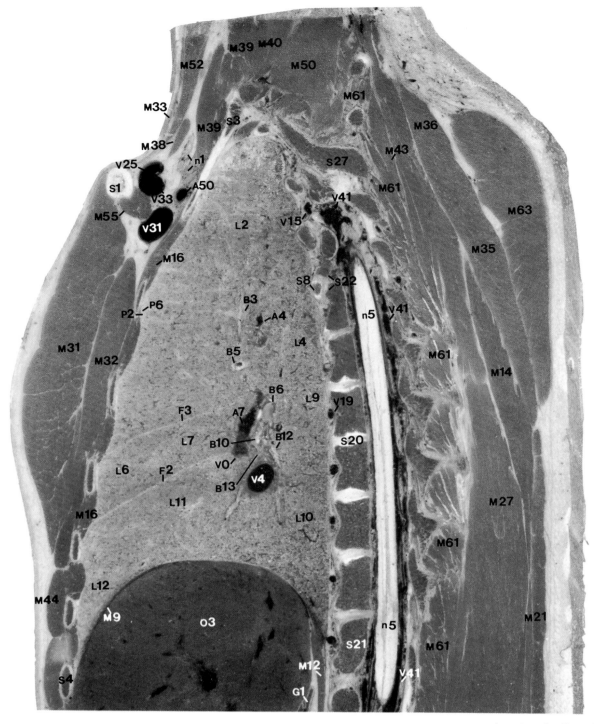

**Anatomic Specimen**

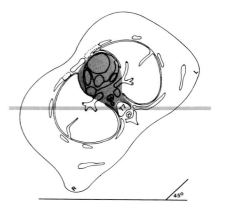

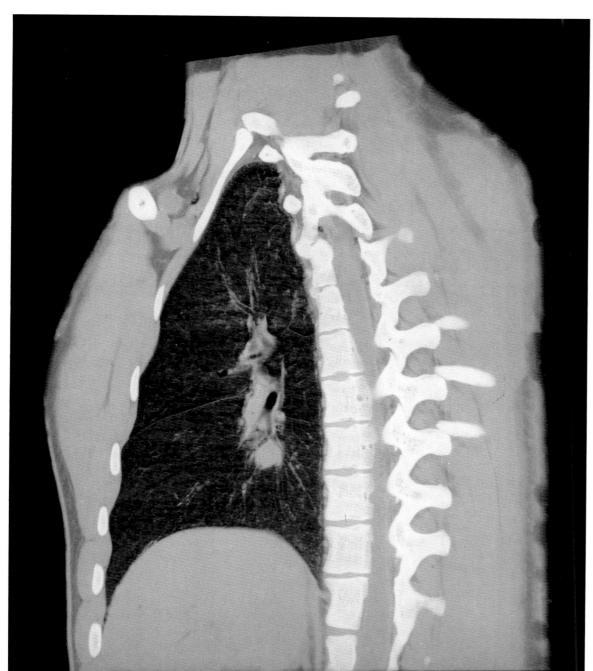

**Specimen Radiograph**

**Magnetic Resonance**

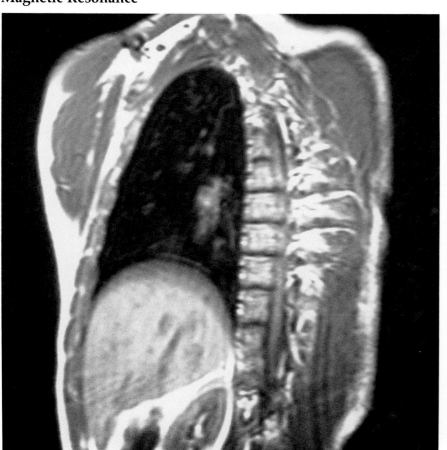

**Trispiral Tomogram**

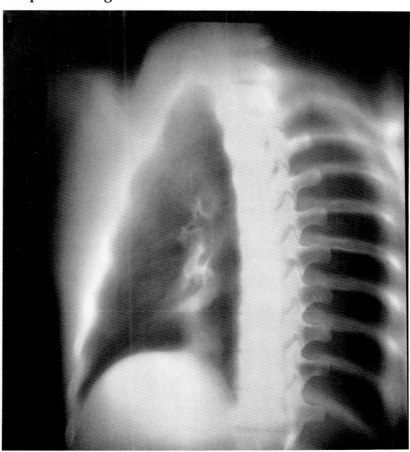

# Right Posterior Oblique

## Section 17

### Anatomic Key

**Arteries**
A 4 RUL, apical branch
A 10 right lower lobe (RLL)
A 33 axillary, right
A 46 intercostal, posterior, right

**Bronchi**
B 3 RUL, apical segment
B 4 RUL, posterior segment
B 5 RUL, anterior segment
B 10 right lower lobe (RLL)
B 12 RLL, medial basal segment
B 13 RLL, anterior basal segment
B 14 RLL, lateral basal segment

**Fissures**
F 2 right major (oblique)
F 3 right minor (horizontal)

**Glands**
G 1 adrenal, right

**Lobes**
L 2 RUL, apical segment
L 3 RUL, posterior segment
L 4 RUL, anterior segment
L 6 RML, lateral segment
L 9 RLL, superior segment
L 10 RLL, medial basal segment
L 11 RLL, anterior basal segment
L 12 RLL, lateral basal segment

**Muscles**
M 9 diaphragm
M 12 diaphragm, crus, right
M 14 iliocostalis, thoracis
M 16 intercostal
M 21 latissimus dorsi
M 23 levator scapulae
M 27 longissimus, thoracis
M 30 omohyoid
M 31 pectoralis, major
M 32 pectoralis, minor
M 35 rhomboid, major
M 36 rhomboid, minor
M 39 scalene, middle
M 40 scalene, posterior
M 43 semispinalis, thoracis
M 44 serratus, anterior
M 50 splenius, capitis
M 52 sternocleidomastoid
M 55 subclavius
M 61 transversospinalis
M 63 trapezius

**Neural Structures**
n 1 brachial plexus
n 5 spinal cord

**Organs**
O 3 liver

**Skeletal Structures**
S 1 clavicle, right
S 3 rib (1st right)
S 4 rib, body, anterior (7th right)
S 21 vertebra, body (T12)

**Veins**
V 4 pulmonary, inferior, RLL
V 6 axillary, right
V 15 intercostal, highest (superior), right
V 19 intercostal, posterior, right
V 25 jugular, external, right
V 33 suprascapular, right
V 36 transverse cervical, left
V 41 venous plexus, vertebral

**Nonanatomic**
NB nylon bolt

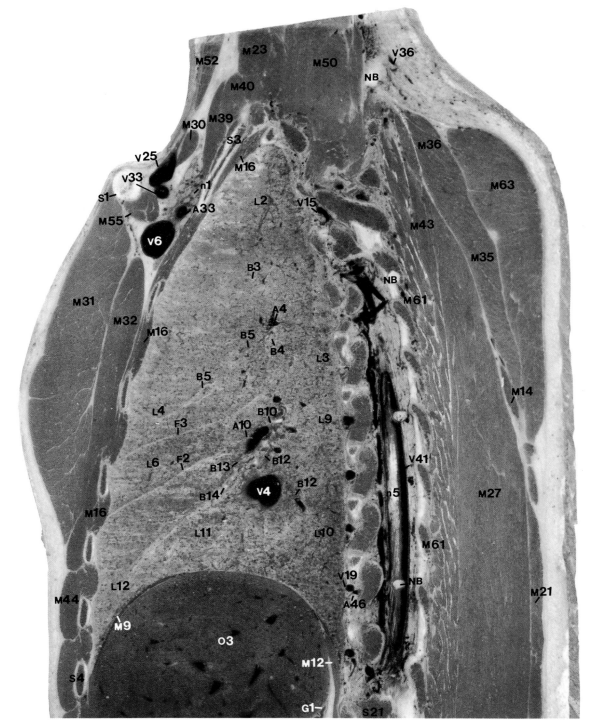

**Anatomic Specimen**

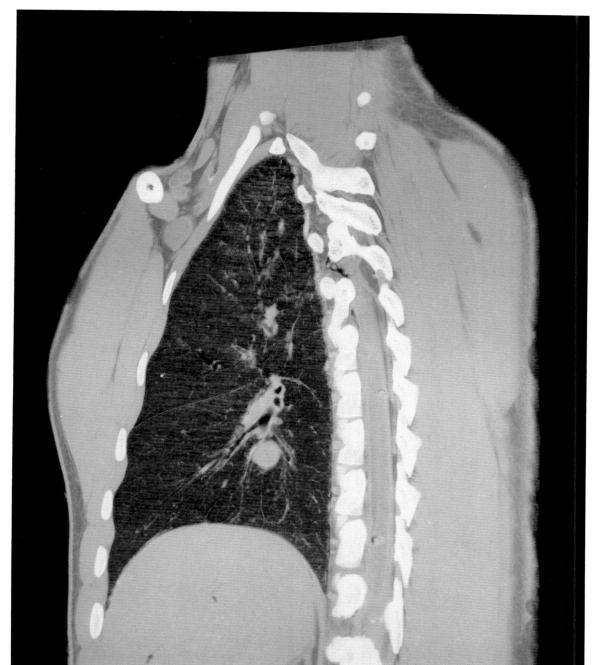

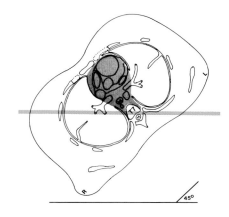

Specimen Radiograph

**Magnetic Resonance**

**Trispiral Tomogram**

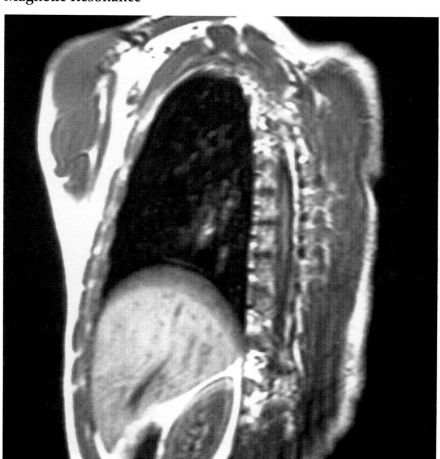

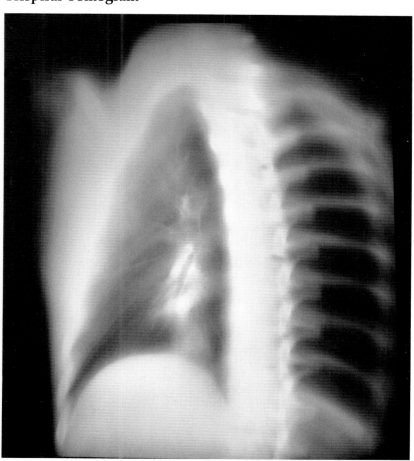

# Right Posterior Oblique

## Section 18

### Anatomic Key

**Arteries**
A 10 right lower lobe (RLL)
A 33 axillary, right
A 46 intercostal, posterior, right
A 58 thoracoacromial, right

**Bronchi**
B 4 RUL, posterior segment
B 5 RUL, anterior segment
B 11 RLL, superior segment
B 14 RLL, lateral basal segment
B 15 RLL, posterior basal segment

**Fat**
f 7 subcutaneous

**Fissures**
F 2 right major (oblique)
F 3 right minor (horizontal)

**Lobes**
L 2 RUL, apical segment
L 3 RUL, posterior segment
L 4 RUL, anterior segment
L 6 RML, lateral segment
L 9 RLL, superior segment
L 11 RLL, anterior basal segment
L 12 RLL, lateral basal segment
L 13 RLL, posterior basal segment

**Muscles**
M 9 diaphragm
M 16 intercostal
M 21 latissimus dorsi
M 23 levator scapulae
M 27 longissimus, thoracis
M 30 omohyoid
M 31 pectoralis, major
M 32 pectoralis, minor
M 35 rhomboid, major
M 36 rhomboid, minor
M 39 scalene, middle
M 40 scalene, posterior
M 43 semispinalis, thoracis
M 44 serratus, anterior
M 50 splenius, capitis
M 55 subclavius
M 61 transversospinalis
M 63 trapezius

**Neural Structures**
n 1 brachial plexus
n 7 spinal nerve

**Organs**
O 3 liver

**Pleural Structure**
P 2 costal (parietal)
P 6 visceral

**Skeletal Structures**
S 1 clavicle, right
S 3 rib (1st right)
S 4 rib, body, anterior (7th right)
S 6 rib, body, posterior (1st right)
S 27 vertebra, spinous process (T10)

**Veins**
v 4 pulmonary, inferior, RLL
v 6 axillary, right
v 11 cephalic, right
v 15 intercostal, highest (superior), right
v 19 intercostal, posterior, right
v 25 jugular, external, right
v 36 transverse cervical, left
v 41 venous plexus, vertebral

**Nonanatomic**
NB nylon bolt

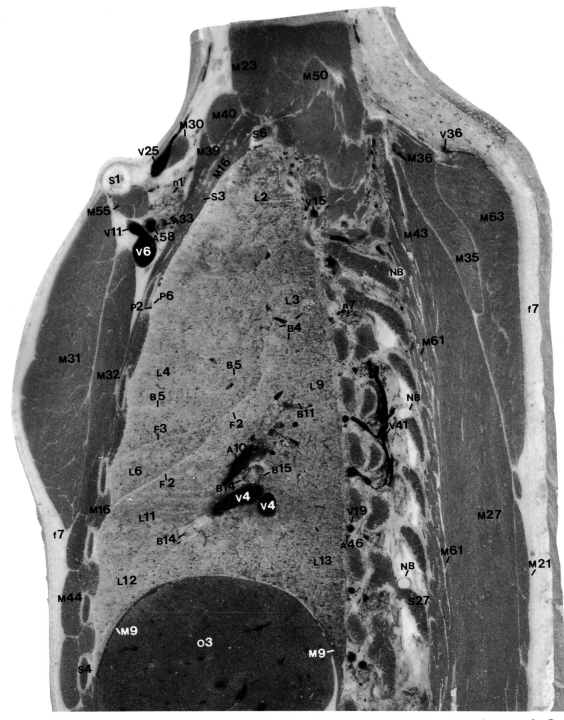

**Anatomic Specimen**

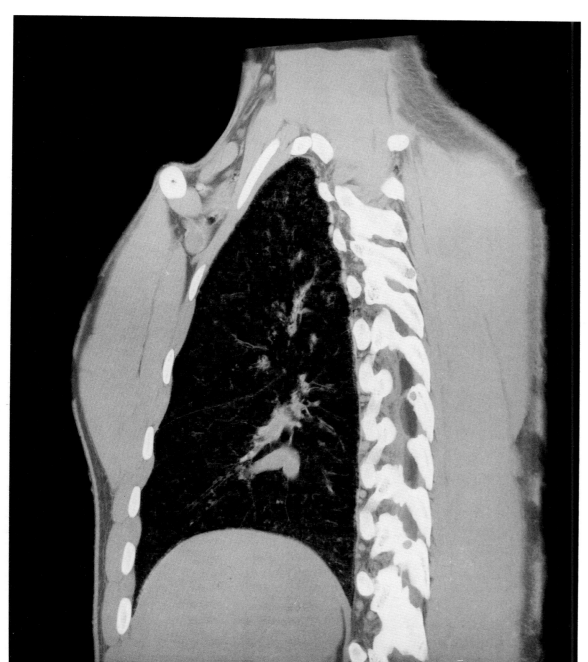

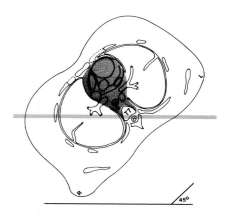

**Specimen Radiograph**

**Magnetic Resonance**

**Trispiral Tomogram**

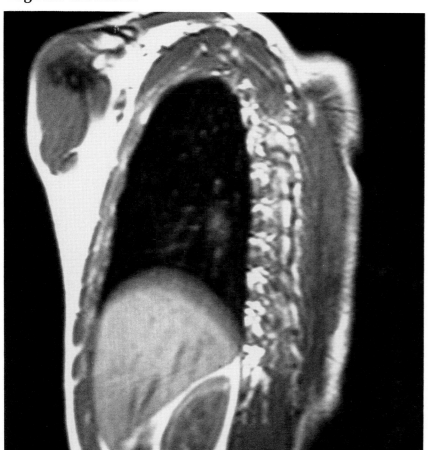

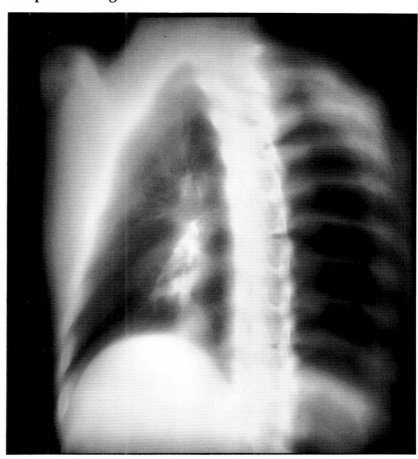

# Right Posterior Oblique

## Section 19

## Anatomic Key

### Arteries
A **33** axillary, right
A **46** intercostal, posterior, right
A **58** thoracoacromial, right

### Bronchi
B **4** RUL, posterior segment
B **5** RUL, anterior segment
B **11** RLL, superior segment
B **14** RLL, lateral basal segment
B **15** RLL, posterior basal segment

### Fat
f **1** axillary

### Fissures
F **2** right major (oblique)
F **3** right minor (horizontal)

### Lobes
L **2** RUL, apical segment
L **3** RUL, posterior segment
L **4** RUL, anterior segment
L **6** RML, lateral segment
L **9** RLL, superior segment
L **12** RLL, lateral basal segment
L **13** RLL, posterior basal segment

### Muscles
M **9** diaphragm
M **13** erector spinae
M **16** intercostal
M **21** latissimus dorsi
M **23** levator scapulae
M **27** longissimus, thoracis
M **30** omohyoid
M **31** pectoralis, major
M **32** pectoralis, minor
M **33** platysma
M **40** scalene, posterior
M **44** serratus, anterior
M **50** splenius, capitis
M **55** subclavius
M **61** transversospinalis
M **63** trapezius

### Neural Structures
n **1** brachial plexus

### Organs
O **3** liver

### Skeletal Structures
S **1** clavicle, right
S **4** rib, body, anterior (7th right)
S **6** rib, body, posterior (1st right)
S **27** vertebra, spinous process (T6)

### Veins
V **4** pulmonary, inferior, RLL
V **6** axillary, right
V **11** cephalic, right
V **19** intercostal, posterior, right
V **33** suprascapular, right
V **37** transverse cervical, right

### Nonanatomic
NB nylon bolt

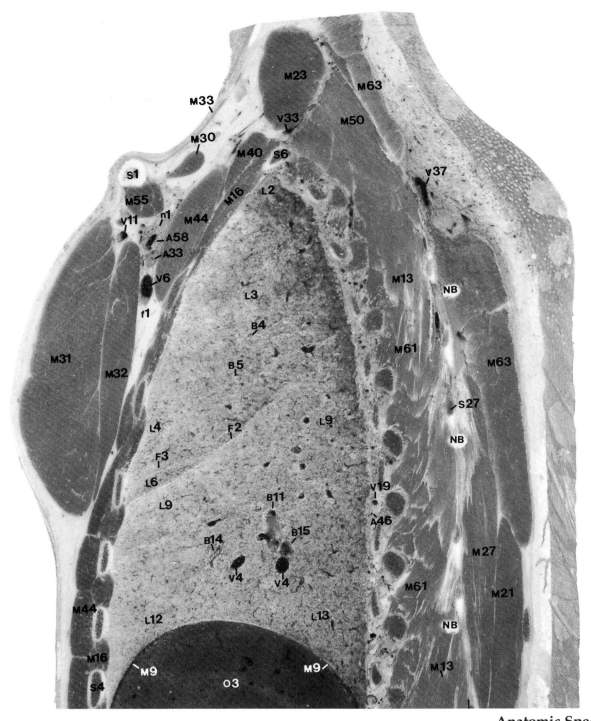

**Anatomic Specimen**

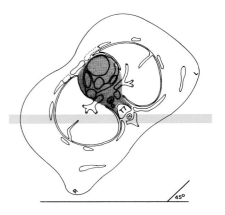

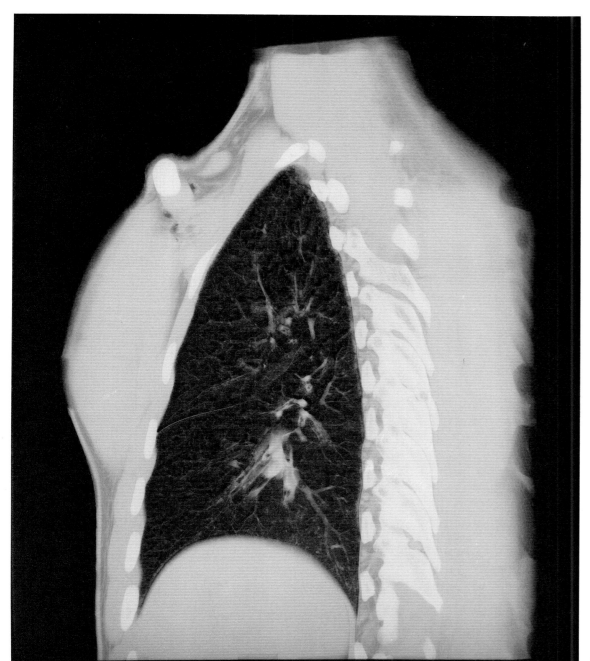

**Specimen Radiograph**

**Magnetic Resonance**

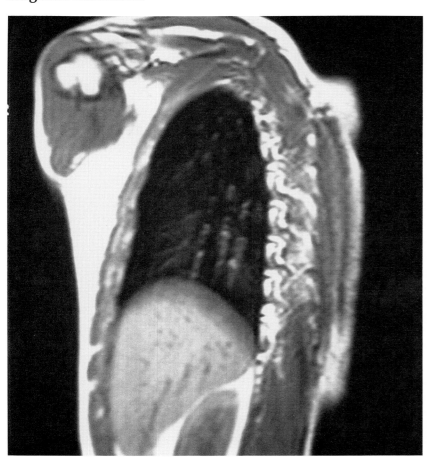

**Trispiral Tomogram**

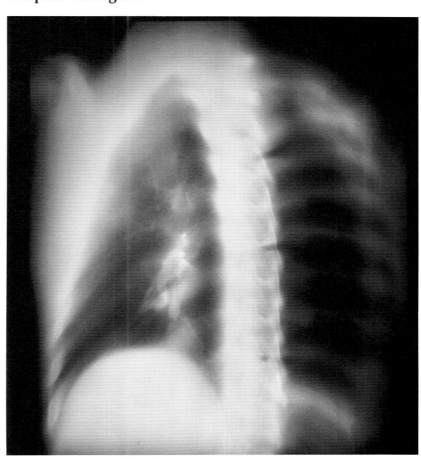

# Right Posterior Oblique

## Section 20

## Anatomic Key

### Arteries
A 33 axillary, right
A 46 intercostal, posterior, right
A 58 thoracoacromial, right

### Bronchi
B 11 RLL, superior segment
B 14 RLL, lateral basal segment
B 15 RLL, posterior basal segment

### Fat
f 1 axillary

### Fissures
F 2 right major (oblique)

### Lobes
L 3 RUL, posterior segment
L 9 RLL, superior segment
L 12 RLL, lateral basal segment
L 13 RLL, posterior basal segment

### Muscles
M 9 diaphragm
M 13 erector spinae
M 16 intercostal
M 21 latissimus dorsi
M 23 levator scapulae
M 27 longissimus, thoracis
M 30 omohyoid
M 31 pectoralis, major
M 32 pectoralis, minor
M 33 platysma
M 35 rhomboid, major
M 44 serratus, anterior
M 55 subclavius
M 63 trapezius

### Neural Structures
n 1 brachial plexus

### Organs
O 3 liver

### Pleural Structures
P 2 costal (parietal)
P 6 visceral

### Skeletal Structures
S 1 clavicle, right
S 3 rib (1st right)
S 4 rib, body, anterior (7th right)
S 8 rib, head and neck (9th right)

### Veins
V 4 pulmonary, inferior, RLL
V 6 axillary, right
V 11 cephalic, right
V 19 intercostal, posterior, right
V 33 suprascapular, right

### Nonanatomic
NB nylon bolt

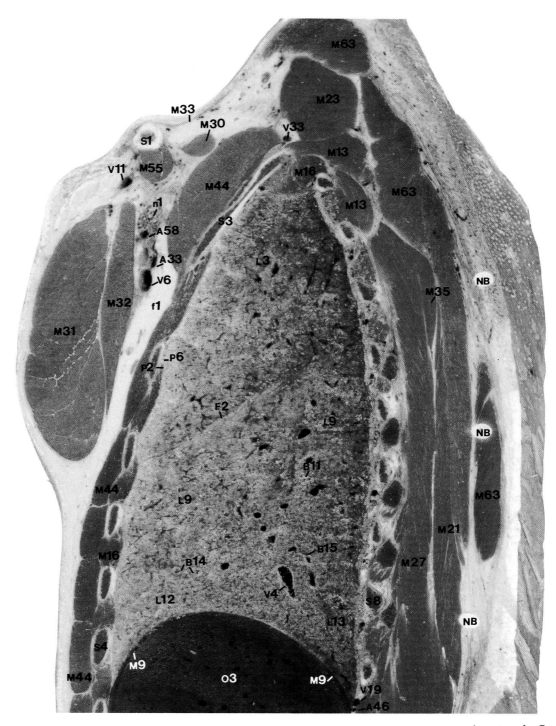

**Anatomic Specimen**

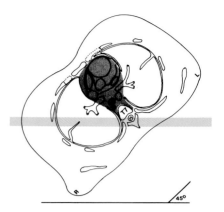

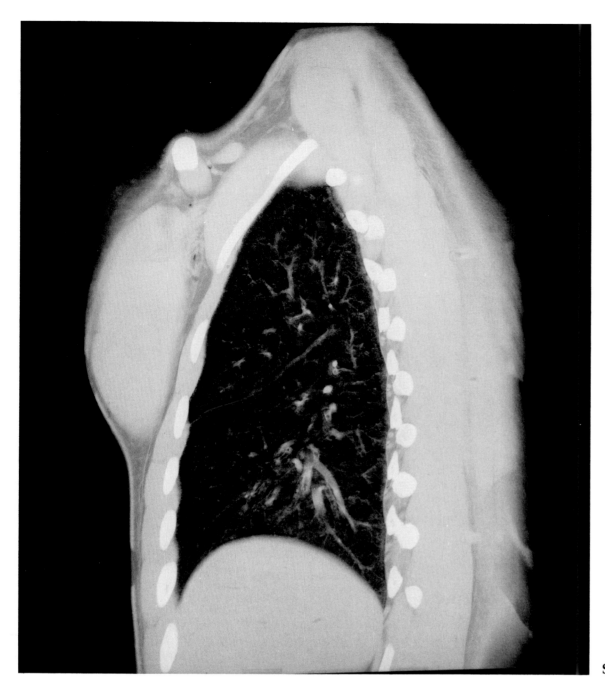

Specimen Radiograph

**Magnetic Resonance**

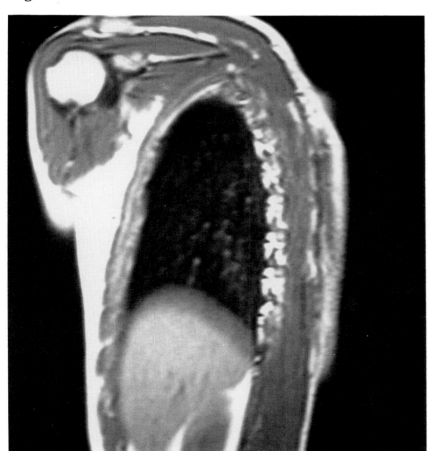

**Trispiral Tomogram**

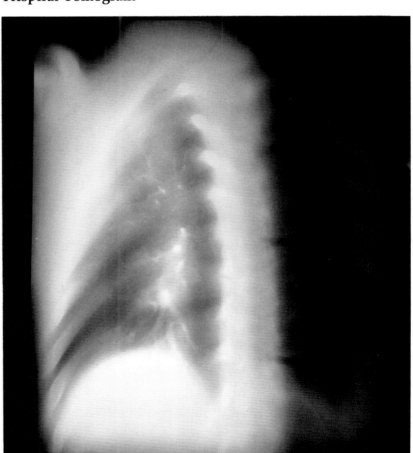

# References

1. *Grant's Atlas of Anatomy*, 9th edition (1991) Agur AM (ed). Williams & Wilkins, Baltimore

2. Barrett CB, Poliakoff SJ, Folder LE (1990) *Primer of Sectional Anatomy with MRI and CT Correlation*. Williams & Wilkins, Baltimore

3. Bergman RA, Afifi AK, Jew JY, Reimann PC (1991) *Atlas of Human Anatomy in Cross Section*. Williams & Wilkins, Baltimore

4. Bo WJ, Meschan I, Krueger WA (1980) *Basic Atlas of Cross-Sectional Anatomy*. W.B. Saunders, Philadelphia

5. Bo WJ, Wolfman ND, Kreuger WA, Meshcan I (1990) *Basic Atlas of Sectional Anatomy with Correlated Imaging*. W.B. Saunders, Philadelphia

6. Cahill DR, Orland MJ (1984) *Atlas of Human Cross-Sectional Anatomy*. Lea & Febiger, Philadelphia

7. Carter BL, Morehead J, Wolpert SM, Hammerschlag SB, Griffiths HJ, Kahn PC (1977) *Cross-Sectional Anatomy. Computed Tomography and Ultrasound Correlation*. Appleton-Century-Crofts, New York

8. Chacko AK, Katzberg RW, MacKay A (1991) *MRI Atlas of Normal Anatomy*. McGraw-Hill, New York

9. Clemente CD (1975) *Anatomy. A Regional Atlas of the Human Body*. Lea & Febiger, Philadelphia

10. El-Khoury GY, Bergman RA, Montgomery WJ (1990) *Sectional Anatomy by MRI/CT*. Churchill Livingstone, New York

11. Ellis H, Logan BM, Dixon A (1991) *Human Cross-Sectional Anatomy: Atlas of Body Sections and CT Images*. Butterworth-Heinemann

12. Eycleshymer AC, Schoemaker DM (1970) *A Cross-Section Anatomy*, 3rd edition. Appleton-Century-Crofts, New York

13. Friedman SM (1970) *Visual Anatomy. Vol. II: Thorax and Abdomen*. Harper and Row, New York Hagerstown London

14. Gerhardt P, Frommhold W (1988) *Atlas of Anatomic Correlations in CT and MRI*. Thieme, Stuttgart New York

15. Gosling JA, Harris PF, Humpherson JR, Whitmore I, Willan PLT (1985) *Atlas of Human Anatomy with Integrated Text*. J.B. Lippincott, Philadelphia

16. Gray H. *Anatomy of the Human Body*, 23rd edition (1936) Lewis WH (ed). Lea & Febiger, Philadelphia

17. Han M-C, Kim C-W (1985) *Sectional Human Anatomy Correlated with CT and MRI*. Ilchokak, Seoul, Korea

18. Heitzman ER (1977) *The Mediastinum. Radiologic Correlations with Anatomy and Pathology*. C.V. Mosby Co., Saint Louis

19. Hollinshead WH (1974) *Textbook of Anatomy*, 3rd edition. Harper and Row, New York Hagerstown London

20. Kieffer SA, Heitzman ER (1979) *An Atlas of Cross-Sectional Anatomy: Computed Tomography, Ultrasound, Radiography, Gross Anatomy*. Harper and Row, New York Hagerstown

21. Koritke JG, Sick H (1983) *Atlas of Sectional Human Anatomy: Frontal, Saqittal, and Horizontal Planes. Vol. I: Head, Neck, Thorax*. Urban & Schwarzenberg, Baltimore Munich

22. Ledley RS, Huang HK, Mazziotta JC (1977) *Cross-Sectional Anatomy: An Atlas for Computerized Tomography*. Williams & Wilkins, Baltimore

23. Littleton JT, Reynolds HC (1964) A new illuminator for copying roentgenograms. *Am J Roentgenol* 91:1368–1370

24. Littleton JT, Brogdon BG, Durizch ML, Callahan WP (1982) The human chest phantom. *Radiology* 145:829–831

25. Littleton JT (1976) *Tomography: Physical Principles and Clinical Applications*. Chapt 11, The chest. Williams & Wilkins, Baltimore

26. Luzsa G (1974) *X-Ray Anatomy of the Vascular System*. J.B. Lippincott, Philadelphia Toronto, and Akademiai Kiado, Budapest

27. Lyons EA (1978) *A Color Atlas of Sectional Anatomy: Chest, Abdomen, and Pelvis*. C.V. Mosby Co., Saint Louis

28. Lyons EA (1990) *Practical Color Atlas of Sectional Anatomy*. Raven Press, New York

29. Marchand, P (1951) The anatomy and applied anatomy of the mediastinal fascia. *Thorax* 6:359–368

30. Medlar EM (1947) Variations in interlobar fissures. *Am J Roentgenol* 57:723–725

31. Moore KL (1985) *Clinically Oriented Anatomy*, 2nd edition. Williams & Wilkins, Baltimore

32. Netter FH (1981) *The CIBA Collection of Medical Illustrations. Vol. V: Heart*. CIBA Pharmaceutical Co., Summit, NJ

33. Netter FH (1980) *The CIBA Collection of Medical Illustrations. Vol. VII: Respiratory System*. CIBA Pharmaceutical Co., Summit, NJ

34. Netter FH (1989) *Atlas of Human Anatomy*. Ciba-Geigy Corp., Summit, NJ

35. *Nomina Anatomica*, 6th edition (1989) Brookes M, Warwick R (eds). Churchill Livingstone, New York Edinburgh London Melbourne

36. Peirce CB, Stocking BW (1937) The oblique projection of the thorax: An anatomic and roentgenologic study. *Am J Roentgenol* 38: 245–267

37. *Pernkopf Anatomy*, 3rd edition (1987) Platzer W (ed). *Vol. II: Atlas of Topographic and Applied Human Anatomy*. Urban and Schwarzenberg, Baltimore Munich

38. Peterson RR (1980) *A Cross-Sectional Approach to Anatomy*. Year Book Medical Publishers, Chicago New York

39. Schnitzlein HN, Murtagh FR (1985) *Imaging Anatomy of the Head and Spine. A Photographic Color Atlas of MRI, CT, Gross and Microscopic Anatomy in Axial, Coronal, and Sagittal Planes*. Urban and Schwarzenberg, Baltimore Munich

40. Sobotta J (1939) McMurrich JP, Watt JC (eds). *Atlas of Human Anatomy. Vol. I: The Bones, Ligaments, Joints, Muscles and Regions of the Human Body*. 5th revised English edition. G.E. Stechert & Co., New York

41. Sobotta J (1939) McMurrich JP, Watt JC (eds). *Atlas of Human Anatomy. Vol. II: The Digestion, Respiratory, and Urogenital Systems of the Human Body*. The vascular system, Part I: The heart. 5th revised

English edition. G.E. Stechert & Co., New York

42. Sobotta J (1939) McMurrich JP, Watt JC (eds). *Atlas of Human Anatomy. Vol. III: The Nervous and Blood Vascular Systems and the Sense Organs of the Human Body.* 5th revised English edition. G.E. Stechert & Co., New York

43. Von Hagens G, Romrell LJ, Ross MH, Tiedemann K (1991) *The Visible Human Body. An Atlas of Sectional Anatomy.* Lea & Febiger, Philadelphia London

44. Wagner M, Lawson TL (1982) *Segmental Anatomy: Applications to Clinical Medicine.* MacMillan Publishing Co., New York

45. Webb R, Muller NL, Naidich DP (1992) *High-Resolution CT of the Lung.* Raven Press, New York

46. Wechsler RJ (1989) *Cross-Sectional Analysis of the Chest and Abdominal Wall.* Mosby-Year Book, Saint Louis

47. Wyman AC, Lawson TL, Goodman LR (1978) *Transverse Anatomy of the Human Thorax, Abdomen, and Pelvis: An Atlas of Anatomic Radiologic Computed Tomographic and Ultrasonic Correlation.* Little, Brown and Co., Boston

48. Yamashita H (1978) *Roentgenologic Anatomy of the Lung.* Igaku-Shoin, Tokyo New York

49. Zylak CL, Littleton JT, Durizch ML (1988) Illusory consolidation of the left lower lobe: A pitfall of portable radiology. *Radiology* 167:653–655

# Appendix

## Direct Acquisition Multiplanar CT

The quality of clinical sectional imaging has been progressively improving, and with computer input will continue to do so. Resolution, contrast sensitivity, technical operability of the equipment, lower dose, and all of the other parameters important to improved sectional imaging are under continuous research by manufacturers and other investigators. The distinctly effective images of high-resolution CT (HRCT), MR, ultrasound, and nuclear emission will continue to improve, and we can predict with reasonable assurance that new sectional images with new and yet unexplored energies will be developed.

Computed tomography has overcome the two prime limitations of conventional pluridirectional tomography (CPT): unwanted marginal blur and limited soft tissue discrimination. The CT images are "clean," i.e., unencumbered by structure blur above and below the plane of focus. Computer applications make CT images easier to obtain technically with excellent quality control.

In spite of these technological advances, *confinement to axial orientation is still a prime limitation of CT*. Secondary CT reconstruction of coronal or sagittal planes from data acquired in the axial plane produces clinically inferior images. The wealth of additional information that would result if direct acquisition *multiplanar CT images* could be attained is obvious from a review of the sectional anatomy presented in the other four planes in this atlas. It is not our province to suggest the type of redesign of hardware and/or software that would permit multiplanar direct acquisition CT, but we do encourage active research in this direction. We feel that such an advance will significantly enhance our diagnostic capabilities.

The concept of multiplanar CT is not new, being most commonly used for investigation of the skull. Wolf et al. (18) and Hammerschlag et al. (7) described techniques to obtain direct coronal CT images of the skull and alluded to the probable clinical applications of this technique in 1976. Van Waes and Zonneveld (17) in 1982 lucidly described multiplanar direct CT studies of the petrous bone, techniques which have now become state-of-the-art. They also described patient positioning for direct coronal CT of the entire body. Lee et al. (10), also in 1982, described a supporting device to obtain direct coronal CT scans of the abdomen and pelvis. Bar-Ziv and Solomon (3, 4) described their clinical experiences with direct coronal CT of thoraco-abdominal problems and recommended general use of this method.

These early attempts to promote and develop multiplanar CT have not met with general acceptance except in the skull, where the required patient positioning is readily adaptable to the design of present CT scanners, and where related anatomic reference support for these studies has already been done in multiple planes. The coronal images of the chest and abdomen shown by Bar-Ziv and Solomon are of good technical quality but were all made with gantry/object angles between 30° and 80°. The resulting sectional images are consequently distorted. They depict anatomic planes that are not familiar to radiologists and for which there are no available anatomic references. Multiplanar CT of the entire body has not generated much interest, despite the fact that growing discontent is being expressed toward the inflexibility of only axial imaging in CT. A frequently cited advantage of MR imaging is its multiplanar capability.

The development of this atlas provided access to suitable anatomical material needed to explore the imaging capabilities of *true* coronal and sagittal direct acquisition CT of the chest. The content of these experimental images was compared to the anatomic detail of the specimen radiographs made at similar levels in the coronal and sagittal projections of this atlas (Chapters 3 and 4).

To obtain direct acquisition scans in coronal and sagittal planes, a frozen chest torso (chapter 2 cadaver) was oriented in an erect position in the gantry of a Siemens Somatom Plus CT scanner. High-resolution CT images were made in both coronal and sagittal planes. Representative sections are reproduced as Figures 1–4 to illustrate the improvement in imaging potential that can be experienced from direct acquisition, high-resolution multiplanar CT. Special attention is drawn to bronchial and parenchymal imaging, and to the superb display of the spine in the sagittal plane with mediastinal window settings.

## Digital Tomography and Tomosynthesis

Although the usefulness of conventional pluridirectional tomography is being lost in the tide of the popularity wave of CT and MR, a significant imaging potential still exists (as shown by a careful inspection of the trispiral tomogram images in this atlas). Perhaps this technique will be "rediscovered" with computer augmentation to make advances toward our ultimate images.

Investigations in this direction have been in motion for some

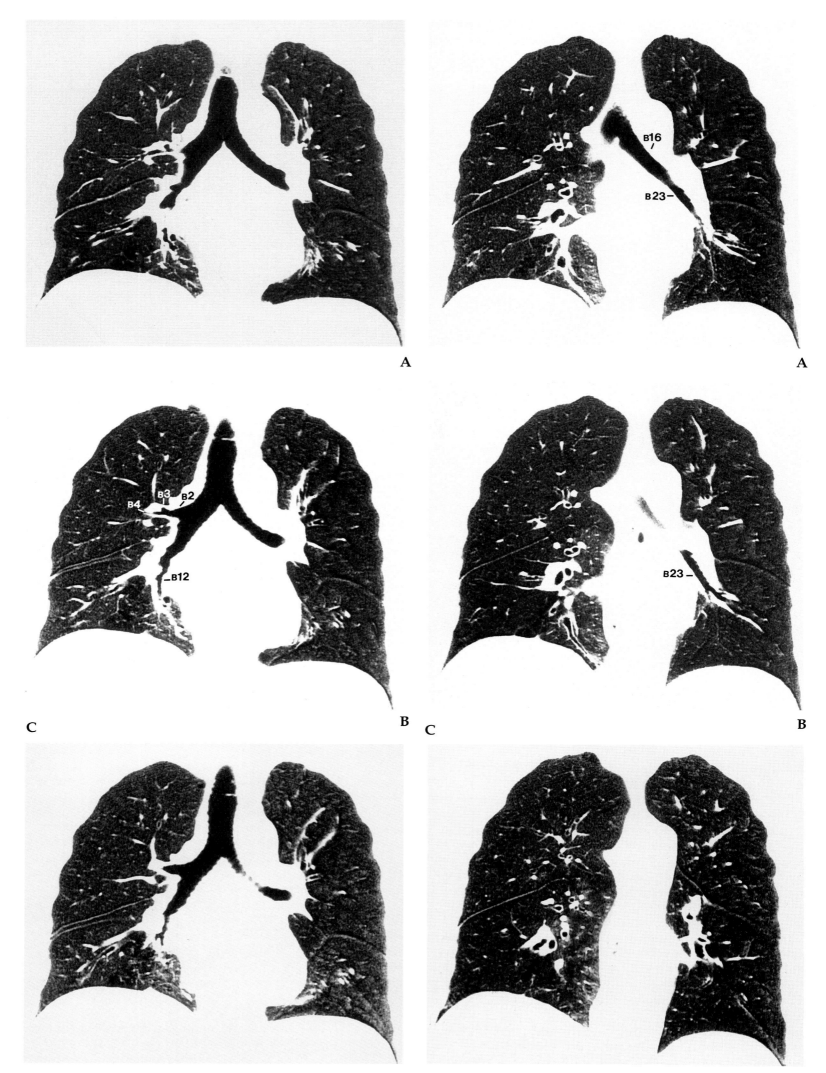

**Figure 1.** (**A**) Coronal CT of isolated thorax, data directly acquired in AP projection. Note tracheal and bronchial detail, fissures, and high-resolution detail of lung parenchyma. (**B**) Coronal CT section just posterior to A. Note RUL bronchus (B2) and origin of apical (B3) and posterior (B4) segmental bronchi to RUL. Note also medial basal segmental bronchus to RLL (B12). (**C**) Coronal CT section posterior to B. Compare to specimen radiograph, Sections 12, 13 and 14, Chapter 3.

**Figure 2.** (**A**) Coronal CT section posterior to carina. Note continuous imaging of left mainstem bronchus (B16), and LLL bronchus (B23) with extension into basilar segmental bronchi LLL. (**B**) Coronal CT section posterior to A showing detail of lung fields and LLL bronchus (B23). (**C**) Coronal CT posterior to B. Compare to specimen radiographs, Sections 15 and 16, Chapter 3.

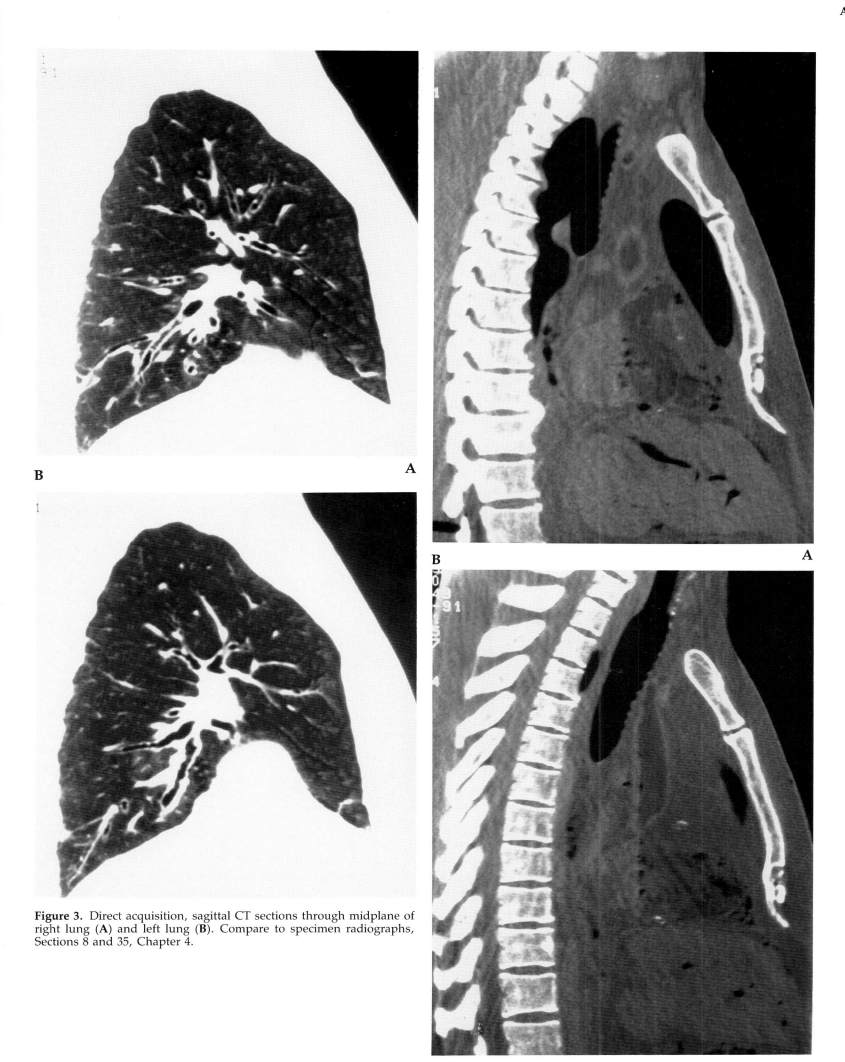

**Figure 3.** Direct acquisition, sagittal CT sections through midplane of right lung (**A**) and left lung (**B**). Compare to specimen radiographs, Sections 8 and 35, Chapter 4.

**Figure 4.** Direct acquisition sagittal CT sections mediastinum technique. (**A**) Lateral margin of thoracic spine well seen. (**B**) Midline sagittal CT section. Bone detail would be significantly enhanced by the use of CT spine technique. Compare to specimen radiographs, Sections 19 and 20, Chapter 4.

time using a combination of tomographic and electronic imaging principles to result in digital tomograms and/or digital tomosynthesis (1, 2, 5, 6, 8, 9, 11–16). Digital tomograms can take advantage of edge enhancement, special filters in processing, and gray scale windowing. Tomosynthesis offers the attractive possibility of an infinite number of tomograms in multiple orientations developed from a single tomographic exposure. The tomosynthesis process, which was formerly done optically, can now be done with digital computer algorithms.

Tomographic Section 12, Chapter 5, was subjected to laser film digitization and digital image processing with selective filters and edge enhancement (15, 16). The obvious improvement in image resolution and contrast can be seen in Figure 5.

## Conclusion

It is hoped that this brief look into the future will stimulate efforts by inventive radiologists in cooperation with industrial scientists to continue the admirable advances in sectional imaging that have appeared in the last two decades. However, an interesting paradox exists: Manufacturers need to be motivated by radiologists, but unfortunately most radiologists are not sufficiently aware of possible new developments in CT to initiate the motivation. Hopefully, this small contribution will serve to arouse both sides.

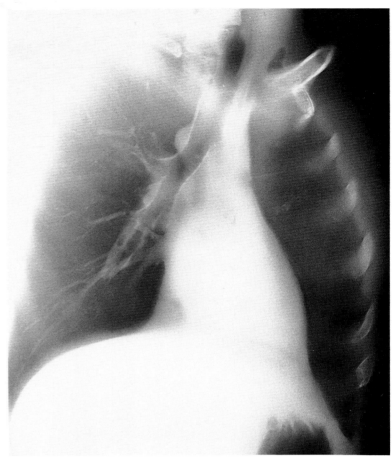

**Figure 5.** Laser digitization (Konica laser digitizer) of trispiral tomogram, Section 12, Chapter 5. Spatial frequency filtration done with a combination of three special filters on the image processing unit of the Shimazu PACS system. Note improved detail, contrast, and marginal enhancement by reduction of marginal blur. Other similar digital techniques by Shimazu have produced pulmonary interstitial detail comparable to high-resolution CT.

## References for Appendix

1. Baily NA, Crepeau RL, Lasser EC (1974) Fluoroscopic tomography. *Invest Radiol* 9:94–103

2. Baily NA, Kampp TD (1981) Digitized longitudinal tomography. *Invest Radiol* 16:126–132

3. Bar-Ziv J, Solomon A (1989) The use of a modified direct coronal tomographic technique for assessing thoraco-abdominal problems. *Gastroenterology* 14:205–208

4. Bar-Ziv J, Solomon A (1990) Direct coronal CT scanning of tracheo bronchial pulmonary and thoraco abdominal lesions in children. *Pediatr Radiol* 20:245–248

5. deVries N, Miller FJ, Wojtowycz MM, Brown PR, Yandow DR, Nelson JA, Kruger RA (1985) Tomographic digital subtraction angiography: Initial clinical studies using tomosynthesis. *Radiology* 157:239–241

6. Friedenberg RM, Lightfoote JB, Wang SP, Smolin MF (1985) Digital tomography: Description and preliminary clinical experience. *Am J Roentgenol* 144:639–643

7. Hammerschlag SB, Wolpert SM, Carter BL (1976) Computed coronal tomography. *Radiology* 120:219–220

8. Kampp TD (1986) The background method applied to classical tomography. *Med Phys* 13:329–333

9. Kruger RA, Sedaghati M, Roy DG, Lue P, Nelson JA, Kubal W, Del Rio P (1984) Tomosynthesis applied to digital subtraction angiography. *Radiology* 152:805–808

10. Lee JKT, Barbier JY, McClennan BL, Stanley RJ (1982) A support device for obtaining direct coronal computed tomographic scans of the pelvis and lower abdomen. *Radiology* 145:209–210

11. Maravilla KR, Murry RC, Horner S (1983) Digital tomosynthesis: Technique for electronic reconstructive tomography. *Am J Neuroradiol* 4:883–888

12. Maravilla KR, Murry RC, Diehl J, Suss R, Allen L, Chang K, Crawford J, McCoy R (1984) Digital tomosynthesis: Technique modifications and clinical applications for neurovascular anatomy. *Radiology* 152:719–724

13. Nadjmi M, Weiss H, Klotz E, Linde R (1980) Flashing tomosynthesis: First clinical results. *Medicamundi* 25:9–17

14. Sklebitz H, Haendle J (1983) Tomoscopy: Dynamic layer imaging without mechanical movement. *Am J Roentgenol* 140:1247–1252

15. Sone S, Kasuga T, Sakai F, Aoki J, Izuno I, Tanezaki Y, Skigita H, Shibata K (1991) Development of a high-resolution digital tomosynthesis system and its clinical application. *RadioGraphics 1991* 11:807–822

16. Sone S, Kasuga T, Sakai F, Izuno I, Oguchi M (1991) Digital image processing to remove blur from linear tomography of the lung. *Acta Radiologica* 32:Fasc 5

17. Van Waes TFGM, Zonneveld FW (1982) Direct coronal body computed tomography. *J Comput Assist Tomogr* 6:58–66

18. Wolf BS, Nakagawa H, Staulcup PH (1976) Feasibility of coronal views in computed scanning of the head. *Radiology* 120:217–218

# Index